PRINCIPLES OF BIOCHEMISTRY

Edited by
Olivia Berthoumieu

ArclerPress

Principles of Biochemistry

Edited by Olivia Berthoumieu

Edition 2018

ArclerPress

Published by
Arcler Press LLC
708 3rd Avenue, 6th Floor
New York
NY 10017
United States of America
www.arclerpress.com

ISBN: 978-1-68094-532-4 (Hardback)

Printed in United States of America

Library of Congress Control Number: 2017933104

About the Editor

Olivia Berthoumieu

Olivia graduated from Oxford University, where she received a PhD in Biochemistry in 2012. Her research interests span from the use of single molecule scale methods on membrane proteins for bionanotechnological applications to the characterisation of potential inhibitors of the amyloid beta peptide aggregation – one of the hallmarks of Alzheimer's disease. A postdoctoral fellow and then research engineer at CNRS (French national centre for scientific research) for three years, she now works as a scientific and medical translator.

Table of Contents

3 DNA STRUCTURE AND REPLICATION

4 RNA SYNTHESIS AND PROCESING

5 PROTEIN BIOSYNTHESIS

6 ENZYMES: STRUCTURE AND MECHANISM

7 ANTIBODIES

8 BIOLOGICAL MEMBRANE

List of Abbreviations

POMC	polyprotein pro-opiomelanocortin
PET	positron emission tomography
PTM	Post-translational modification
PIGS	Prediction of Immunoglobulin Structure
PCNA	Proliferating Cell Nuclear Antigen
PG	prostaglandins
Sec	selenocysteine
SHM	somatic hypermutation
TdT	Terminal deoxynucleotidyl transferase
TIMPs	tissue inhibitors of metalloproteinases
rDNA	Recombinant DNA
RT-PCR	reverse transcription polymerase chain reaction
RhD	Rhesus D
RdDp	RNA-dependent DNA polymerase
RCR	rolling circle replication
SHM	somatic hypermutation
TIMPs	tissue inhibitors of metalloproteinases
WAM	Web Antibody Modeling
XPS	Xeroderma Pigmentosum Variant

Preface

Biochemistry is the application of chemistry to the study of biological processes at the cellular and molecular level. It emerged as a distinct discipline around the beginning of the 20th century when scientists combined chemistry, physiology and biology to investigate the chemistry of living systems. In the mid 1900s, biochemistry and its presence in the scientific world increased as feats in medicine, microbiology, and nutrition, not to mention war time efforts, dictated the need to understand the underlying processes behind life. This meant the study of how large chemical molecules like carbohydrates, lipids, and amino acids function in terms of structure and interactions within a living system. This led to studies in protein function and structure, which would later give scientists an understanding of nucleic acids in the later part of the 20th century allowing for the discovery of deoxyribonucleic acid, or DNA, discovered by James Watson, Francis Crick, Rosalind Franklin and Maurice Wilkins. Biochemistry has become the foundation for understanding all biological processes. In general, biochemistry explains the processes of life and how chemical interactions support life. Biochemistry is everywhere. When discussing topics in biochemistry, it is important to understand that if you have an interest in any 'ology,' it has roots directly or indirectly in biochemistry. General topics in biochemistry include: medicine, nutrition, molecular biology and plant and animal biology. Medicine is a broad category but relates to biochemistry on many levels. Doctors and nurses give drugs to patients to help cure a disease or prevent it. This is due to the fact that when that drug is added to the human system, it alters how other chemicals in the body function, resulting in disease prevention or recovery. Every time a vaccination is injected into a human it prevents disease because biochemically the body already thinks that the disease has been there. Nutrition is a category of biochemistry that is very important to all life on the planet. All life needs a certain amount of amino acids, carbohydrates, fats, lipids and proteins to survive. These compounds are either consumed or fabricated by the organism to provide cellular structure, energy and other chemicals to survive. It has provided explanations for the causes of many diseases in humans, animals and plants. It can frequently suggest ways by which such diseases may be treated or cured.

Content Coverage

Chapters One and *Two* present an overview of biochemistry and amino acids essentials, respectively. In biochemistry, amino acids having both the amine and the carboxylic acid groups attached to the first (alpha-) carbon atom have particular importance.

Chapter Third Membranes serve diverse functions in eukaryotic and prokaryotic cells, while DNA replication occurs in all living organisms and is the basis for biological inheritance.

Chapters Four & Five clarifies process of RNA synthesis and protein synthesis, separately. Ribonucleic acid (RNA) is a polymeric molecule essential in various biological roles in coding, decoding, regulation, and expression of genes.

Chapters Six focus on biological membranes and DNA structure & replication, correspondingly.

Chapters Seven and *Eight* provide mechanism of enzymes and antibodies, separately. Enzymes are macromolecular biological catalysts. Enzymes accelerate, or catalyze, chemical reactions, whereas antibodies are glycoproteins belonging to the immunoglobulin superfamily.

CHAPTER 1

Biochemistry: An Overview

Chapter Contents

- **Introduction**
- **History**
- **Metabolism**
- **Mineral (Nutrient)**

Biochemistry is the study of the chemical processes and reactions that take place within living organisms. It can be considered a subdivision of both chemistry and biology, although the skills and techniques used within it place great emphasis on traditional chemistry.

INTRODUCTION

Biochemistry, sometimes called biological chemistry, is the study of chemical processes within and relating to living organisms. By controlling information flow through biochemical signaling and the flow of chemical energy through metabolism, biochemical processes give rise to the complexity of life. Over the last decades of the 20th century, biochemistry has become so successful at explaining living processes that now almost all areas of the life sciences from botany to medicine to genetics are engaged in biochemical research. Today, the main focus of pure biochemistry is on understanding how biological molecules give rise to the processes that occur within living cells, which in turn relates greatly to the study and understanding of tissues, organs, and whole organisms—that is, all of biology.

Biochemistry is closely related to molecular biology, the study of the molecular mechanisms by which genetic information encoded in DNA is able to result in the processes of life. Depending on the exact definition of the terms used, molecular biology can be thought of as a branch of biochemistry, or biochemistry as a tool with which to investigate and study molecular biology.

Much of biochemistry deals with the structures, functions and interactions of biological macromolecules, such as proteins, nucleic acids, carbohydrates and lipids, which provide the structure of cells and perform many of the functions associated with life. The chemistry of the cell also depends on the reactions of smaller molecules and ions. These can be inorganic, for example water and metal ions, or organic, for example the amino acids, which are used to synthesize proteins. The mechanisms by which cells harness energy from their environment via chemical reactions are known as metabolism. The findings of biochemistry are applied primarily in medicine, nutrition, and agriculture. In medicine, biochemists investigate the causes and cures of diseases. In nutrition, they study how to maintain health and study the effects of nutritional deficiencies. In agriculture, biochemists investigate soil and fertilizers, and try to discover ways to improve crop cultivation, crop storage and pest control.

HISTORY

Gerty Cori and Carl Cori jointly won the Nobel Prize in 1947 for their discovery of the Cori cycle at RPMI.

At its broadest definition, biochemistry can be seen as a study of the components and composition of living things and how they come together to become life, and the history of biochemistry may therefore go back as far as the ancient Greeks. However, biochemistry as a specific scientific discipline has its beginning sometime in the 19th century, or a little earlier, depending on which aspect of biochemistry is being focused on. Some argued that the beginning of biochemistry may have been the discovery of the first enzyme, diastase (today called amylase), in 1833 by Anselme Payen, while others considered Eduard Buchner's first demonstration of a complex biochemical process alcoholic fermentation in cell-free extracts in 1897 to be the birth of biochemistry. Some might also point as its beginning to the influential 1842 work by Justus von Liebig, *Animal chemistry, or, Organic chemistry in its applications to physiology and pathology*, which presented a chemical theory of metabolism, or even earlier to the 18th century studies on fermentation and respiration by Antoine Lavoisier. Many other pioneers in the field who helped to uncover the layers of complexity of biochemistry have been proclaimed founders of modern biochemistry, for example Emil Fischer for his work on the chemistry of proteins, and F. Gowland Hopkins on enzymes and the dynamic nature of biochemistry.

The term "biochemistry" itself is derived from a combination of biology and chemistry. In 1877, Felix Hoppe-Seyler used the term (*biochemie* in German) as a synonym for physiological chemistry in the foreword to the first issue of *Zeitschrift für Physiologische Chemie* (Journal of Physiological Chemistry) where he argued for the setting up of institutes dedicated to this field of study. The German chemist Carl Neuberg however is often cited to have coined the word in 1903, while some credited it to Franz Hofmeister.

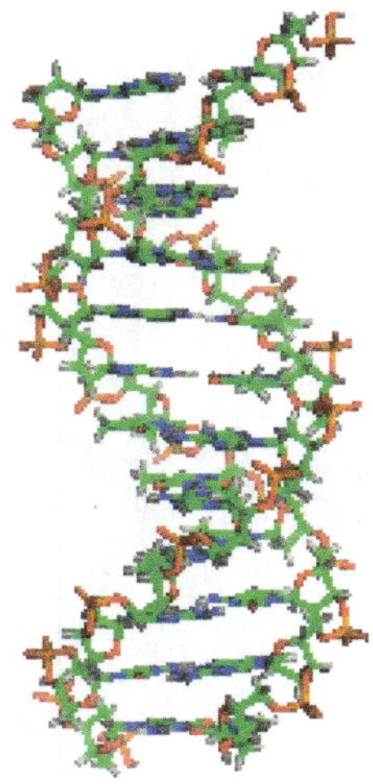

DNA structure (1D65)

It was once generally believed that life and its materials had some essential property or substance (often referred to as the "vital principle") distinct from any found in non-living matter, and it was thought that only living beings could produce the molecules of life. Then, in 1828, Friedrich Wöhler published a paper on the synthesis of urea, proving that organic compounds can be created artificially. Since then, biochemistry has advanced, especially since the mid-20th century, with the development of new techniques such as chromatography, X-ray diffraction, dual polarisation interferometry, NMR spectroscopy, radioisotopic labeling, electron microscopy, and molecular dynamics simulations. These techniques allowed for the discovery and detailed analysis of many molecules and metabolic pathways of the cell, such as glycolysis and the Krebs cycle (citric acid cycle).

Another significant historic event in biochemistry is the discovery of the gene and its role in the transfer of information in the cell. This part of biochemistry is often called molecular biology. In the 1950s, James D. Watson, Francis Crick, Rosalind Franklin, and Maurice Wilkins were instrumental in solving DNA structure and suggesting its relationship with genetic transfer of information. In 1958, George Beadle and Edward Tatum received the Nobel Prize for work in fungi showing that one gene produces one enzyme. In 1988, Colin Pitchfork was the first person convicted of murder with DNA evidence, which led to the growth of forensic science. More recently, Andrew Z. Fire and Craig C. Mello received the 2006 Nobel Prize for discovering the role of RNA interference (RNAi), in the silencing of gene expression.

Starting Materials: the Chemical Elements of Life

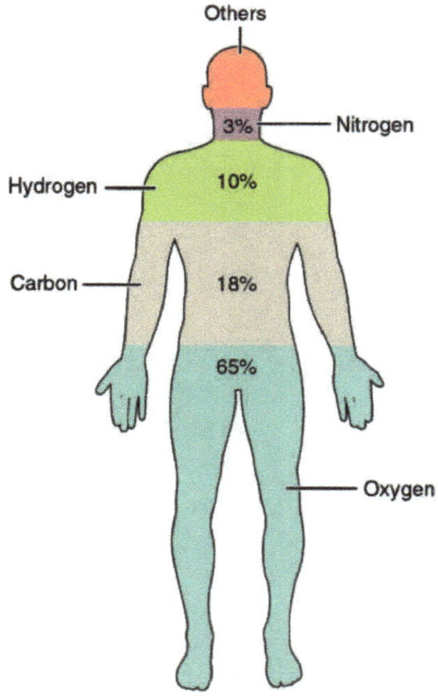

The main elements that compose the human body are shown from most abundant (by mass) to least abundant.

Around two dozen of the 92 naturally occurring chemical elements are essential to various kinds of biological life. Most rare elements on Earth are not needed by life (exceptions being selenium and iodine), while a few common ones (aluminum and titanium) are not used. Most organisms share element needs, but there are a few differences between plants and animals. For example, ocean algae use bromine, but land plants and animals seem to need none. All animals require sodium, but some plants do not. Plants need boron and silicon, but animals may not (or may need ultra-small amounts).

Just six elements—carbon, hydrogen, nitrogen, oxygen, calcium, and phosphorus—make up almost 99% of the mass of living cells, including those in the human body (see composition of the human body for a complete list). In addition to the six major elements that compose most of the human body, humans require smaller amounts of possibly 18 more.

Biomolecules

The four main classes of molecules in biochemistry (often called biomolecules) are carbohydrates, lipids, proteins, and nucleic acids. Many biological molecules are polymers: in this terminology, monomers are relatively small micromolecules that are linked together to create large macromolecules known as polymers. When monomers are linked together to synthesize a biological polymer, they undergo a process called dehydration synthesis. Different macromolecules can assemble in larger complexes, often needed for biological activity.

Carbohydrates

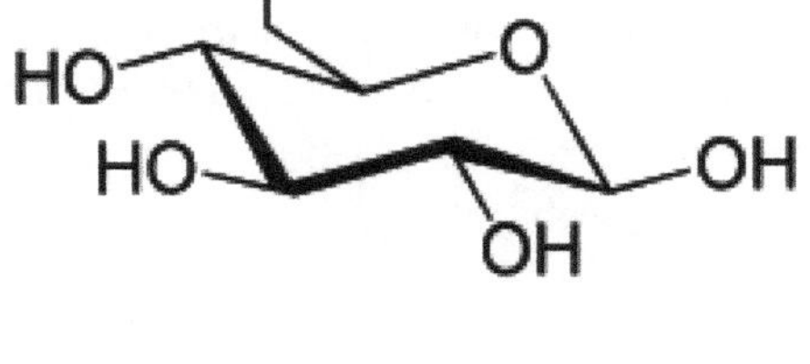

Glucose, a monosaccharide

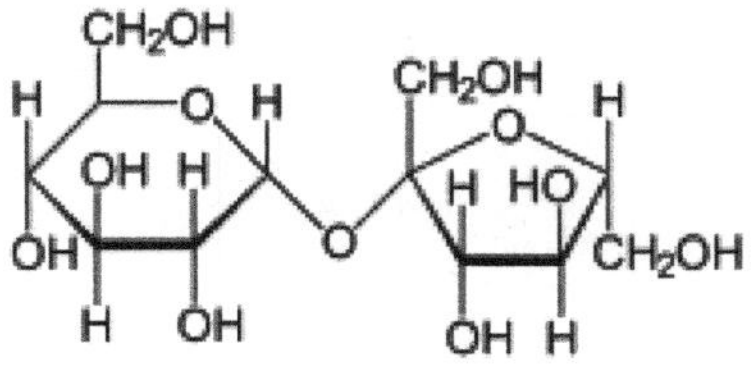

A molecule of sucrose (glucose + fructose), a disaccharide

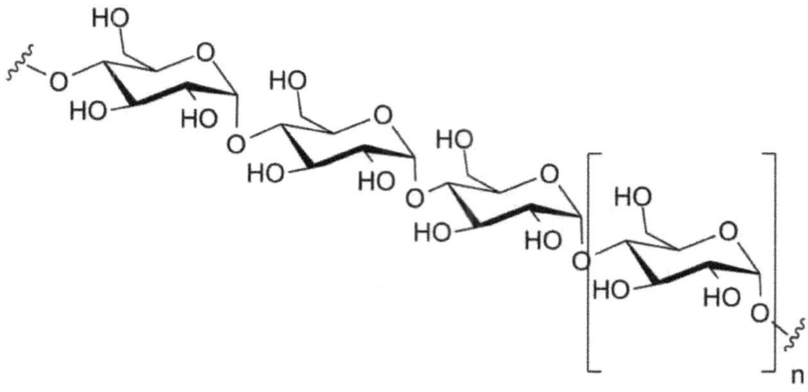

Amylose, a polysaccharide made up of several thousand glucose units

The function of carbohydrates includes energy storage and providing structure. Sugars are carbohydrates, but not all carbohydrates are sugars. There are more carbohydrates on Earth than any other known type of biomolecule; they are used to store energy and genetic information, as well as play important roles in cell to cell interactions and communications.

The simplest type of carbohydrate is a monosaccharide, which among other properties contains carbon, hydrogen, and oxygen, mostly in a ratio of 1:2:1 (generalized formula $C_nH_{2n}O_n$, where n is at least 3). Glucose ($C_6H_{12}O_6$) is one of the most important carbohydrates, others include fructose ($C_6H_{12}O_6$), the sugar commonly associated with the sweet taste of fruits,[a] and deoxyribose ($C_5H_{10}O_4$).

A monosaccharide can switch from the acyclic (open-chain) form to a cyclic form, through a nucleophilic addition reaction between the carbonyl group and one of the hydroxyls of the same molecule. The reaction creates a ring of carbon atoms closed by one bridging oxygen atom. The resulting molecule has an hemiacetal or hemiketal group, depending on whether the linear form was an aldose or a ketose. The reaction is easily reversed, yielding the original open-chain form.

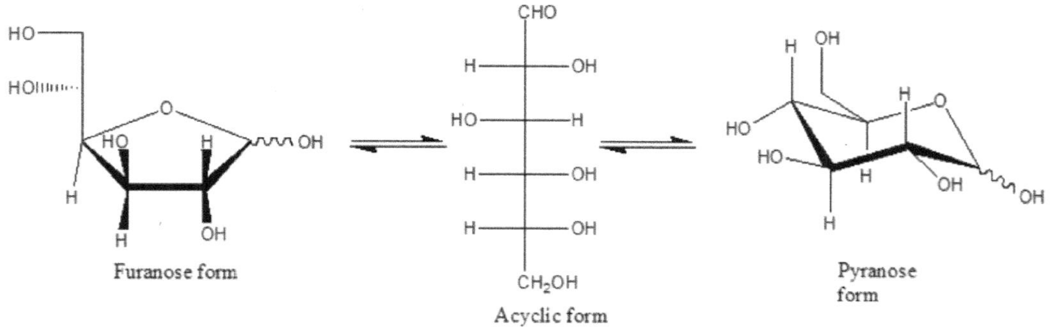

Conversion between the furanose, acyclic, and pyranose forms of D-glucose.

In these cyclic forms, the ring usually has **5** or **6** atoms. These forms are called furanoses and pyranoses, respectively — by analogy with furan and pyran, the simplest compounds with the same carbon-oxygen ring (although they lack the double bonds of these two molecules). For example, the aldohexose glucose may form a hemiacetal linkage between the hydroxyl on carbon 1 and the oxygen on carbon 4, yielding a molecule with a 5-membered ring, called glucofuranose. The same reaction can take place between carbons 1 and 5 to form a molecule with a 6-membered ring, called glucopyranose. Cyclic forms with a 7-atom ring (the same of oxepane), rarely encountered, are called heptoses.

When two monosaccharides undergo dehydration synthesis whereby a molecule of water is released, as two hydrogen atoms and one oxygen atom are lost from the two monosaccharides. The new molecule, consisting of two monosaccharides, is called a *disaccharide* and is conjoined together by a glycosidic or ether bond. The reverse reaction can also occur, using a molecule of water to split up a disaccharide and break the glycosidic bond; this is termed *hydrolysis*. The most well-known disaccharide is sucrose, ordinary sugar (in scientific contexts, called *table sugar* or *cane sugar* to differentiate it from other sugars). Sucrose consists of a glucose molecule and a fructose molecule joined together. Another important disaccharide is lactose, consisting of a glucose molecule and a galactose molecule. As most humans age, the production of lactase, the enzyme that hydrolyzes lactose back into glucose and galactose, typically decreases. This results in lactase deficiency, also called *lactose intolerance*.

When a few (around three to six) monosaccharides are joined, it is called an *oligosaccharide* (*oligo-* meaning "few"). These molecules tend to be used as markers and signals, as well as having some other uses. Many monosaccharides joined together make a polysaccharide. They can be joined together in one long linear chain, or they may be branched. Two of the most common polysaccharides are cellulose and glycogen, both consisting of repeating glucose monomers. Examples are *Cellulose* which is an important structural component of plant's cell walls, and *glycogen*, used as a form of energy storage in animals.

Sugar can be characterized by having reducing or non-reducing ends. A reducing end of a carbohydrate is a carbon atom that can be in equilibrium with the open-chain aldehyde (aldose) or keto form (ketose). If the joining of monomers takes place at such a carbon atom, the free hydroxy group of the pyranose or furanose form is exchanged with an OH-side-chain of another sugar, yielding a full acetal. This prevents opening of the chain to the aldehyde or keto form and renders the modified residue non-reducing. Lactose contains a reducing end at its glucose moiety, whereas the galactose moiety form a full acetal with the C4-OH group of glucose. Saccharose does not have a reducing end because of full acetal formation between the aldehyde carbon of glucose (C1) and the keto carbon of fructose (C2).

Lipids

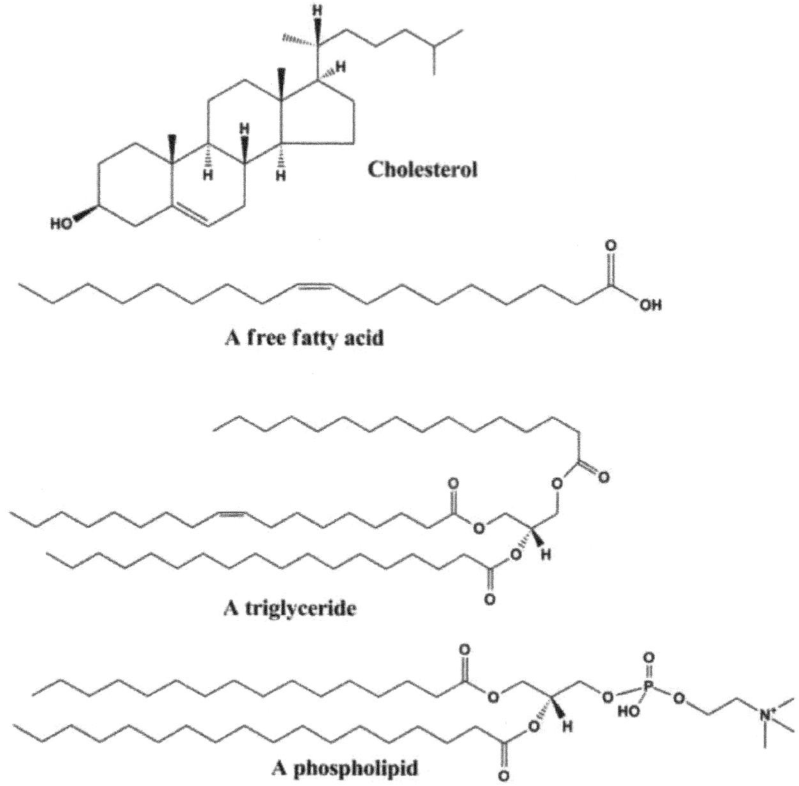

Structures of some common lipids. At the top are cholesterol and oleic acid. The middle structure is a triglyceride composed of oleoyl, stearoyl, and palmitoyl chains attached to a glycerol backbone. At the bottom is the common phospholipid, phosphatidylcholine.

Lipids comprises a diverse range of molecules and to some extent is a catchall for relatively water-insoluble or nonpolar compounds of biological origin, including waxes, fatty acids, fatty-acid derived phospholipids, sphingolipids, glycolipids, and terpenoids (e.g., retinoids and steroids). Some lipids are linear aliphatic molecules, while others have ring structures. Some are aromatic, while others are not. Some are flexible, while others are rigid.

Lipids are usually made from one molecule of glycerol combined with other molecules. In triglycerides, the main group of bulk lipids, there is one molecule of glycerol and three fatty acids. Fatty acids are considered the monomer in that case, and may be saturated (no double bonds in the carbon chain) or unsaturated (one or more double bonds in the carbon chain).

Most lipids have some polar character in addition to being largely nonpolar. In general, the bulk of their structure is nonpolar or hydrophobic ("water-fearing"), meaning that it does not interact well with polar solvents like water. Another part of their structure is polar or hydrophilic ("water-loving") and will tend to associate with polar solvents like water. This makes them amphiphilic molecules (having both hydrophobic and hydrophilic portions). In the case of cholesterol, the polar group is a mere -OH (hydroxyl or alcohol). In the case of phospholipids, the polar groups are considerably larger and more polar, as described below.

Lipids are an integral part of our daily diet. Most oils and milk products that we use for cooking and eating like butter, cheese, ghee etc., are composed of fats. Vegetable oils are rich in various polyunsaturated fatty acids (PUFA). Lipid-containing foods undergo digestion within the body and are broken into fatty acids and glycerol, which are the final degradation products of fats and lipids. Lipids, especially phospholipids, are also used in various pharmaceutical products, either as co-solubilisers (e.g., in parenteral infusions) or else as drug carrier components (e.g., in a liposome or transfersome).

Proteins

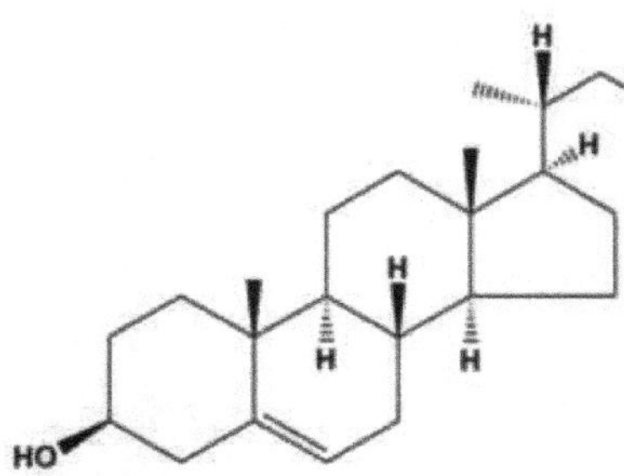

Cholesterol

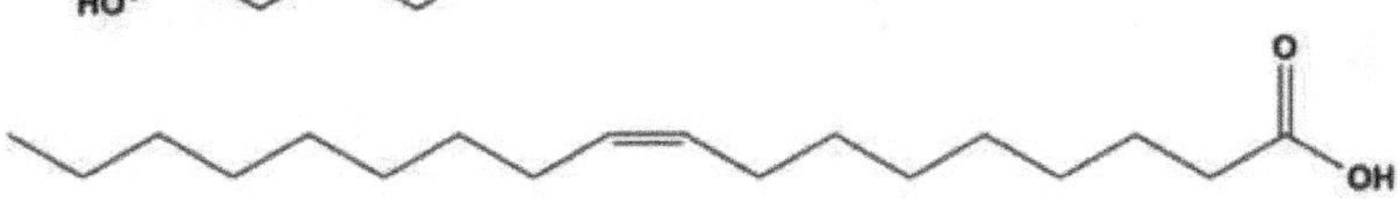

A free fatty acid

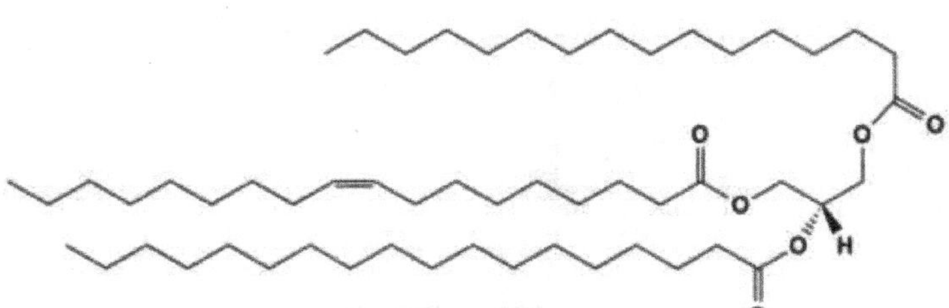

A triglyceride

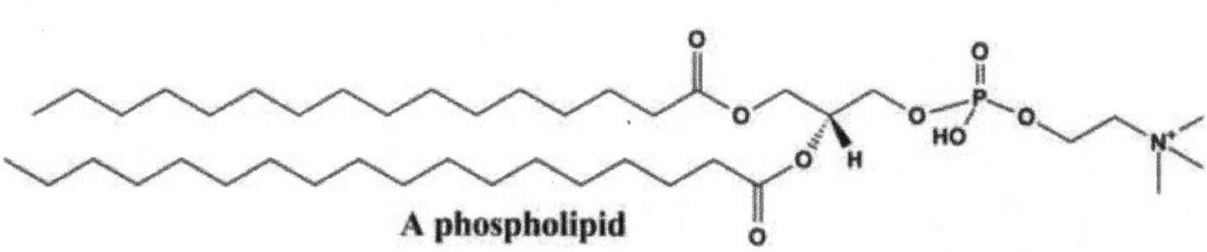

A phospholipid

The general structure of an α-amino acid, with the amino group on the left and the carboxyl group on the right.

Proteins are very large molecules – macro-biopolymers – made from monomers called amino acids. An amino acid consists of a carbon atom bound to four groups. One is an amino group, —NH_2, and one is a carboxylic acid group, —COOH (although these exist as —NH_3^+ and —COO^- under physiologic conditions). The third is a simple hydrogen atom. The fourth is commonly denoted "—R" and is different for each amino acid. There are 20 standard amino acids, each containing a carboxyl group, an amino group, and a side-chain (known as an "R" group). The "R" group is what makes each amino acid different, and the properties of the side-chains greatly influence the overall three-dimensional conformation of a protein. Some amino acids have functions by themselves or in a modified form; for instance, glutamate functions as an important neurotransmitter. Amino acids can be joined via a peptide bond. In this dehydration synthesis, a water molecule is removed and the peptide bond connects the nitrogen of one amino acid's amino group to the carbon of the other's carboxylic acid group. The resulting molecule is called a *dipeptide*, and short stretches of amino acids (usually, fewer than thirty) are called *peptides* or polypeptides. Longer stretches merit the title *proteins*. As an example, the important blood serum protein albumin contains 585 amino acid residues.

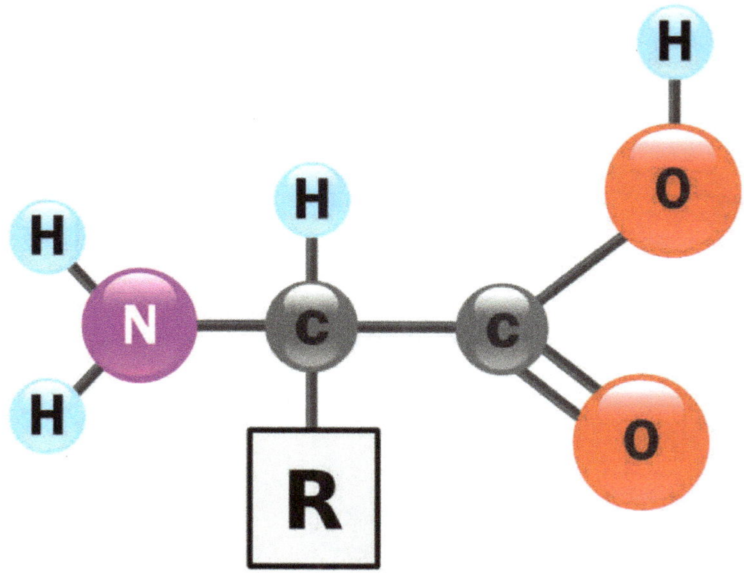

Generic amino acids (1) in neutral form, (2) as they exist physiologically, and (3) joined together as a dipeptide.

Some proteins perform largely structural roles. For instance, movements of the proteins actin and myosin ultimately are responsible for the contraction of skeletal muscle. One property many proteins have is that they specifically bind to a certain molecule or class of molecules—they may be *extremely* selective in what they bind. Antibodies are an example of proteins that attach to one specific type of molecule. In fact, the enzyme-linked immunosorbent assay (ELISA), which uses antibodies, is one of the most sensitive tests modern medicine uses to detect various biomolecules. Probably the most important proteins, however, are the enzymes. Virtually every reaction in a living cell requires an enzyme to lower the activation energy of the reaction. These molecules recognize specific reactant molecules called *substrates*; they then catalyze the reaction between them. By lowering the activation energy, the enzyme speeds up that reaction by a rate of 10^{11} or more; a reaction that would normally take over 3,000 years to complete spontaneously might take less than a second with an enzyme. The enzyme itself is not used up in the process, and is free to catalyze the same reaction with a new set of substrates. Using various modifiers, the activity of the enzyme can be regulated, enabling control of the biochemistry of the cell as a whole.

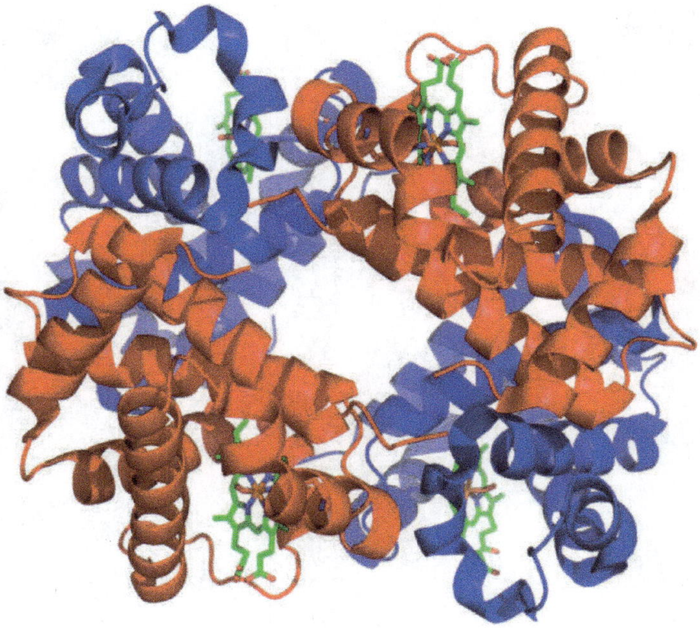

A schematic of hemoglobin. The red and blue ribbons represent the protein globin; the green structures are the heme groups.

The structure of proteins is traditionally described in a hierarchy of four levels. The primary structure of a protein simply consists of its linear sequence of amino acids; for instance, "alanine-glycine-tryptophan-serine-glutamate-asparagine-glycine-lysine-…». Secondary structure is concerned with local morphology (morphology being the study of structure). Some combinations of amino acids will tend to curl up in a coil called an α-helix or into a sheet called a β-sheet; some α-helixes can be seen in the hemoglobin schematic above. Tertiary structure is the entire three-dimensional shape of the protein. This shape is determined by the sequence of amino acids. In fact, a single change can change the entire structure. The alpha chain of hemoglobin contains 146 amino acid residues; substitution of the glutamate residue at position 6 with a valine residue changes the behavior of hemoglobin so much that it results in sickle-cell disease. Finally, quaternary structure is concerned with the structure of a protein with multiple peptide subunits, like hemoglobin with its four subunits. Not all proteins have more than one subunit.

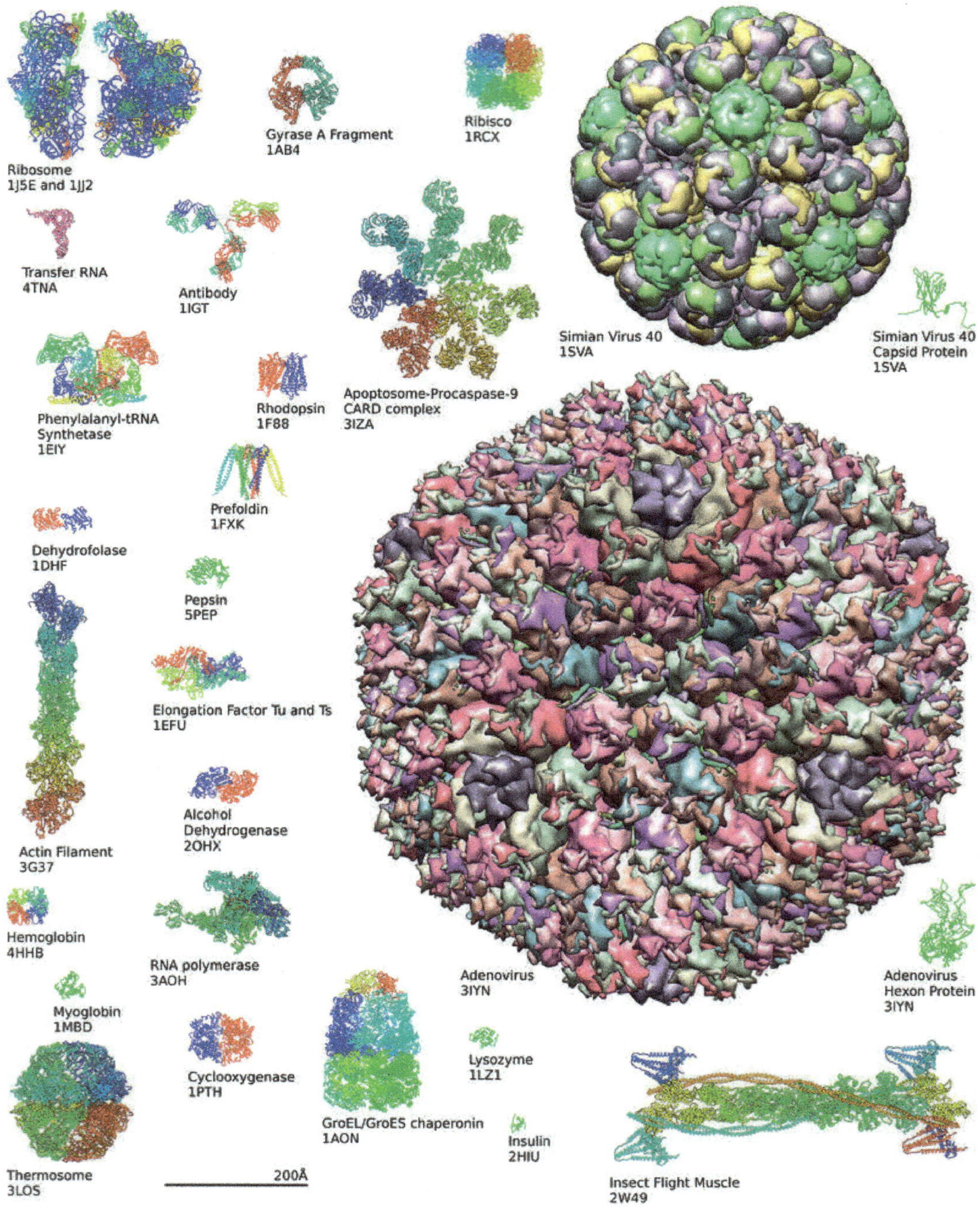

Examples of protein structures from the Protein Data Bank

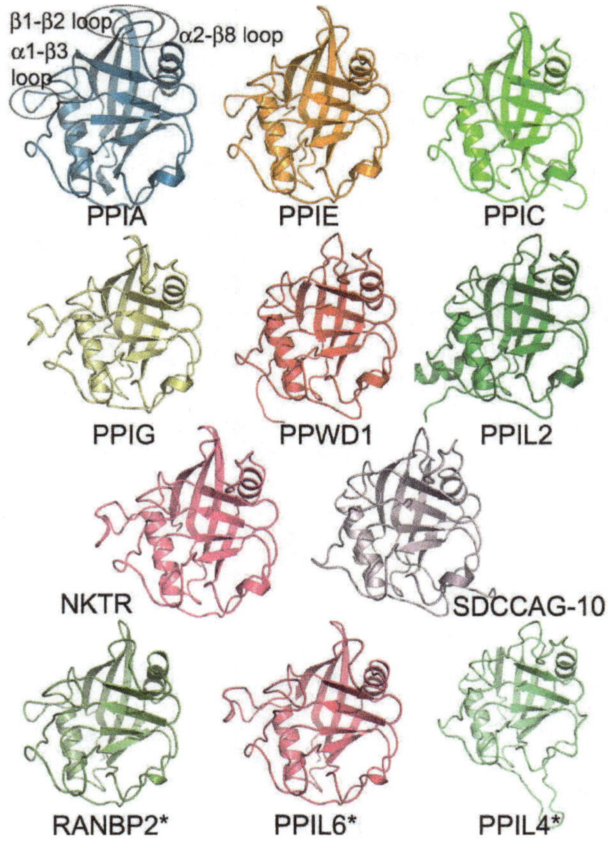

Members of a protein family, as represented by the structures of the isomerase domains.

Ingested proteins are usually broken up into single amino acids or dipeptides in the small intestine, and then absorbed. They can then be joined to make new proteins. Intermediate products of glycolysis, the citric acid cycle, and the pentose phosphate pathway can be used to make all twenty amino acids, and most bacteria and plants possess all the necessary enzymes to synthesize them. Humans and other mammals, however, can synthesize only half of them. They cannot synthesize isoleucine, leucine, lysine, methionine, phenylalanine, threonine, tryptophan, and valine. These are the essential amino acids, since it is essential to ingest them. Mammals do possess the enzymes to synthesize alanine, asparagine, aspartate, cysteine, glutamate, glutamine, glycine, proline, serine, and tyrosine, the nonessential amino acids. While they can synthesize arginine and histidine, they cannot produce it in sufficient amounts for young, growing animals, and so these are often considered essential amino acids.

If the amino group is removed from an amino acid, it leaves behind a carbon skeleton called an α-keto acid. Enzymes called transaminases can easily transfer the amino group from one amino acid (making it an α-keto acid) to another α-keto acid (making it an amino acid). This is important in the biosynthesis of amino acids, as for many of the pathways, intermediates from other biochemical pathways are converted to the α-keto acid skeleton, and then an amino group is added, often via transamination. The amino acids may then be linked together to make a protein.

A similar process is used to break down proteins. It is first hydrolyzed into its component amino acids. Free ammonia (NH_3), existing as the ammonium ion (NH_4^+) in blood, is toxic to life forms. A suitable method for excreting it must therefore exist. Different tactics have evolved in different animals, depending on the animals' needs. Unicellular organisms, of course, simply release the ammonia into the environment. Likewise, bony fish can release the ammonia into the water where it is quickly diluted. In general, mammals convert the ammonia into urea, via the urea cycle.

In order to determine whether two proteins are related, or in other words to decide whether they are homologous or not, scientists use sequence-comparison methods. Methods like sequence alignments and structural alignments are powerful tools that help scientists identify homologies between related molecules. The relevance of finding homologies among proteins goes beyond forming an evolutionary pattern of protein families. By finding how similar two protein sequences are, we acquire knowledge about their structure and therefore their function.

Nucleic Acids

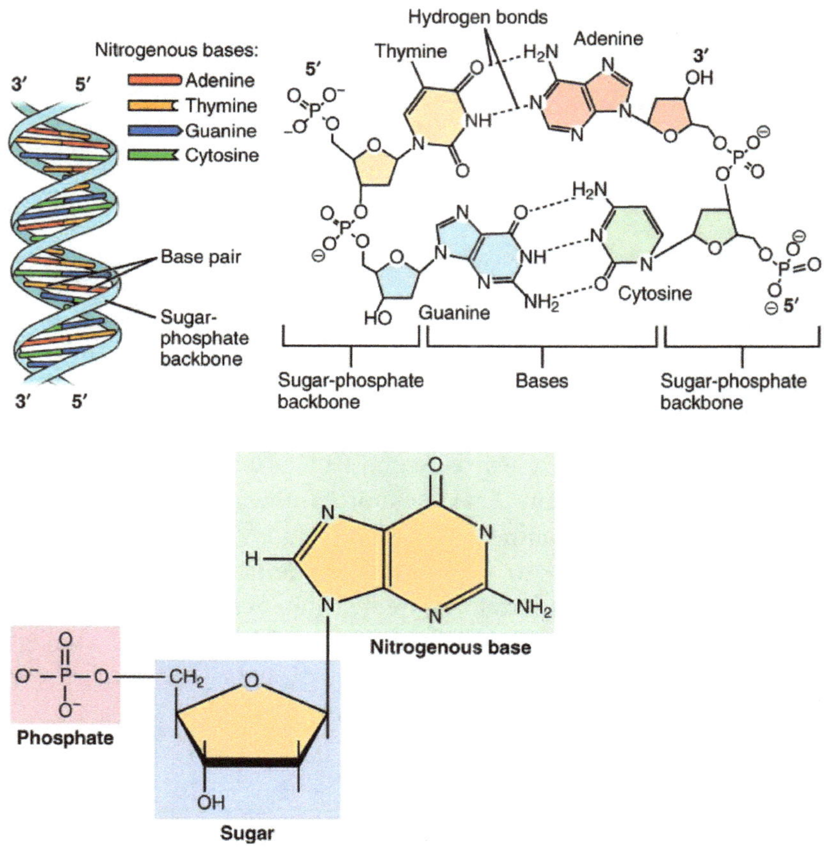

The structure of deoxyribonucleic acid (DNA), the picture shows the monomers being put together.

Nucleic acids, so called because of its prevalence in cellular nuclei, is the generic name of the family of biopolymers. They are complex, high-molecular-weight biochemical macromolecules that can convey genetic information in all living cells and viruses. The monomers are called nucleotides, and each consists of three components: a nitrogenous heterocyclic base (either a purine or a pyrimidine), a pentose sugar, and a phosphate group.

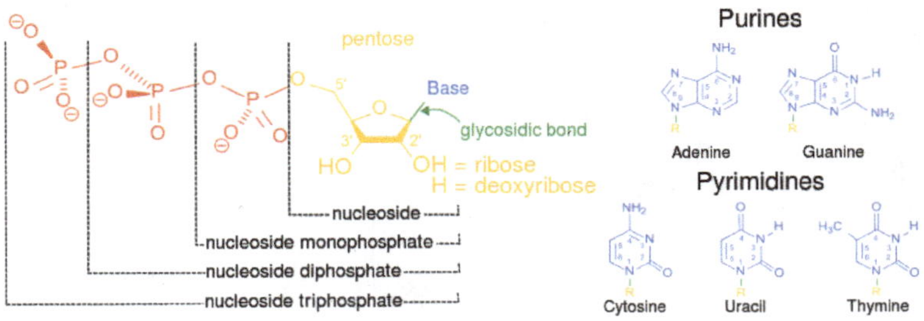

Structural elements of common nucleic acid constituents. Because they contain at least one phosphate group, the compounds marked *nucleoside monophosphate*, *nucleoside diphosphate* and *nucleoside triphosphate* are all nucleotides (not simply phosphate-lacking nucleosides).

The most common nucleic acids are deoxyribonucleic acid (DNA) and ribonucleic acid (RNA). The phosphate group and the sugar of each nucleotide bond with each other to form the backbone of the nucleic acid, while the sequence of nitrogenous bases stores the information. The most common nitrogenous bases are adenine, cytosine, guanine, thymine, and uracil. The nitrogenous bases of each strand of a nucleic acid will form hydrogen bonds with certain other nitrogenous bases in a complementary strand of nucleic acid (similar to a zipper). Adenine binds with thymine and uracil; Thymine binds only with adenine; and cytosine and guanine can bind only with one another.

Aside from the genetic material of the cell, nucleic acids often play a role as second messengers, as well as forming the base molecule for adenosine triphosphate (ATP), the primary energy-carrier molecule found in all living organisms. Also, the nitrogenous bases possible in the two nucleic acids are different: adenine, cytosine, and guanine occur in both RNA and DNA, while thymine occurs only in DNA and uracil occurs in RNA.

METABOLISM

Carbohydrates as Energy Source

Glucose is the major energy source in most life forms. For instance, polysaccharides are broken down into their monomers (glycogen phosphorylase removes glucose residues from glycogen). Disaccharides like lactose or sucrose are cleaved into their two component monosaccharides.

Glycolysis (Anaerobic)

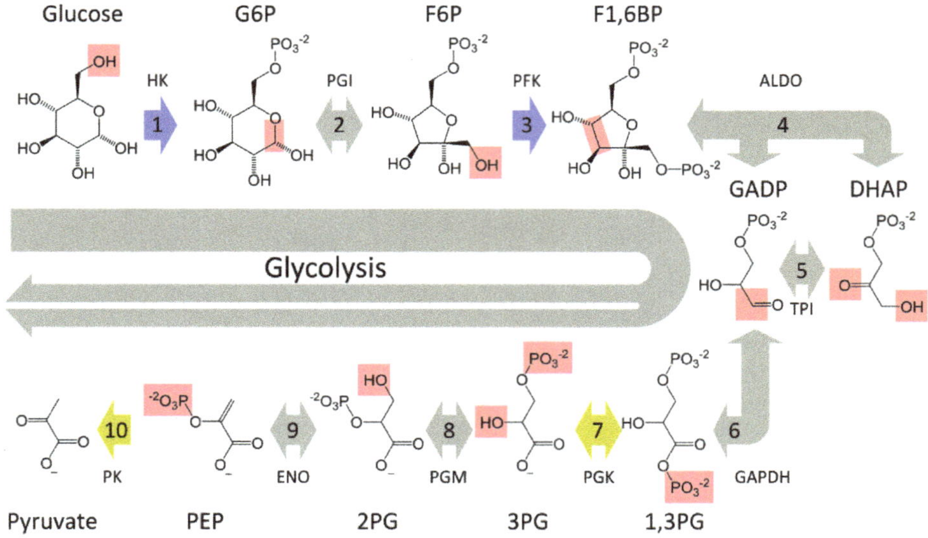

The metabolic pathway of glycolysis converts glucose to pyruvate by via a series of intermediate metabolites. Each chemical modification (red box) is performed by a different enzyme. Steps 1 and 3 consume ATP (blue) and steps 7 and 10 produce ATP (yellow). Since steps 6-10 occur twice per glucose molecule, this leads to a net production of ATP.

Aerobic

In aerobic cells with sufficient oxygen, as in most human cells, the pyruvate is further metabolized. It is irreversibly converted to acetyl-CoA, giving off one carbon atom as the waste product carbon dioxide, generating another reducing equivalent as NADH. The two molecules acetyl-CoA (from one molecule of glucose) then enter the citric acid cycle, producing two more molecules of ATP, six more NADH molecules and two reduced (ubi) quinones (via $FADH_2$ as enzyme-bound cofactor), and releasing the remaining carbon atoms as carbon dioxide. The produced NADH and quinol molecules then feed into the enzyme complexes of the respiratory chain, an electron transport system transferring the electrons ultimately to oxygen and conserving the released energy in the form of a proton gradient over a membrane (inner mitochondrial membrane in eukaryotes). Thus, oxygen is reduced to water and the original electron acceptors NAD^+ and quinone are regenerated. This is why humans breathe in oxygen and breathe out carbon dioxide. The energy released from transferring the electrons from high-energy states in NADH and quinol is conserved first as proton gradient and converted to ATP via ATP synthase. This generates an additional *28* molecules of ATP (24 from the 8 NADH + 4 from the 2 quinols), totaling to 32 molecules of ATP conserved per degraded glucose (two from glycolysis + two from the citrate cycle). It is clear that using oxygen to completely oxidize glucose provides an organism with far more energy than any oxygen-independent metabolic feature, and this is thought to be the reason why complex life appeared only after Earth's atmosphere accumulated large amounts of oxygen.

Gluconeogenesis

In vertebrates, vigorously contracting skeletal muscles (during weightlifting or sprinting, for example) do not receive enough oxygen to meet the energy demand, and so they shift to anaerobic metabolism, converting glucose to lactate. The liver regenerates the glucose, using a process called gluconeogenesis. This process is not quite the opposite of glycolysis, and actually requires three times the amount of energy gained from glycolysis (six molecules of ATP are used, compared to the two gained in glycolysis). Analogous to the above reactions, the glucose produced can then undergo glycolysis in tissues that need energy, be stored as glycogen (or starch in plants), or be converted to other monosaccharides or joined into di- or oligosaccharides. The combined pathways of glycolysis during exercise, lactate's crossing via the bloodstream to the liver, subsequent gluconeogenesis and release of glucose into the bloodstream is called the Cori cycle.

Relationship to Other "Molecular-Scale" Biological Sciences

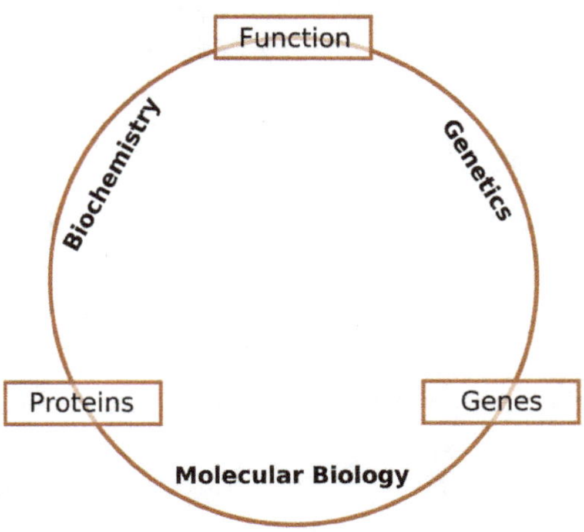

Schematic relationship between biochemistry, genetics, and molecular biology.

Researchers in biochemistry use specific techniques native to biochemistry, but increasingly combine these with techniques and ideas developed in the fields of genetics, molecular biology and biophysics. There has never been a hard-line among these disciplines in terms of content and technique. Today, the terms *molecular biology* and *biochemistry* are nearly interchangeable. The following figure is a schematic that depicts one possible view of the relationship between the fields:

- *Biochemistry* is the study of the chemical substances and vital processes occurring in living organisms. Biochemists focus heavily on the role, function, and structure of biomolecules. The study of the chemistry behind biological processes and the synthesis of biologically active molecules are examples of biochemistry.
- *Genetics* is the study of the effect of genetic differences on organisms. Often this can be inferred by the absence of a normal component (e.g., one gene). The study of "mutants" – organisms with a changed gene that leads to the organism being different with respect to the so-called "wild type" or normal phenotype. Genetic interactions (epistasis) can often confound simple interpretations of such "knock-out" or "knock-in" studies.
- *Molecular biology* is the study of molecular underpinnings of the process of replication, transcription and translation of the genetic material. The central dogma of molecular biology where genetic material is transcribed into RNA and then translated into protein, despite being an oversimplified picture of molecular biology, still provides a good starting point for understanding the field. This picture, however, is undergoing revision in light of emerging novel roles for RNA.
- *Chemical biology* seeks to develop new tools based on small molecules that allow minimal perturbation of biological systems while providing detailed information about their function. Further, chemical biology employs biological systems to create non-natural hybrids between biomolecules and synthetic devices (for example emptied viral capsids that can deliver gene therapy or drug molecules).

MINERAL (NUTRIENT)

A **mineral** is a chemical element required as an essential nutrient by organisms, other than carbon, hydrogen, nitrogen, oxygen and sulfur present in common organic molecules. The remaining elements are classed as *minerals* in the four groups of essential nutrients (the others are vitamins, essential fatty acids, and essential amino acids).

The six major minerals in the human body are calcium, phosphorus, potassium, sodium, chlorine, and magnesium, in order of abundance. Calcium and phosphorus, as skeletal minerals comprising about 1.6 kg of an adult, make up 99% of the mass of the body along with the five non-mineral elements carbon, hydrogen, oxygen, nitrogen, and sulfur. The last four major mineral elements (potassium, sodium, chlorine and magnesium) make up only about 0.85% of the weight of the body. Together these eleven chemical elements make up 99.85% of the body. All of the remaining functional elements in a human body are sometimes known as "trace elements". All of the trace elements together total less than 10 grams, less than the weight of the least common major mineral, magnesium. The trace elements necessary for mammalian life include iron, cobalt, copper, zinc, manganese, molybdenum, iodine, and selenium. Because inorganic mineral content of foods do not form volatile combustion products, nutrition analysis methods involving combustion may report the total mineral content of food as "crude ash".

Over twenty minerals are necessary for mammals, and several more for various other types of life. The total number of chemical elements that are absolutely needed is not known for any organism. Ultratrace elements of some minerals such as silicon and boron are known to have a role but the exact biochemical nature is unknown, and others such as arsenic and chromium are suspected to have a role in health, but with weaker evidence. Bromine has recently been found to have a biochemical role in the body in the synthesis of collagen, but these findings have not yet been confirmed.

Most chemical elements that are ingested by organisms are in the form of simple compounds. Larger chemical compounds of elements need to be broken down for absorption. Plants absorb dissolved elements in soils, which are subsequently picked up by the herbivores that eat them, and the elements move up the food chain. Larger organisms may also consume soil (geophagia) and visit salt licks to obtain limiting minerals they are unable to acquire through other components of their diet.

Bacteria and fungi play an essential role in the weathering of primary elements that results in the release of nutrients for their own nutrition and for the nutrition of others in the ecological food chain. One element, cobalt, is available for use by animals only after having been processed into complex molecules (e.g., vitamin B12) by bacteria. Scientists are only recently starting to appreciate the magnitude and role that microorganisms have in the global cycling and formation of biominerals.

Essential Chemical Elements for Humans

At least twenty chemical elements are known to be *required* to support human biochemical processes by serving structural and functional roles as well as electrolytes. However, as many as twenty-nine elements in total (including the common hydrogen, carbon, nitrogen and oxygen) are suggested to be used by mammals, as a result of studies of biochemical, special uptake, and metabolic handling studies. Many of these additional elements have no well-defined biochemical function known at present. Most of the known and suggested mineral nutrients are of relatively low atomic weight, and are reasonably common on land, or, at least, common in the ocean (iodine, sodium):

Nutritional elements in the periodic table

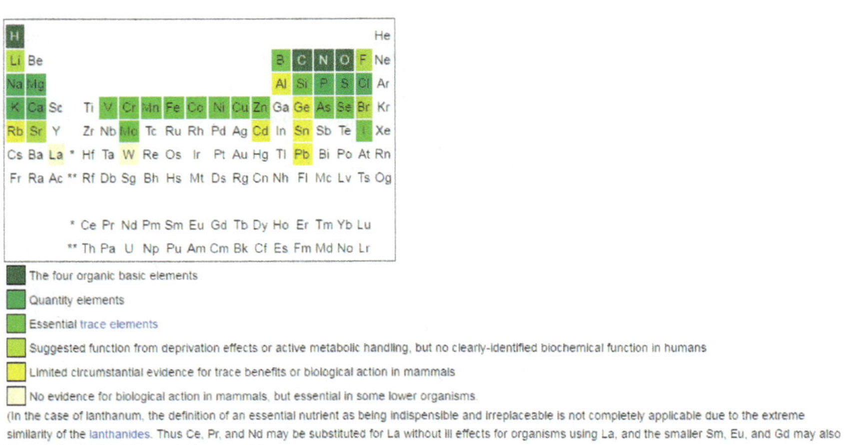

The following play important roles in biological processes:

DV = Daily Value (formerly Recommended Daily Allowance); **UL** = Tolerable Upper Intake Level (figures shown are for adults age 31-50, male or female neither pregnant nor lactating)

Dietary element	DV [mg]	UL [mg]	Amount	Category	High nutrient density dietary sources	Insufficiency	Excess
Potassium	3500	NE	Quantity	A systemic electrolyte and is essential in coregulating ATP with sodium	Sweet potato, tomato, potato, beans, lentils, dairy products, seafood, banana, prune, carrot, orange	hypokalemia	hyperkalemia
Chlorine	3400	3600	Quantity	Needed for production of hydrochloric acid in the stomach and in cellular pump functions	Table salt (sodium chloride) is the main dietary source.	hypochloremia	hyperchloremia
Sodium	2400	2300	Quantity	A systemic electrolyte and is essential in coregulating ATP with potassium	Table salt (sodium chloride, the main source), sea vegetables, milk, and spinach.	hyponatremia	hypernatremia
Calcium	1000	2500	Quantity	Needed for muscle, heart and digestive system health, builds bone, supports synthesis and function of blood cells	Dairy products, eggs, canned fish with bones (salmon, sardines), green leafy vegetables, nuts, seeds, tofu, thyme, oregano, dill, cinnamon.	hypocalcaemia	hypercalcaemia
Phosphorus	1000	4000	Quantity	A component of bones (see apatite), cells, in energy processing, in DNA and ATP (as phosphate) and many other functions	Red meat, dairy foods, fish, poultry, bread, rice, oats.In biological contexts, usually seen as phosphate	hypophosphatemia	hyperphosphatemia
Magnesium	400	350	Quantity	Required for processing ATP and for bones	Spinach, legumes, nuts, seeds, whole grains, peanut butter, avocado	hypomagnesemia, magnesium deficiency	hypermagnesemia
Iron	18	45	Trace	Required for many proteins and enzymes, notably hemoglobin to prevent anemia	Meat, seafood, nuts, beans, dark chocolate	iron deficiency	iron overload disorder
Zinc	15	40	Trace	Pervasive and required for several enzymes such as carboxypeptidase, liver alcohol dehydrogenase, and carbonic anhydrase	Oysters*, red meat, poultry, nuts, whole grains, dairy products	zinc deficiency	zinc toxicity
Manganese	2	350	Trace	A cofactor in enzyme functions	Grains, legumes, seeds, nuts, leafy vegetables, tea, coffee	manganese deficiency	manganism
Copper Copper in health	2	11	Trace	Required component of many redox enzymes, including cytochrome c oxidase	Liver, seafood, oysters, nuts, seeds; some: whole grains, legumes	copper deficiency	copper toxicity
Iodine	0.150	1.1	Trace	Required for synthesis of thyroid hormones, thyroxine and triiodothyronine and to prevent goiter: Iodine in biology	Seaweed (kelp or kombu)*, grains, eggs, iodized salt	iodine deficiency	iodismHyperthyroidism

Dietary element	DV [mg]	UL [mg]	Amount	Category	High nutrient density dietary sources	Insufficiency	Excess
Chromium	0.120	NE	Trace	Involved in glucose and lipid metabolism, although its mechanisms of action in the body and the amounts needed for optimal health are not well-defined	Broccoli, grape juice (especially red), meat, whole grain products	Chromium deficiency	Chromium toxicity
Molybdenum	0.075	2	Trace	The oxidases xanthine oxidase, aldehyde oxidase, and sulfite oxidase	Legumes, whole grains, nuts	molybdenum deficiency	molybdenum toxicity
Selenium	0.070	0.4	Trace	Essential to activity of antioxidant enzymes like glutathione peroxidase	Brazil nuts, seafoods, organ meats, meats, grains, dairy products, eggs	selenium deficiency	selenosis
Cobalt	none	NE	Trace	Required in the synthesis of vitamin B12, but because bacteria are required to synthesize the vitamin, it is usually considered part of vitamin B12			Cobalt poisoning

Blood Concentrations of Minerals

Minerals are present in a healthy human being's blood at certain mass and molar concentrations. The figure below presents the concentrations of each of the chemical elements discussed in this article, from center-right to the right. Depending on the concentrations, some are in upper part of the picture, while others are in the lower part. The figure includes the relative values of other constituents of blood such as hormones. In the figure, minerals are color highlighted in purple.

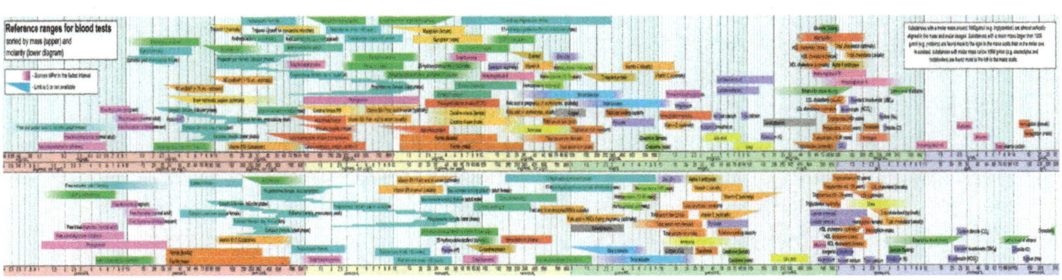

Dietary Nutrition

Dietitians may recommend that minerals are best supplied by ingesting specific foods rich with the chemical element(s) of interest. The elements may be naturally present in the food (e.g., calcium in dairy milk) or added to the food (e.g., orange juice fortified with calcium; iodized salt, salt fortified with iodine). Dietary supplements can be formulated to contain several different chemical elements (as compounds), a combination of vitamins and/or other chemical compounds, or a single element (as a compound or mixture of compounds), such as calcium (as calcium carbonate, calcium citrate, etc.) or magnesium (as magnesium oxide, etc.), chromium (usually as chromium(III) picolinate), or iron (as iron bis-glycinate). The dietary focus on chemical elements derives from an interest in supporting the biochemical reactions of metabolism with the required elemental components. Appropriate intake levels of certain chemical elements have been demonstrated to be required to maintain optimal health. Diet can meet all the body's chemical element requirements, although supplements can be used when some requirements (e.g., calcium, which is found mainly in dairy products) are not adequately met by the diet, or when chronic or acute deficiencies arise from pathology, injury, etc. Research has supported that altering inorganic mineral compounds (carbonates, oxides, etc.) by reacting them with organic ligands (amino acids, organic acids, etc.) improves the bioavailability of the supplemented mineral.

Elements Considered Possibly Essential but not Confirmed

Many ultratrace elements have been suggested as essential, but such claims have usually not been confirmed. Definitive evidence for efficacy comes from the characterization of a biomolecule containing the element with an identifiable and testable function. One problem with identifying efficacy is that some elements are innocuous at low concentrations and are pervasive (examples: silicon and nickel in solid and dust), so proof of efficacy is lacking because deficiencies are difficult to reproduce.

Element	Description	Excess
Bromine	Possibly important to basement membrane architecture and tissue development, as a needed catalyst to make collagen IV.	bromism
Arsenic	Essential in rat, hamster, goat and chicken models, but no biochemical mechanism known in humans.	arsenic poisoning
Nickel	Nickel is an essential component of several enzymes, including urease and hydrogenase. Although not required by humans, some are thought to be required by gut bacteria, such as urease required by some varieties of Bifidobacterium. In humans, nickel may be a cofactor or structural component of certain metalloenzymes involved in hydrolysis, redox rections, and gene expression. Nickel deficiency depressed growth in goats, pigs, and sheep, and diminished circulating thyroid hormone concentration in rats.	Nickel toxicity
Fluorine	Fluorine (as fluoride) is not generally considered an essential element because humans do not require it for growth or to sustain life. However, if one considers the prevention of dental cavities an important criterion in determining essentiality, then fluoride might well be considered an essential trace element. However, recent research indicates that the primary action of fluoride occurs topically (at the surface).	Fluoride poisoning

Boron	Boron is an essential plant nutrient, required primarily for maintaining the integrity of cell walls. Boron has been shown to be essential to complete the life cycle in representatives of all phylogenetic kingdoms, including the model species danio rerio (zebrafish) and Xenopus laevis (African clawed frog). In animals, supplemental boron has been shown to reduce calcium excretion and activate vitamin D.	Nontoxic
Lithium	It is not known whether lithium has a physiological role in any species, but nutritional studies in mammals have indicated its importance to health, leading to a suggestion that it be classed as an essential trace element.	Lithium toxicity
Strontium	Strontium has been found to be involved in the utilization of calcium in the body. It has promoting action on calcium uptake into bone at moderate dietary strontium levels, but a rachitogenic (rickets-producing) action at higher dietary levels.	Rachitogenic (causing Rickets)
Other	Silicon and vanadium have established, albeit specialized, biochemical roles as structural or functional cofactors in other organisms, and are possibly, even probably, used by mammals (including humans). By contrast, tungsten, lanthanum, and cadmium have specialized biochemical uses in certain lower organisms, but these elements appear not to be utilized by humans. Other elements considered to be possibly essential include aluminium, germanium, lead, rubidium, and tin.	Multiple

Mineral Ecology

Recent studies have shown a tight linkage between living organisms and chemical elements on this planet. This led to the redefinition of minerals as "an element or compound, amorphous or crystalline, formed through 'biogeochemical' processes. The addition of 'bio' reflects a greater appreciation, although an incomplete understanding, of the processes of mineral formation by living forms." Biologists and geologists have only recently started to appreciate the magnitude of mineral biogeoengineering. Bacteria have contributed to the formation of minerals for billions of years and critically define the biogeochemical mineral cycles on this planet. Microorganisms can precipitate metals from solution contributing to the formation of ore deposits in addition to their ability to catalyze mineral dissolution, to respire, precipitate, and form minerals.

Most minerals are inorganic in nature. Mineral nutrients refers to the smaller class of minerals that are metabolized for growth, development, and vitality of living organisms. Mineral nutrients are recycled by bacteria that are freely suspended in the vast water columns of the worlds oceans. They absorb dissolved organic matter containing mineral nutrients as they scavenge through the dying individuals that fall out of large phytoplankton blooms. Flagellates are effective bacteriovores and are also commonly found in the marine water column. The flagellates are preyed upon by zooplankton while the phytoplankton concentrates on the larger particulate matter that is suspended in the water column as they are consumed by larger zooplankton, with fish as the top predator. Mineral nutrients cycle through this marine food chain, from bacteria and phytoplankton to flagellates and zooplankton which are then eaten by fish. The bacteria are important in this chain because only they have the physiological ability to absorb the dissolved mineral nutrients from the sea. These recycling principals from marine environments apply to many soil and freshwater ecosystems as well. In terrestrial ecosystems, fungi play similar roles as bacteria: they mobilize nutritional elements composing matter that is inaccessible for other organisms and transport acquired nutrients to nutritionally scarce patches of ecosystem.

REFERENCES

1. "Definition". fig.net. Retrieved 2016-02-17.
2. Johnson, Anthony, Solving Stonehenge: The New Key to an Ancient Enigma. (Thames & Hudson, 2008) ISBN 978-0-500-05155-9
3. Hong-Sen Yan & Marco Ceccarelli (2009), International Symposium on History of Machines and Mechanisms: Proceedings of HMM 2008, Springer, p. 107, ISBN 1-4020-9484-1
4. Lewis, M. J. T. (2001-04-23). Surveying Instruments of Greece and Rome. Cambridge University Press. ISBN 9780521792974. Retrieved 30 August 2012.
5. Turner, Gerard L'E. Nineteenth Century Scientific Instruments, Sotheby Publications, 1983, ISBN 0-85667-170-3
6. Sturman, Brian; Wright, Alan. "The History of the Tellurometer" (PDF). International Federation of Surveyors. Retrieved 20 July 2014.
7. Cheves, Marc. "Geodimeter-The First Name in EDM". Archived from the original on March 10, 2014. Retrieved 2014-07-20.
8. Mahun, Jerry. "Electronic Distance Measurement". Jerrymahun.com. Retrieved 2014-07-20.
9. National Cooperative Highway Research Program: Collecting, Processing and Integrating GPS data into GIS, p. 40. Published by Transportation Research Board, 2002 ISBN 0-309-06916-5 ISBN 978-0-309-06916-8
10. Toni Schenk, Suyoung Seo, Beata Csatho:Accuracy Study of Airborne Laser Scanning Data with Photogrammetry, p. 118 ArchivedMarch 25, 2009, at the Wayback Machine.
11. "View DigitalGlobe Imagery Solutions @ Geospatial Forum".
12. Ronalds, B.F. (2016). Sir Francis Ronalds: Father of the Electric Telegraph. London: Imperial College Press. ISBN 978-1-78326-917-4.
13. Kahmen, Heribert; Faig, Wolfgang (1988). Surveying. Berlin: de Gruyter. p. 9. ISBN 3-11-008303-5. Retrieved 2014-08-10.
14. BC Punmia. Surveying by BC Punmia. p. 2. Retrieved 9 December 2014h.

15. N N Basak. Surveying and Levelling. p. 542. Retrieved 28 July 2016.
16. BC Punmia. Surveying by BC Punmia. p. 2. Retrieved 9 December 2014.
17. A History of the Rectangular Survey System by C. Albert White, 1983, Pub: Washington, D.C. : U.S. Dept. of the Interior, Bureau of Land Management : For sale by Supt. of Docs., U.S. G.P.O.,
18. Richards, D., & Hermansen, K. (1995). Use of extrinsic evidence to aid interpretation of deeds. Journal of Surveying Engineering, (121), 178.
19. Nelson, David L.; Michael M. Cox (2000-02-15). Lehninger Principles of Biochemistry, Third Edition (3 Har/Com ed.). W. H. Freeman. p. 1200. ISBN 1-57259-931-6.
20. Ultratrace minerals. Authors: Nielsen, Forrest H. USDA, ARS Source: Modern nutrition in health and disease / editors, Maurice E. Shils ... et al.. Baltimore : Williams & Wilkins, c1999., p. 283-303. Issue Date: 1999 URI:
21. U.S. Food and Drug Administration 14. Appendix F(mg)
22. Dietary Reference Intakes (DRIs): Elements Food and Nutrition Board, Institute of Medicine, National Academies (2011)(mg)
23. Dietary Reference Intakes : Electrolytes and Water The National Academies (2004)
24. "Dietary Guidelines for Americans 2005: Appendix B-1. Food Sources of Potassium". United States Department of Agriculture. 2005.
25. Adam Drewnowski (2010). "The Nutrient Rich Foods Index helps to identify healthy, affordable foods" (PDF). The American Journal of Clinical Nutrition. 91(suppl): 1095S–1101S.
26. "NHS Choices:Vitamins and minerals – Others". Retrieved November 8, 2011.
27. Corbridge, D. E. C. (1995-02-01). Phosphorus: An Outline of Its Chemistry, Biochemistry, and Technology (5th ed.). Amsterdam: Elsevier Science Pub Co. p. 1220. ISBN 0-444-89307-5.
28. "Linus Pauling Institute at Oregon State University". Retrieved 2008-11-29.
29. "Magnesium—Fact Sheet for Health Professionals". National Institutes of Health. 2016.
30. "Iron—Dietary Supplement Fact Sheet". National Institutes of Health. 2016.
31. "Zinc—Fact Sheet for Health Professionals". National Institutes of Health. 2016.
32. Schlenker, Eleanor; Gilbert, Joyce Ann (28 August 2014). Williams' Essentials of Nutrition and Diet Therapy. Elsevier Health Sciences. pp. 162–3. ISBN 978-0-323-29401-0. Retrieved 15 July 2016.
33. "Iodine—Fact Sheet for Health Professionals". National Institutes of Health. 2016.
34. Jameson, J. Larry; De Groot, Leslie J. (25 February 2015). Endocrinology: Adult and Pediatric. Elsevier Health Sciences. p. 1510. ISBN 978-0-323-32195-2. Retrieved 14 July 2016.
35. Kim, Myoung Jin; Anderson, John; Mallory, Caroline (1 February 2014). Human Nutrition. Jones & Bartlett Publishers. p. 241. ISBN 978-1-4496-4742-1. Retrieved

10 July 2016.

36. Gropper, Sareen S.; Smith, Jack L. (1 June 2012). Advanced Nutrition and Human Metabolism. Cengage Learning. pp. 527–8. ISBN 1-133-10405-3. Retrieved 10 July 2016.
37. "Chromium". Office of Dietary Supplements, US National Institutes of Health. 2016. Retrieved 10 July 2016.
38. Sardesai VM (December 1993). "Molybdenum: an essential trace element". Nutr Clin Pract. 8 (6): 277–81. doi:10.1177/011542659300800627 7. PMID 8302261.
39. Momcilović, B. (September 1999). "A case report of acute human molybdenum toxicity from a dietary molybdenum supplement—a new member of the "Lucor metallicum" family.". Archives of Industrial Hygiene and Toxicology. De Gruyter. 50 (3): 289–97. PMID 10649845.
40. "Selenium—Fact Sheet for Health Professionals". National Institutes of Health. 2016.
41. Lippard, Stephen J.; Jeremy M. Berg (1994). Principles of Bioinorganic Chemistry. Mill Valley, CA: University Science Books. p. 411. ISBN 0-935702-72-5.
42. Ashmead, H. DeWayne (1993). The Roles of Amino Acid Chelates in Animal Nutrition. Westwood: Noyes Publications.
43. "USDA Table of Nutrient Retention Factors, Release 6" (PDF). USDA. USDA. Dec 2007.
44. A. Scott McCall; Christopher F. Cummings; Gautam Bhave; Roberto Vanacore; Andrea Page-McCaw; Billy G. Hudson (5 June 2014). "Bromine Is an Essential Trace Element for Assembly of Collagen IV Scaffolds in Tissue Development and Architecture". Cell. 157 (6): 1380–1392. doi:10.1016/j.cell.2014.05.009. PMID 24906154.
45. Anke M. Arsenic. In: Mertz W. ed., Trace elements in human and Animal Nutrition, 5th ed. Orlando, FL: Academic Press, 1986, 347–372; Uthus E.O., Evidency for arsenical essentiality, Environ. Geochem. Health, 1992, 14:54–56; Uthus E.O., Arsenic essentiality and factors affecting its importance. In: Chappell W.R, Abernathy C.O, Cothern C.R. eds., Arsenic Exposure and Health. Northwood, UK: Science and Technology Letters, 1994, 199–208.
46. Berdanier, Carolyn D.; Dwyer, Johanna T.; Heber, David (19 April 2016). Handbook of Nutrition and Food, Third Edition. CRC Press. pp. 211–26. ISBN 978-1-4665-0572-8. Retrieved 3 July 2016.
47. Sigel, Astrid; Sigel, Helmut; Sigel, Roland K. O. (27 January 2014). Interrelations between Essential Metal Ions and Human Diseases. Springer Science & Business Media. p. 349. ISBN 978-94-007-7500-8. Retrieved 4 July 2016.
48. Institute of Medicine (29 September 2006). Dietary Reference Intakes: The Essential Guide to Nutrient Requirements. National Academies Press. pp. 313–19, 415–22. ISBN 978-0-309-15742-1. Retrieved 21 June 2016.
49. Cerklewski FL (May 1998). "Fluoride—essential or just beneficial". Nutrition.

14 (5): 475–6. doi:10.1016/S0899-9007(98)00023-9. PMID 9614319.

50. "Linus Pauling Institute at Oregon State University". Retrieved 2008-11-29.
51. Mahler, R. L. "Essential Plant Micronutrients. Boron in Idaho" (PDF). University of Idaho. Archived from the original (PDF) on 1 October 2009. Retrieved 2009-05-05.
52. "Functions of Boron in Plant Nutrition" (PDF). U.S. Borax Inc. Archived from the original (PDF) on 20 March 2009.
53. Blevins, Dale G.; Lukaszewski, KM (1998). "Functions of Boron in Plant Nutrition". Annual Review of Plant Physiology and Plant Molecular Biology. 49 (1): 481–500. doi:10.1146/annurev.arplant.49.1.481. PMID 15012243.
54. Erdman, John W., Jr.; MacDonald, Ian A.; Zeisel, Steven H. (30 May 2012). Present Knowledge in Nutrition. John Wiley & Sons. p. 1324. ISBN 978-0-470-96310-4. Retrieved 4 July 2016.
55. Nielsen, Forrest H. (1997). "Boron in human and animal nutrition". Plant and Soil. 193 (2): 199–208. doi:10.1023/A:1004276311956. ISSN 0032-079X.
56. "Some Facts about Lithium". ENC Labs. Retrieved 2010-10-15.
57. "The biological role of strontium". Retrieved 2010-10-06.
58. Gottschlich, Michele M. (2001). The Science and Practice of Nutrition Support: A Case-based Core Curriculum. Kendall Hunt. p. 98. ISBN 978-0-7872-7680-5. Retrieved 9 July 2016.
59. Insel, Paul M.; Turner, R. Elaine; Ross, Don (2004). Nutrition. Jones & Bartlett Learning. p. 499. ISBN 978-0-7637-0765-1. Retrieved 10 July 2016.
60. Skinner, H. C. W. (2005). "Biominerals". Mineralogical Magazine. 69 (5): 621–641. doi:10.1180/0026461056950275.
61. Newman, D. K.; Banfield, J. F. (2002). "Geomicrobiology: How Molecular-Scale Interactions Underpin Biogeochemical Systems". Science. 296 (5570): 1071–7. doi:10.1126/science.1010716. PMID 12004119.
62. Warren, L. A.; Kauffman, M. E. (2003). "Microbial geoengineers". Science. 299 (5609): 1027–9. doi:10.1126/science.1072076. JSTOR 3833546. PMID 12586932.
63. González-Muñoz, M. T.; Rodriguez-Navarro, C.; Martinez-Ruiz, F.; Arias, J. M.; Merroun, M. L.; Rodriguez-Gallego, M. (2010). "Bacterial biomineralization: new insights from Myxococcus-induced mineral precipitation". Geological Society, London, Special Publications. 336 (1): 31–50. doi:10.1144/SP336.3.
64. Kirkby, H.; Kirkby, E. A.; Cakmak, I. (1996). "Effect of mineral nutritional status on shoot-root partitioning of photoassimilates and cycling of mineral nutrients" (PDF). Journal of Experimental Biology. 47 (S1255): 1255. doi:10.1093/jxb/47.Special_Issue.1255.
65. Adame, L. (2002). "Leaf absorption of mineral nutrients in carnivorous plants stimulates root nutrient uptake" (PDF). New Phytologist. 155: 89–100. doi:10.1046/j.1469-8137.2002.00441.x.
66. Azam, F.; Fenchel, T.; Field, J. G.; Gray, J. S.; Meyer-Reil, L. A.; Thingstad,

F. (1983). "The ecological role of water-column microbes in the sea" (PDF). Mar. Ecol. Prog. Ser. 10: 257–263. doi:10.3354/meps010257.

67. Uroz, S.; Calvaruso, C.; Turpault, M.; Frey-Klett, Pascale (2009). "Mineral weathering by bacteria: ecology, actors and mechanisms"(PDF). Trends in Microbiology. 17 (8): 378–87. doi:10.1016/j.tim.2009.05.004. PMID 19660952.

68. J. Dighton (2007). "Nutrient Cycling by Saprotrophic Fungi in Terrestrial Habitats". In Kubicek, Christian P.; Druzhinina, Irina S. Environmental and microbial relationships (2nd ed. ed.). Berlin: Springer. pp. 287–300. ISBN 978-3-540-71840-6.

CHAPTER 2

Amino Acids: Structure and Functions

Chapter Contents

Amino acids are the building blocks of proteins. They band together in chains to form the stuff from which life is born. This is a two-step process: first, they get together and form peptides or polypeptides, and it is from these groupings that proteins are made.

INTRODUCTION

Amino acids are biologically important organic compounds containing amine (-NH_2) and carboxyl (-COOH) functional groups, along with a side-chain (R group) specific to each amino acid. The key elements of an amino acid are carbon, hydrogen, oxygen, and nitrogen, though other elements are found in the side-chains of certain amino acids. About 500 amino acids are known (though only 20 appear in the genetic code) and can be classified in many ways. They can be classified according to the core structural functional groups' locations as alpha- (α-), beta- (β-), gamma- (γ-) or delta- (δ-) amino acids; other categories relate to polarity, pH level, and side-chain group type (aliphatic, acyclic, aromatic, containing hydroxyl or sulfur, etc.). In the form of proteins, amino acids comprise the second-largest component (water is the largest) of human muscles, cells and other tissues. Outside proteins, amino acids perform critical roles in processes such as neurotransmitter transport and biosynthesis.

In biochemistry, amino acids having both the amine and the carboxylic acid groups attached to the first (alpha-) carbon atom have particular importance. They are known as 2-, alpha-, or **α-amino acids** (generic formula $H_2NCHRCOOH$ in most cases, where R is an organicsubstituent known as a "side-chain"); often the term "amino acid" is used to refer specifically to these. They include the 23 proteinogenic ("protein-building") amino acids, which combine into peptide chains ("polypeptides") to form the building-blocks of a vast array of proteins. These are all L-stereoisomers ("left-handed" isomers), although a few D-amino acids ("right-handed") occur in bacterial envelopes, as a neuromodulator (D-serine), and in some antibiotics.Twenty of the proteinogenic amino acids are encoded directly by triplet codons in the genetic code and are known as "standard" amino acids. The other three ("non-standard" or "non-canonical") are selenocysteine (present in many noneukaryotes as well as most eukaryotes, but not coded directly by DNA), pyrrolysine (found only in some archea and one bacterium) and *N*-formylmethionine (which is often the initial amino acid of proteins in bacteria, mitochondria, and chloroplasts). Pyrrolysine and selenocysteine are encoded via variant codons; for example, selenocysteine is encoded by stop codon and SECIS element.Codon–tRNA combinations not found in nature can also be used to "expand" the genetic code and create novel proteins known as alloproteins incorporating non-proteinogenic amino acids.

Many important proteinogenic and non-proteinogenic amino acids also play critical non-protein roles within the body. For example, in the human brain, glutamate (standard glutamic acid) and gamma-amino-butyric acid ("GABA", non-standard gamma-amino acid) are, respectively, the main excitatory and inhibitory neurotransmitters; hydroxyproline (a major component of the connective tissue collagen) is synthesised from proline; the standard amino acid glycine is used to synthesise porphyrins used in red blood cells; and the non-standard carnitine is used in lipid transport.

Nine proteinogenic amino acids are called "essential" for humans because they cannot be created from other compounds by the human body and so must be taken in as food. Others may be conditionally essential for certain ages or medical conditions. Essential amino acids may also differ between species.

Because of their biological significance, amino acids are important in nutrition and are commonly used in nutritional supplements, fertilizers, and food technology. Industrial uses include the production of drugs, biodegradable plastics, and chiral catalysts.

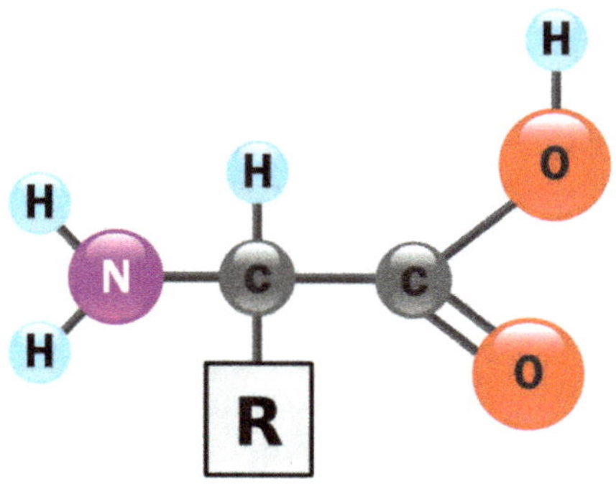

The structure of an alpha amino acid in its un-ionized form

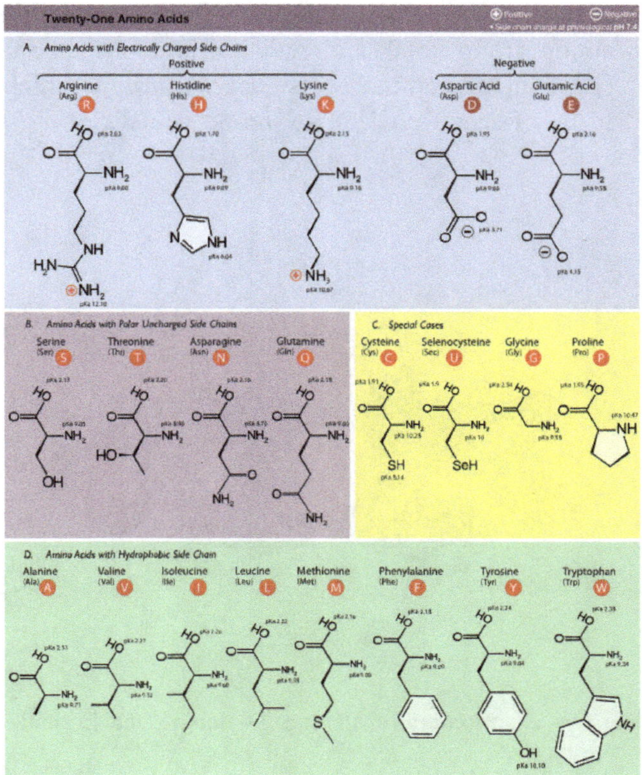

The 21 proteinogenic α-amino acids found in eukaryotes, grouped according to their side-chains' pK_a values and charges carried at physiological pH 7.4

The first few amino acids were discovered in the early 19th century. In 1806, French chemists Louis-Nicolas Vauquelin and Pierre Jean Robiquet isolated a compound in asparagus that was subsequently named asparagine, the first amino acid to be discovered. Cystine was discovered in 1810, although its monomer, cysteine, remained undiscovered until 1884. Glycine and leucine were discovered in 1820. The last of the 20 common amino acids to be discovered was threonine in 1935 by William Cumming Rose, who also determined the essential amino acids and established the minimum daily requirements of all amino acids for optimal growth.

Usage of the term *amino acid* in the English language is from 1898. Proteins were found to yield amino acids after enzymatic digestion or acid hydrolysis. In 1902, Emil Fischer and Franz Hofmeister proposed that proteins are the result of the formation of bonds between the amino group of one amino acid with the carboxyl group of another, in a linear structure that Fischer termed "peptide".

GENERAL STRUCTURE

In the structure shown at the top of the page, **R** represents a side-chain specific to each amino acid. The carbon atom next to the carboxyl group (which is therefore numbered

2 in the carbon chain starting from that functional group) is called the α–carbon. Amino acids containing an amino group bonded directly to the alpha carbon are referred to as *alpha amino acids*. These include amino acids such as proline which contain secondary amines, which used to be often referred to as "imino acids".

Isomerism

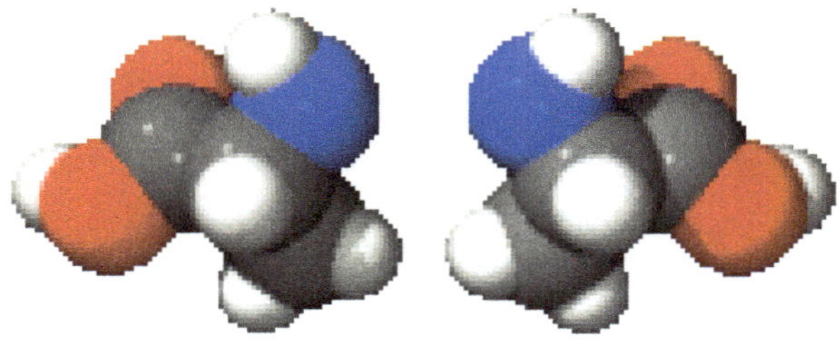

The two enantiomers of alanine, D-alanine and L-alanine

The alpha amino acids are the most common form found in nature, but only when occurring in the L-isomer. The alpha carbon is a chiral carbon atom, with the exception of glycine which has two indistinguishable hydrogen atoms on the alpha carbon. Therefore, all alpha amino acids but glycine can exist in either of two enantiomers, called L or D amino acids, which are mirror images of each other. While L-amino acids represent all of the amino acids found in proteins during translation in the ribosome, D-amino acids are found in some proteins produced by enzyme posttranslational modifications after translation and translocation to the endoplasmic reticulum, as in exotic sea-dwelling organisms such as cone snails. They are also abundant components of the peptidoglycan cell walls of bacteria, and D-serine may act as a neurotransmitter in the brain. D-amino acids are used in racemic crystallography to create centrosymmetric crystals, which (depending on the protein) may allow for easier and more robust protein structure determination. The L and D convention for amino acid configuration refers not to the optical activity of the amino acid itself but rather to the optical activity of the isomer of glyceraldehyde from which that amino acid can, in theory, be synthesized (D-glyceraldehyde is dextrorotatory; L-glyceraldehyde is levorotatory). In alternative fashion, the *(S)* and *(R)* designators are used to indicate the absolute stereochemistry. Almost all of the amino acids in proteins are *(S)* at the α carbon, with cysteine being *(R)* and glycine non-chiral. Cysteine has its side-chain in the same geometric position as the other amino acids, but the *R/S* terminology is reversed because of the higher atomic number of sulfur compared to the carboxyl oxygen gives the side-chain a higher priority, whereas the atoms in most other side-chains give them lower priority.

Side Chains

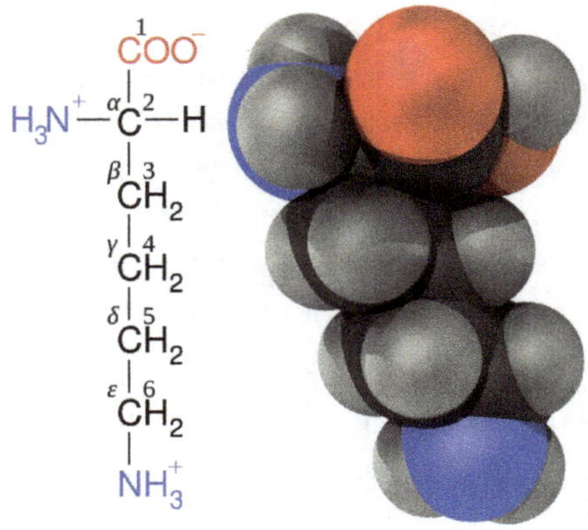

Lysine with carbon atoms labeled

In amino acids that have a carbon chain attached to the α–carbon (such as lysine, shown to the right) the carbons are labeled in order as α, β, γ, δ, and so on. In some amino acids, the amine group is attached to the β or γ-carbon, and these are therefore referred to as *beta* or *gamma amino acids*.

Amino acids are usually classified by the properties of their side-chain into four groups. The side-chain can make an amino acid a weak acid or a weak base, and a hydrophile if the side-chain is polar or a hydrophobe if it is nonpolar. The chemical structures of the 22 standard amino acids, along with their chemical properties, are described more fully in the article on these proteinogenic amino acids.

The phrase "branched-chain amino acids" or BCAA refers to the amino acids having aliphatic side-chains that are non-linear; these are leucine, isoleucine, and valine. Proline is the only proteinogenic amino acid whose side-group links to the α-amino group and, thus, is also the only proteinogenic amino acid containing a secondary amine at this position. In chemical terms, proline is, therefore, an imino acid, since it lacks a primary amino group, although it is still classed as an amino acid in the current biochemical nomenclature, and may also be called an "N-alkylated alpha-amino acid".

Zwitterions

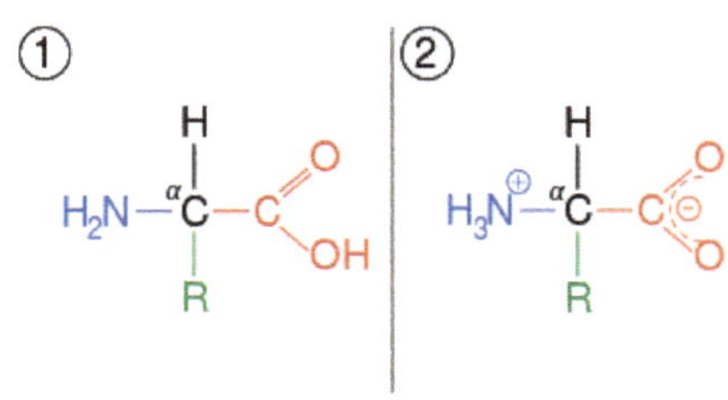

An amino acid in its (1) un-ionized and (2) zwitterionic forms

The α-carboxylic acid group of amino acids is a weak acid, meaning that it releases a hydron (such as a proton) at moderate pH values. In other words, carboxylic acid groups ($-CO_2H$) can be deprotonated to become negative carboxylates ($-CO_2^-$). The negatively charged carboxylate ion predominates at pH values greater than the pKa of the carboxylic acid group (mean for the 20 common amino acids is about 2.2, see the table of amino acid structures above). In a complementary fashion, the α-amine of amino acids is a weak base, meaning that it accepts a proton at moderate pH values. In other words, α-amino groups (NH_2-) can be protonated to become positive α-ammonium groups ($^+NH_3-$). The positively charged α-ammonium group predominates at pH values less than the pKa of the α-ammonium group (mean for the 20 common α-amino acids is about 9.4).

Because all amino acids contain amine and carboxylic acid functional groups, they share amphiprotic properties. Below pH 2.2, the predominant form will have a neutral carboxylic acid group and a positive α-ammonium ion (net charge +1), and above pH 9.4, a negative carboxylate and neutral α-amino group (net charge −1). But at pH between 2.2 and 9.4, an amino acid usually contains both a negative carboxylate and a positive α-ammonium group, as shown in structure (2) on the right, so has net zero charge. This molecular state is known as a zwitterion, from the German **Zwitter** meaning *hermaphrodite* or *hybrid*. The fully neutral form (structure (1) on the right) is a very minor species in aqueous solution throughout the pH range (less than 1 part in 10^7). Amino acids exist as zwitterions also in the solid phase, and crystallize with salt-like properties unlike typical organic acids or amines.

Isoelectric Point

The variation in titration curves when the amino acids are grouped by category can be seen here. With the exception of tyrosine, using titration to differentiate between hydrophobic amino acids is problematic.

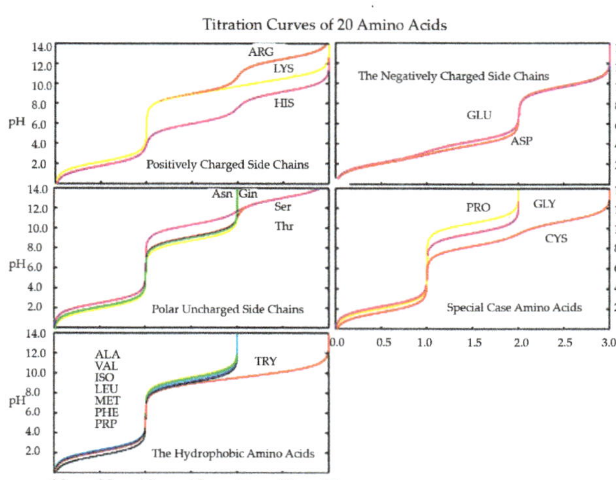

Composite of titration curves of twenty proteinogenic amino acids grouped by side chain category.

At pH values between the two pKa values, the zwitterion predominates, but coexists in dynamic equilibrium with small amounts of net negative and net positive ions. At the exact midpoint between the two pKa values, the trace amount of net negative and trace of net positive ions exactly balance, so that average net charge of all forms present is zero. This pH is known as the isoelectric point pI, so pI = $\frac{1}{2}(pKa_1 + pKa_2)$. The individual amino acids all have slightly different pKa values, so have different isoelectric points. For amino acids with charged side-chains, the pKa of the side-chain is involved. Thus for Asp, Glu with negative side-chains, pI = $\frac{1}{2}(pKa_1 + pKa_R)$, where pKa_R is the side-chain pKa. Cysteine also has potentially negative side-chain with $pKa_R = 8.14$, so pI should be calculated as for Asp and Glu, even though the side-chain is not significantly charged at neutral pH. For His, Lys, and Arg with positive side-chains, pI = $\frac{1}{2}(pKa_R + pKa_2)$. Amino acids have zero mobility in electrophoresis at their isoelectric point, although this behaviour is more usually exploited for peptides and proteins than single amino acids. Zwitterions have minimum solubility at their isoelectric point and some amino acids (in particular, with non-polar side-chains) can be isolated by precipitation from water by adjusting the pH to the required isoelectric point.

OCCURRENCE AND FUNCTIONS IN BIOCHEMISTRY

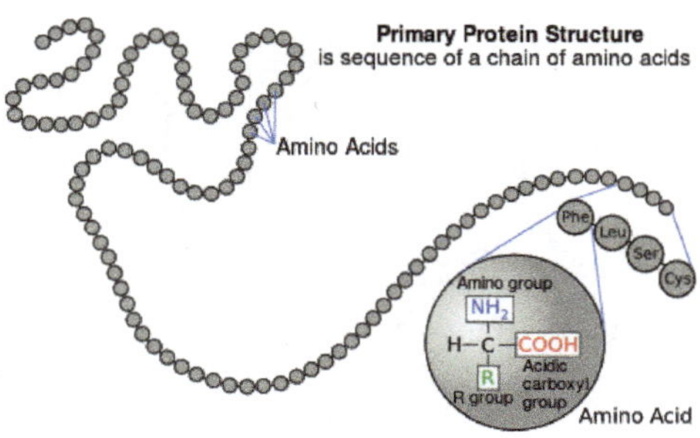

A polypeptide is an unbranched chain of amino acids.

Proteinogenic Amino Acids

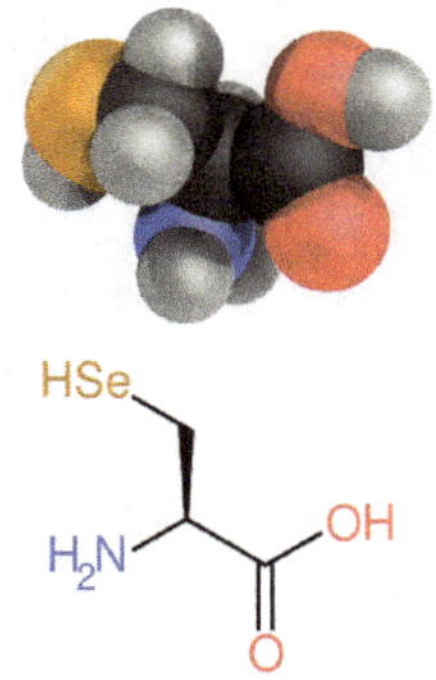

The amino acid selenocysteine

Amino acids are the structural units (monomers) that make up proteins. They join together to form short polymer chains called peptides or longer chains called either polypeptides or proteins. These polymers are linear and unbranched, with each amino acid within the chain attached to two neighboring amino acids. The process of making proteins is called *translation* and involves the step-by-step addition of amino acids to a growing protein chain by a ribozyme that is called a ribosome. The order in which the amino acids are added is read through the genetic code from an mRNA template, which is a RNA copy of one of the organism's genes.

Twenty-two amino acids are naturally incorporated into polypeptides and are called proteinogenic or natural amino acids. Of these, 20 are encoded by the universal genetic code. The remaining 2, selenocysteine and pyrrolysine, are incorporated into proteins by unique synthetic mechanisms. Selenocysteine is incorporated when the mRNA being translated includes a SECIS element, which causes the UGA codon to encode selenocysteine instead of a stop codon. Pyrrolysine is used by some methanogenic archaea in enzymes that they use to produce methane. It is coded for with the codon UAG, which is normally a stop codon in other organisms. This UAG codon is followed by a PYLIS downstream sequence.

Non-Proteinogenic Amino Acids

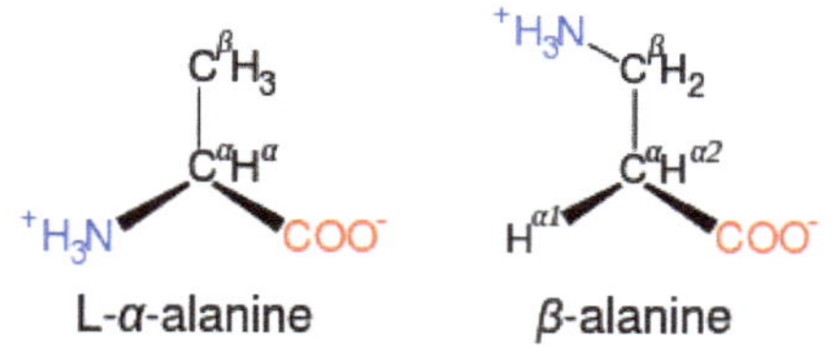

β-alanine and its α-alanine isomer

Aside from the 22 proteinogenic amino acids, there are many other amino acids that are called *non-proteinogenic*. Those either are not found in proteins (for example carnitine, GABA) or are not produced directly and in isolation by standard cellular machinery (for example, hydroxyproline and selenomethionine).

Non-proteinogenic amino acids that are found in proteins are formed by post-translational modification, which is modification after translation during protein synthesis. These modifications are often essential for the function or regulation of a protein; for example, the carboxylation of glutamate allows for better binding of calcium cations, and the hydroxylation of proline is critical for maintaining connective tissues.Another example is the formation of hypusine in the translation initiation factor EIF5A, through modification of a lysine residue. Such modifications can also determine the localization of the protein, e.g., the addition of long hydrophobic groups can cause a protein to bind to a phospholipid membrane.

Some non-proteinogenic amino acids are not found in proteins. Examples include 2-aminoisobutyric acid and the neurotransmitter gamma-aminobutyric acid. Non-proteinogenic amino acids often occur as intermediates in the metabolic pathways for standard amino acids – for example, ornithine and citrulline occur in the urea cycle, part of amino acid catabolism (see below). A rare exception to the dominance of α-amino acids in biology is the β-amino acid beta alanine (3-aminopropanoic acid), which is used in plants and microorganisms in the synthesis of pantothenic acid (vitamin B_5), a component of coenzyme A.

D-Amino Acid Natural Abundance

d-isomers are uncommon in live organisms. For instance, gramicidin is a polypeptide made up from mixture of d- and l-amino acids. Other compounds containing D-amino acid are tyrocidine and valinomycin. These compounds disrupt bacterial cell walls, particularly in Gram-positive bacteria. Only 837 d-amino acids were found in Swiss-Prot database (187 million amino acids analysed).

Non-Standard Amino Acids

The 20 amino acids that are encoded directly by the codons of the universal genetic code are called *standard* or *canonical* amino acids. The others are called *non-standard* or *non-canonical*. Most of the non-standard amino acids are also non-proteinogenic (i.e. they cannot be used to build proteins), but three of them are proteinogenic, as they can be used to build proteins by exploiting information not encoded in the universal genetic code.

The three non-standard proteinogenic amino acids are selenocysteine (present in many non-eukaryotes as well as most eukaryotes, but not coded directly by DNA), pyrrolysine (found only in some archaea and one bacterium), and *N*-formylmethionine (which is often the initial amino acid of proteins in bacteria, mitochondria, and chloroplasts). For example, 25 human proteins include selenocysteine (Sec) in their primary structure,and the structurally characterized enzymes (selenoenzymes) employ Sec as the catalytic moiety in their active sites. Pyrrolysine and selenocysteine are encoded via variant codons. For example, selenocysteine is encoded by stop codon and SECIS element.

In Human Nutrition

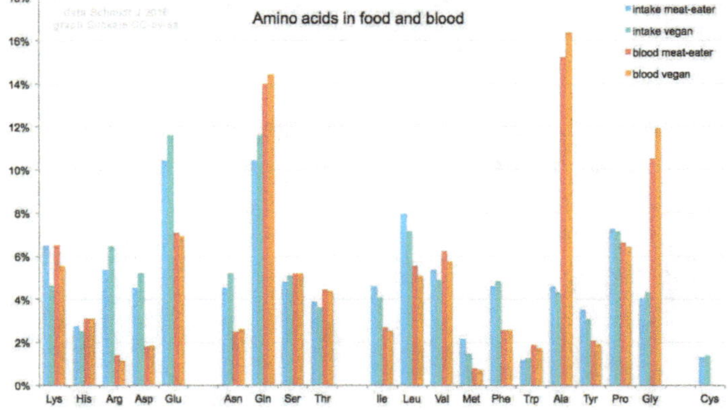

Share of amino acid in different human diets and the resulting mix of amino acids in human blood serum. Glutamate and glutamine are the most frequent in food at over 10%, while alanine, glutamine, and glycine are the most common in blood.

When taken up into the human body from the diet, the 20 standard amino acids either are used to synthesize proteins and other biomolecules or are oxidized to urea and carbon dioxide as a source of energy. The oxidation pathway starts with the removal of the amino group by a transaminase; the amino group is then fed into the urea cycle. The other product of transamidation is a keto acid that enters the citric acid cycle. Glucogenic amino acids can also be converted into glucose, through gluconeogenesis. Of the 20 standard amino acids, nine (His, Ile, Leu, Lys, Met, Phe, Thr, Trp and Val), are called essential amino acids because the human body cannot synthesize them from other compounds at the level needed for normal growth, so they must be obtained from food. In addition, cysteine, taurine, tyrosine, and arginine are considered semiessential amino-acids in children (though taurine is not technically an amino acid), because the metabolic pathways that synthesize these amino acids are not fully developed. The amounts required also depend on the age and health of the individual, so it is hard to make general statements about the dietary requirement for some amino acids. Dietary exposure to the non-standard amino acid BMAA has been linked to human neurodegenerative diseases, including ALS.

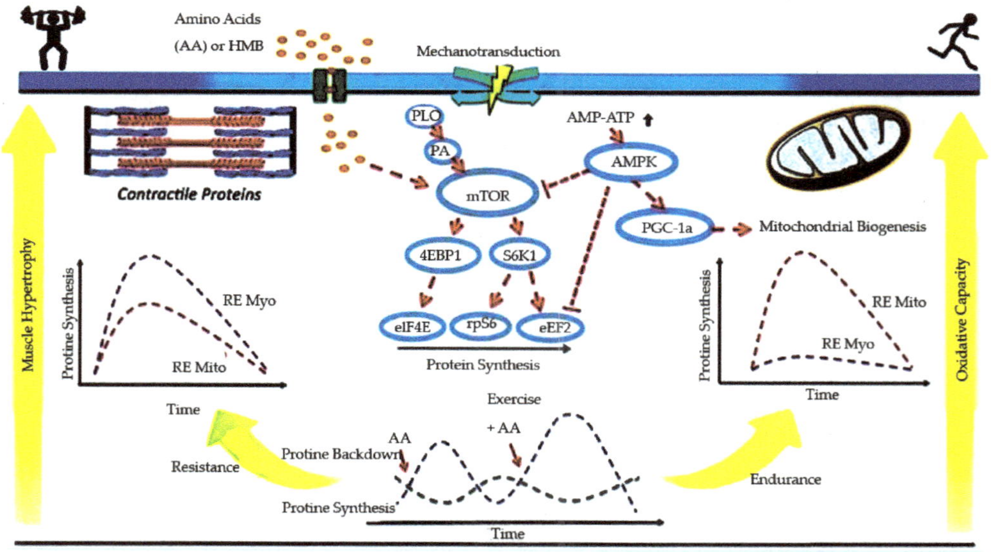

Diagram of the molecular signaling cascades that are involved in myofibrillar muscle protein synthesis and mitochondrial biogenesis in response to physical exercise and specific amino acids or their derivatives (primarily l-leucine and HMB). Many amino acids derived from food protein promote the activation of mTORC1 and increase protein synthesis by signaling through Rag GTPases.

Abbreviations and Representations:

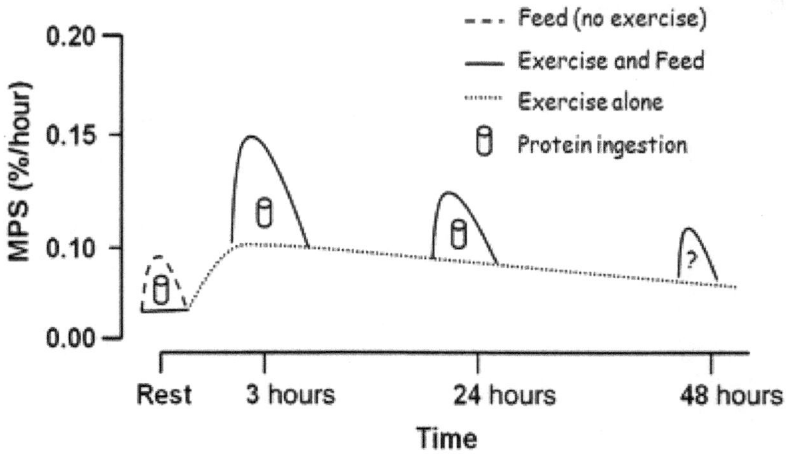

Resistance training stimulates muscle protein synthesis (MPS) for a period of up to 48 hours following exercise (shown by dotted line). Ingestion of a protein-rich meal at any point during this period will augment the exercise-induced increase in muscle protein synthesis (shown by solid lines).

Non-Protein Functions

In humans, non-protein amino acids also have important roles as metabolic intermediates, such as in the biosynthesis of the neurotransmitter gamma-amino-butyric acid (GABA). Many amino acids are used to synthesize other molecules, for example:

- Tryptophan is a precursor of the neurotransmitter serotonin.
- Tyrosine (and its precursor phenylalanine) are precursors of the catecholamineneurotransmitters dopamine, epinephrine and norepinephrine and various trace amines.
- Phenylalanine is a precursor of phenethylamine and tyrosine in humans. In plants, it is a precursor of various phenylpropanoids, which are important in plant metabolism.
- Glycine is a precursor of porphyrins such as heme.
- Arginine is a precursor of nitric oxide.
- Ornithine and S-adenosylmethionine are precursors of polyamines.
- Aspartate, glycine, and glutamine are precursors of nucleotides.

However, not all of the functions of other abundant non-standard amino acids are known.

Some non-standard amino acids are used as defenses against herbivores in plants. For example, canavanine is an analogue of arginine that is found in many legumes, and in particularly large amounts in *Canavalia gladiata* (sword bean). This amino acid protects the plants from predators such as insects and can cause illness in people if some types of legumes are eaten without processing. The non-protein amino acid mimosine is found in other species of legume, in particular *Leucaena leucocephala*. This compound is an analogue of tyrosine and can poison animals that graze on these plants.

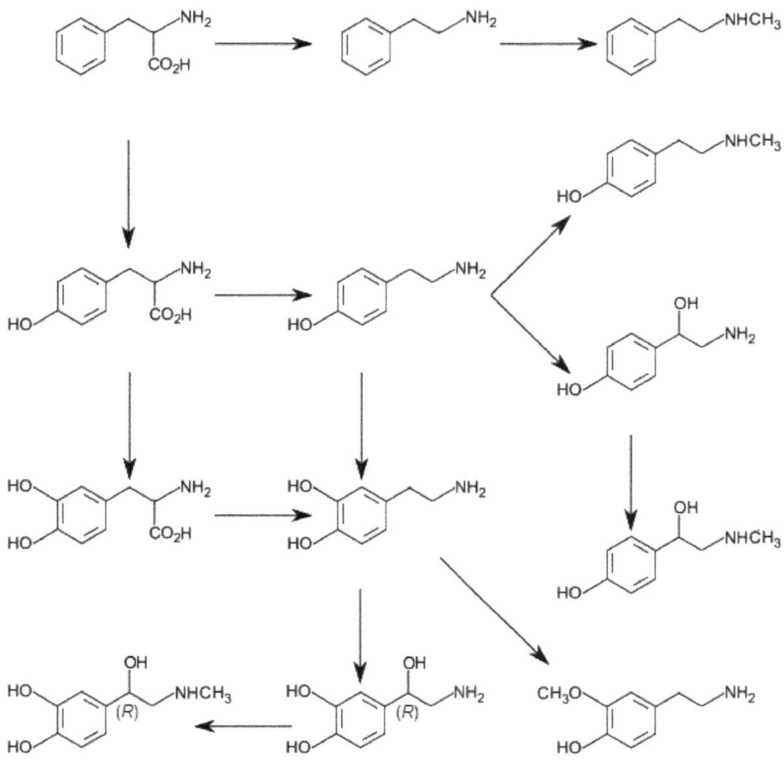

Catecholamines and trace amines are synthesized from phenylalanine and tyrosine in humans.

USES IN INDUSTRY

Amino acids are used for a variety of applications in industry, but their main use is as additives to animal feed. This is necessary, since many of the bulk components of these feeds, such as soybeans, either have low levels or lack some of the essential amino acids: lysine, methionine, threonine, and tryptophan are most important in the production of these feeds. In this industry, amino acids are also used to chelate metal cations in order to improve the absorption of minerals from supplements, which may be required to improve the health or production of these animals.

The food industry is also a major consumer of amino acids, in particular, glutamic acid, which is used as a flavor enhancer, and aspartame (aspartyl-phenylalanine-1-methyl ester) as a low-calorie artificial sweetener. Similar technology to that used for animal nutrition is employed in the human nutrition industry to alleviate symptoms of mineral deficiencies, such as anemia, by improving mineral absorption and reducing negative side effects from inorganic mineral supplementation.

The chelating ability of amino acids has been used in fertilizers for agriculture to facilitate the delivery of minerals to plants in order to correct mineral deficiencies, such as iron chlorosis. These fertilizers are also used to prevent deficiencies from occurring and improving the overall health of the plants. The remaining production of amino acids is used in the synthesis of drugs and cosmetics.

Similarly, some amino acids derivatives are used in pharmaceutical industry. They include 5-HTP (5-hydroxytryptophan) used for experimental treatment of depression, L-DOPA (L-dihydroxyphenylalanine) for Parkinson's treatment, and eflornithine drug that inhibits ornithine decarboxylase and used in the treatment of sleeping sickness.

Expanded Genetic Code

Since 2001, 40 non-natural amino acids have been added into protein by creating a unique codon (recoding) and a corresponding transfer-RNA:aminoacyl – tRNA-synthetase pair to encode it with diverse physicochemical and biological properties in order to be used as a tool to exploring protein structure and function or to create novel or enhanced proteins.

Nullomers

Nullomers are codons that in theory code for an amino acid, however in nature there is a selective bias against using this codon in favor of another, for example bacteria prefer to use CGA instead of AGA to code for arginine. This creates some sequences that do not appear in the genome. This characteristic can be taken advantage of and used to create new selective cancer-fighting drugs and to prevent cross-contamination of DNA samples from crime-scene investigations.

Chemical Building Blocks

Amino acids are important as low-cost feedstocks. These compounds are used in chiral pool synthesis as enantiomerically pure building-blocks.

Amino acids have been investigated as precursors chiral catalysts, e.g., for asymmetric hydrogenation reactions, although no commercial applications exist.

Biodegradable Plastics

Amino acids are under development as components of a range of biodegradable polymers. These materials have applications as environmentally friendly packaging and in medicine in drug delivery and the construction of prosthetic implants. These polymers include polypeptides, polyamides, polyesters, polysulfides, and polyurethanes with amino acids either forming part of their main chains or bonded as side-chains. These modifications alter the physical properties and reactivities of the polymers. An interesting example of such materials is polyaspartate, a water-soluble biodegradable polymer that may have applications in disposable diapers and agriculture. Due to its solubility and ability to chelate metal ions, polyaspartate is also being used as a biodegradeable anti-scaling agent and a corrosion inhibitor. In addition, the aromatic amino acid tyrosine is being developed as a possible replacement for toxic phenols such as bisphenol A in the manufacture of polycarbonates.

Reactions

As amino acids have both a primary amine group and a primary carboxyl group, these chemicals can undergo most of the reactions associated with these functional groups. These include nucleophilic addition, amide bond formation, and imine formation for the amine group, and esterification, amide bond formation, and decarboxylation for the carboxylic acid group. The combination of these functional groups allow amino acids to be effective polydentate ligands for metal-amino acid chelates. The multiple side-chains of amino acids can also undergo chemical reactions. The types of these reactions are determined by the groups on these side-chains and are, therefore, different between the various types of amino acid.

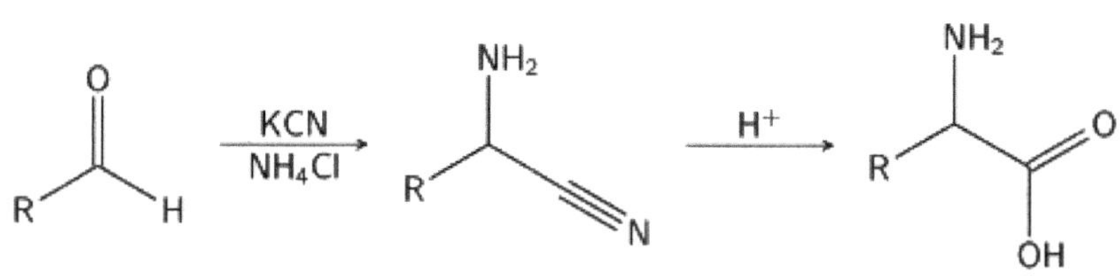

The Strecker amino acid synthesis

Chemical Synthesis

Several methods exist to synthesize amino acids. One of the oldest methods begins with the bromination at the α-carbon of a carboxylic acid. Nucleophilic substitution with ammonia then converts the alkyl bromide to the amino acid. In alternative fashion, the Strecker amino acid synthesis involves the treatment of an aldehyde with potassium cyanide and ammonia, this produces an α-amino nitrile as an intermediate. Hydrolysis of the nitrile in acid then yields a α-amino acid. Using ammonia or ammonium salts in this reaction gives unsubstituted amino acids, whereas substituting primary and secondary amines will yield substituted amino acids. Likewise, using ketones, instead of aldehydes, gives α,α-disubstituted amino acids. The classical synthesis gives racemic mixtures of α-amino acids as products, but several alternative procedures using asymmetric auxiliaries or asymmetric catalysts have been developed.

At the current time, the most-adopted method is an automated synthesis on a solid support (e.g., polystyrene beads), using protecting groups (e.g., Fmoc and t-Boc) and activating groups (e.g., DCC and DIC).

Peptide Bond Formation

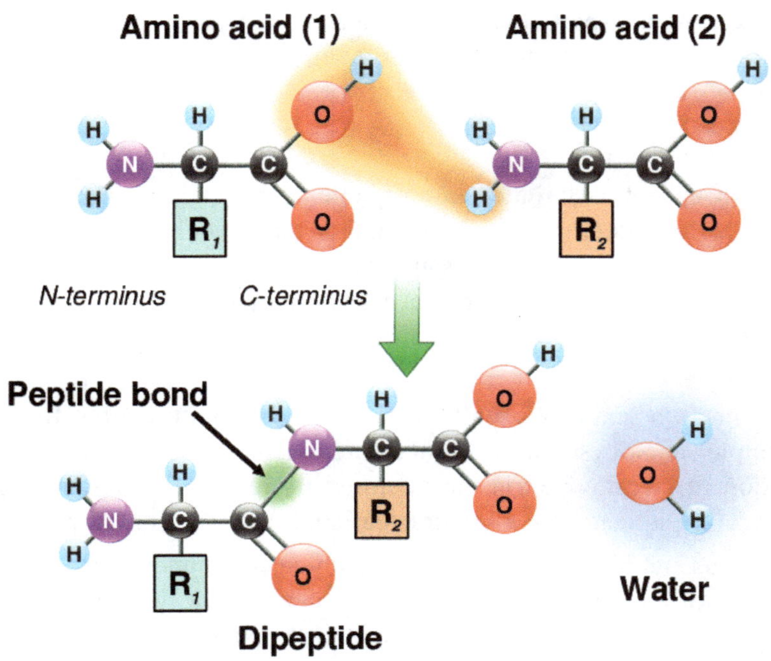

The condensation of two amino acids to form a *dipeptide* through a *peptide bond*

As both the amine and carboxylic acid groups of amino acids can react to form amide bonds, one amino acid molecule can react with another and become joined through an amide linkage. This polymerization of amino acids is what creates proteins. This condensation reaction yields the newly formed peptide bond and a molecule of water. In cells, this reaction does not occur directly; instead, the amino acid is first activated by attachment to a transfer RNA molecule through an ester bond. This aminoacyl-tRNA is produced in an ATP-dependent reaction carried out by an aminoacyl tRNA synthetase. This aminoacyl-tRNA is then a substrate for the ribosome, which catalyzes the attack of the amino group of the elongating protein chain on the ester bond. As a result of this mechanism, all proteins made by ribosomes are synthesized starting at their N-terminus and moving toward their C-terminus.

However, not all peptide bonds are formed in this way. In a few cases, peptides are synthesized by specific enzymes. For example, the tripeptide glutathione is an essential part of the defenses of cells against oxidative stress. This peptide is synthesized in two steps from free amino acids. In the first step, gamma-glutamylcysteine synthetase condenses cysteine and glutamic acid through a peptide bond formed between the side-chain carboxyl of the glutamate (the gamma carbon of this side-chain) and the amino group of the cysteine. This dipeptide is then condensed with glycine by glutathione synthetase to form glutathione.

In chemistry, peptides are synthesized by a variety of reactions. One of the most-used in solid-phase peptide synthesis uses the aromatic oxime derivatives of amino acids as activated units. These are added in sequence onto the growing peptide chain, which is attached to a solid resin support. The ability to easily synthesize vast numbers of different peptides by varying the types and order of amino acids (using combinatorial chemistry) has made peptide synthesis particularly important in creating libraries of peptides for use in drug discovery through high-throughput screening.

Biosynthesis

In plants, nitrogen is first assimilated into organic compounds in the form of glutamate, formed from alpha-ketoglutarate and ammonia in the mitochondrion. In order to form other amino acids, the plant uses transaminases to move the amino group to another alpha-keto carboxylic acid. For example, aspartate aminotransferase converts glutamate and oxaloacetate to alpha-ketoglutarate and aspartate. Other organisms use transaminases for amino acid synthesis, too.

Nonstandard amino acids are usually formed through modifications to standard amino acids. For example, homocysteine is formed through the transsulfuration pathway or by the demethylation of methionine via the intermediate metabolite S-adenosyl methionine, while hydroxyproline is made by a posttranslational modification of proline.

Microorganisms and plants can synthesize many uncommon amino acids. For example, some microbes make 2-aminoisobutyric acid and lanthionine, which is a sulfide-bridged derivative of alanine. Both of these amino acids are found in peptidic lantibiotics such as alamethicin.However, in plants, 1-aminocyclopropane-1-carboxylic acid is a small disubstituted cyclic amino acid that is a key intermediate in the production of the plant hormone ethylene.

Catabolism

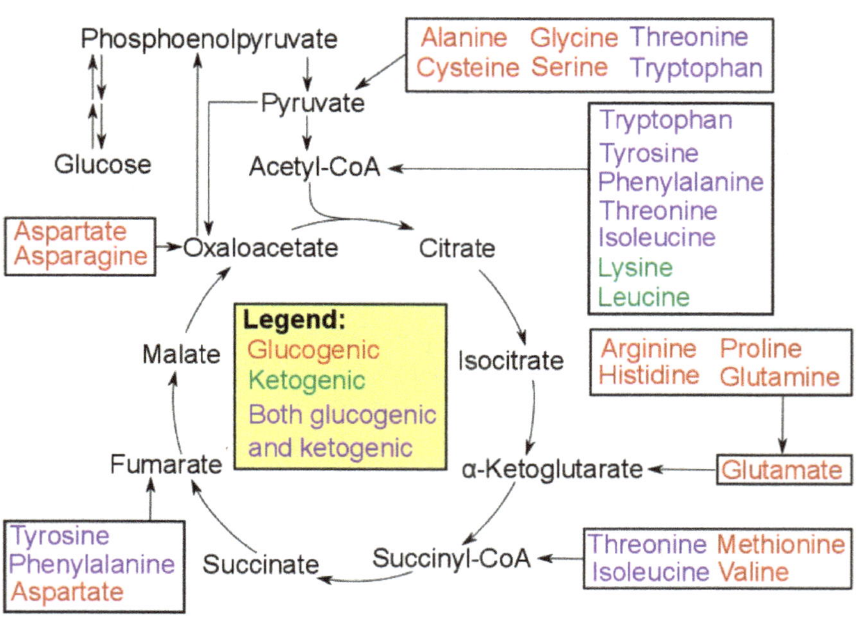

Catabolism of proteinogenic amino acids. Amino acids can be classified according to the properties of their main products as either of the following: *Glucogenic*, with the products having the ability to form glucose by gluconeogenesis *Ketogenic*, with the products not having the ability to form glucose. These products may still be used for ketogenesis or lipid synthesis. Amino acids catabolized into both glucogenic and ketogenic products.

Amino acids must first pass out of organelles and cells into blood circulation via amino acid transporters, since the amine and carboxylic acid groups are typically ionized. Degradation of an amino acid, occurring in the liver and kidneys, often involves deamination by moving its amino group to alpha-ketoglutarate, forming glutamate. This process involves transaminases, often the same as those used in amination during synthesis. In many vertebrates, the amino group is then removed through the urea cycle and is excreted in the form of urea. However, amino acid degradation can produce uric acid or ammonia instead. For example, serine dehydratase converts serine to pyruvate and ammonia. After removal of one or more amino groups, the remainder of the molecule can sometimes be used to synthesize new amino acids, or it can be used for energy by entering glycolysis or the citric acid cycle, as detailed in image at right.

Physicochemical Properties of Amino Acids

The 20 amino acids encoded directly by the genetic code can be divided into several groups based on their properties. Important factors are charge, hydrophilicity or hydrophobicity, size, and functional groups. These properties are important for protein structure and protein–protein interactions. The water-soluble proteins tend to have their hydrophobic residues (Leu, Ile, Val, Phe, and Trp) buried in the middle of the protein, whereas hydrophilic side-chains are exposed to the aqueous solvent. (Note that in biochemistry, a residue refers to a specific monomer within the polymeric chain of a polysaccharide, protein or nucleic acid.) The integral membrane proteins tend to have outer rings of exposed hydrophobic amino acids that anchor them into the lipid bilayer. In the case part-way between these two extremes, some peripheral membrane proteins have a patch of hydrophobic amino acids on their surface that locks onto the membrane. In similar fashion, proteins that have to bind to positively charged molecules have surfaces rich with negatively charged amino acids like glutamate and aspartate, while proteins binding to negatively charged molecules have surfaces rich with positively charged chains like lysine and arginine. There are different hydrophobicity scales of amino acid residues.

Some amino acids have special properties such as cysteine, that can form covalent disulfide bonds to other cysteine residues, proline that forms a cycle to the polypeptide backbone, and glycine that is more flexible than other amino acids.

Many proteins undergo a range of posttranslational modifications, when additional chemical groups are attached to the amino acids in proteins. Some modifications can produce hydrophobic lipoproteins, or hydrophilic glycoproteins. These type of modification allow the reversible targeting of a protein to a membrane. For example, the addition and removal of the fatty acid palmitic acid to cysteine residues in some signaling proteins causes the proteins to attach and then detach from cell membranes.

Able Of Standard Amino Acid Abbreviations and Properties

Amino Acid	3-Letter	1-Letter	Side-chain class	Side-chain polarity	Side-chain charge (pH 7.4)	Hydropathy index	Absorbance λ_{max}(nm)	ε at λ_{max}($mM^{-1}cm^{-1}$)	MW (Weight)	Occurrence in proteins (%)
Alanine	Ala	A	aliphatic	nonpolar	neutral	1.8			89.094	8.76
Arginine	Arg	R	basic	basic polar	positive	−4.5			174.203	5.78
Asparagine	Asn	N	amide	polar	neutral	−3.5			132.119	3.93
Aspartic acid	Asp	D	acid	acidic polar	negative	−3.5			133.104	5.49
Cysteine	Cys	C	sulfur-containing	nonpolar	neutral	2.5	250	0.3	121.154	1.38
Glutamic acid	Glu	E	acid	acidic polar	negative	−3.5			147.131	6.32
Glutamine	Gln	Q	amide	polar	neutral	−3.5			146.146	3.9
Glycine	Gly	G	aliphatic	nonpolar	neutral	−0.4			75.067	7.03
Histidine	His	H	basic aromatic	basic polar	positive(10%) neutral(90%)	−3.2	211	5.9	155.156	2.26
Isoleucine	Ile	I	aliphatic	nonpolar	neutral	4.5			131.175	5.49
Leucine	Leu	L	aliphatic	nonpolar	neutral	3.8			131.175	9.68
Lysine	Lys	K	basic	basic polar	positive	−3.9			146.189	5.19
Methionine	Met	M	sulfur-containing	nonpolar	neutral	1.9			149.208	2.32
Phenylalanine	Phe	F	aromatic	nonpolar	neutral	2.8	257, 206, 188	0.2, 9.3, 60.0	165.192	3.87
Proline	Pro	P	cyclic	nonpolar	neutral	−1.6			115.132	5.02
Serine	Ser	S	hydroxyl-containing	polar	neutral	−0.8			105.093	7.14
Threonine	Thr	T	hydroxyl-containing	polar	neutral	−0.7			119.119	5.53
Tryptophan	Trp	W	aromatic	nonpolar	neutral	−0.9	280, 219	5.6, 47.0	204.228	1.25
Tyrosine	Tyr	Y	aromatic	polar	neutral	−1.3	274, 222, 193	1.4, 8.0, 48.0	181.191	2.91
Valine	Val	V	aliphatic	nonpolar	neutral	4.2			117.148	6.73

Two additional amino acids are in some species coded for by codons that are usually interpreted as stop codons:

21st and 22nd amino acids	3-Letter	1-Letter	MW(Weight)
Selenocysteine	Sec	U	168.064
Pyrrolysine	Pyl	O	255.313

In addition to the specific amino acid codes, placeholders are used in cases where chemical or crystallographic analysis of a peptide or protein cannot conclusively determine the identity of a residue. They are also used to summarise conserved protein sequence motifs. The use of single letters to indicate sets of similar residues is similar to the use of abbreviation codes for degenerate bases.

Ambiguous Amino Acids	3-Letter	1-Letter	
Any / unknown	Xaa	X	All
Asparagine or aspartic acid	Asx	B	D, N
Glutamine or glutamic acid	Glx	Z	E, Q
Leucine or Isoleucine	Xle	J	I, L
Hydrophobic		Φ	V, I, L, F, W, Y, M
Aromatic		Ω	F, W, Y, H
Aliphatic (Non-Aromatic)		Ψ	V, I, L, M
Small		π	P, G, A, S
Hydrophilic		ζ	S, T, H, N, Q, E, D, K, R
Positively charged		+	K, R, H
Negatively charged		−	D, E

Unk is sometimes used instead of **Xaa**, but is less standard.

In addition, many non-standard amino acids have a specific code. For example, several peptide drugs, such as Bortezomib and MG132, are artificially synthesized and retain their protecting groups, which have specific codes. Bortezomib is Pyz-Phe-boroLeu, and MG132 is Z-Leu-Leu-Leu-al. To aid in the analysis of protein structure, photo-reactive amino acid analogs are available. These include photoleucine (**pLeu**) and photomethionine (**pMet**).

REFERENCES

1. "Amino". Dictionary.com. 2015. Retrieved 3 July 2015.
2. "amino acid". Cambridge Dictionaries Online. Cambridge University Press. 2015. Retrieved 3 July 2015.
3. "amino". FreeDictionary.com. Farlex. 2015. Retrieved 3 July 2015.
4. Wagner I, Musso H (November 1983). "New Naturally Occurring Amino Acids". Angewandte Chemie International Edition in English. 22 (11): 816–28. doi:10.1002/anie.198308161.
5. Latham, Michael C. (1997). "Chapter 8. Body composition, the functions of food, metabolism and energy". Human nutrition in the developing world. Food and Nutrition Series – No. 29. Rome: Food and Agriculture Organization of the United Nations.

6. Proline is an exception to this general formula. It lacks the NH2 group because of the cyclization of the side-chain and is known as an imino acid; it falls under the category of special structured amino acids.
7. Clark, Jim (August 2007). "an introduction to amino acids". chemguide. Retrieved 4 July 2015.
8. Jakubke, Hans-Dieter; Sewald, Norbert (2008). "Amino acids". Peptides from A to Z: A Concise Encyclopedia. Germany: Wiley-VCH. p. 20. ISBN 9783527621170 – via Google Books.
9. Pollegioni, Loredano; Servi, Stefano, eds. (2012). Unnatural Amino Acids: Methods and Protocols. Methods in Molecular Biology – Volume 794. Humana Press. p. v. ISBN 978-1-61779-331-8. OCLC 756512314.
10. Hertweck C (2011). "Biosynthesis and Charging of Pyrrolysine, the 22nd Genetically Encoded Amino Acid". Angewandte Chemie International Edition. 50 (41): 9540–1. doi:10.1002/anie.201103769. PMID 21796749.
11. "Chapter 1: Proteins are the Body's Worker Molecules". The Structures of Life. National Institute of General Medical Sciences. 27 October 2011. Retrieved 20 May 2008.
12. Michal, Gerhard; Schomburg, Dietmar, eds. (2012). Biochemical Pathways: An Atlas of Biochemistry and Molecular Biology (2nd ed.). Wiley. p. 5.
13. Modeling Electrostatic Contributions to Protein Folding and Binding – Tjong, p.1 footnote
14. Frontiers in Drug Design and Discovery ed. Atta-Ur-Rahman & others, p.299
15. Elzanowski A, Ostell J (7 April 2008). "The Genetic Codes". National Center for Biotechnology Information (NCBI). Retrieved 10 March 2010.
16. Xie J, Schultz PG (December 2005). "Adding amino acids to the genetic repertoire". Current Opinion in Chemical Biology. 9 (6): 548–54. doi:10.1016/j.cbpa.2005.10.011. PMID 16260173.
17. Wang Q, Parrish AR, Wang L (March 2009). "Expanding the genetic code for biological studies". Chem. Biol. 16 (3): 323–36. doi:10.1016/j.chembiol.2009.03.001. PMC 2696486 . PMID 19318213.
18. Simon M (2005). Emergent computation: emphasizing bioinformatics. New York: AIP Press/Springer Science+Business Media. pp. 105–106. ISBN 0-387-22046-1.
19. Petroff OA (December 2002). "GABA and glutamate in the human brain". Neuroscientist. 8 (6): 562–573. doi:10.1177/1073858402238515. PMID 12467378.
20. For example, ruminants such as cows obtain a number of amino acids via microbes in the first two stomach chambers.
21. Vauquelin LN, Robiquet PJ (1806). "The discovery of a new plant principle in Asparagus sativus". Annales de Chimie. 57: 88–93.
22. Anfinsen CB, Edsall JT, Richards FM (1972). Advances in Protein Chemistry. New York: Academic Press. pp. 99, 103. ISBN 978-0-12-034226-6.

23. Wollaston WH (1810). "On cystic oxide, a new species of urinary calculus". Philosophical Transactions of the Royal Society. 100: 223–30. doi:10.1098/rstl.1810.0015.
24. Baumann E (1884). "Über cystin und cystein". Z Physiol Chem. 8 (4): 299–305. Archived from the original on 14 March 2011. Retrieved 28 March 2011.
25. Braconnot HM (1820). "Sur la conversion des matières animales en nouvelles substances par le moyen de l'acide sulfurique". Annales de Chimie et de Physique. 2nd Series. 13: 113–25.
26. Robert D. Simoni; Robert L. Hill; Martha Vaughan (13 September 2002). "The Discovery of the Amino Acid Threonine: the Work of William C. Rose". Journal of Biological Chemistry. 277 (37): 56–58. PMID 12218068.
27. McCoy, R. H.; Meyer, C. E. & Rose, W. C. (1935). "Feeding Experiments with Mixtures of Highly Purified Amino Acids. VIII. Isolation and Identification of a New Essential Amino Acid". Journal of Biological Chemistry. 112: 283–302.
28. "etymonline.com entry for amino". www.etymonline.com. Retrieved 19 July 2010.
29. Joseph S. Fruton (1990). "Chapter 5- Emil Fischer and Franz Hofmeister". Contrasts in Scientific Style: Research Groups in the Chemical and Biochemical Sciences,. 191. American Philosophical Society. pp. 163–165. ISBN 0-87169-191-4.
30. "Alpha amino acid – Medical definition". Merriam-Webster dictionary.
31. Proline at the US National Library of Medicine Medical Subject Headings (MeSH)
32. http://opbs.okstate.edu/5753/Amino%20Acids.html
33. IUPAC, Compendium of Chemical Terminology, 2nd ed. (the "Gold Book") (1997). Online corrected version: (2006–) "Imino acids".
34. Creighton, Thomas H. (1993). "Chapter 1". Proteins: structures and molecular properties. San Francisco: W. H. Freeman. ISBN 978-0-7167-7030-5.
35. Pisarewicz K, Mora D, Pflueger FC, Fields GB, Marí F (May 2005). "Polypeptide chains containing D-gamma-hydroxyvaline". Journal of the American Chemical Society. 127 (17): 6207–15. doi:10.1021/ja050088m. PMID 15853325.
36. van Heijenoort J (March 2001). "Formation of the glycan chains in the synthesis of bacterial peptidoglycan". Glycobiology. 11 (3): 25R–36R. doi:10.1093/glycob/11.3.25R. PMID 11320055.
37. Wolosker H, Dumin E, Balan L, Foltyn VN (July 2008). "D-amino acids in the brain: D-serine in neurotransmission and neurodegeneration". The FEBS Journal. 275 (14): 3514–26. doi:10.1111/j.1742-4658.2008.06515.x. PMID 18564180.
38. Matthews BW (2009). "Racemic crystallography—easy crystals and easy structures: What's not to like?". Protein Science. 18 (6): 1135–1138. doi:10.1002/pro.125. PMC 2774423 . PMID 19472321.
39. Hatem, Salama Mohamed Ali (2006). "Gas chromatographic determination of Amino Acid Enantiomers in tobacco and bottled wines". University of Giessen. Retrieved 17 November 2008.

40. "Nomenclature and Symbolism for Amino Acids and Peptides". IUPAC-IUB Joint Commission on Biochemical Nomenclature. 1983. Archived from the original on 9 October 2008. Retrieved 17 November 2008.
41. Jodidi, S. L. (1 March 1926). "The Formol Titration of Certain Amino Acids". Journal of the American Chemical Society. 48 (3): 751–753. doi:10.1021/ja01414a033.
42. Liebecq, Claude, ed. (1992). Biochemical Nomenclature and Related Documents (2nd ed.). Portland Press. pp. 39–69. ISBN 978-1-85578-005-7.
43. Smith, Anthony D. (1997). Oxford dictionary of biochemistry and molecular biology. Oxford: Oxford University Press. p. 535. ISBN 978-0-19-854768-6. OCLC 37616711.
44. Simmons, William J.; Gerhard Meisenberg (2006). Principles of medical biochemistry. Mosby Elsevier. p. 19. ISBN 0-323-02942-6.
45. Fennema OR. Food Chemistry 3rd Ed. CRC Press. pp. 327–8. ISBN 0-8247-9691-8.
46. Rodnina MV, Beringer M, Wintermeyer W (January 2007). "How ribosomes make peptide bonds". Trends in Biochemical Sciences. 32 (1): 20–6. doi:10.1016/j.tibs.2006.11.007. PMID 17157507.
47. Driscoll DM, Copeland PR (2003). "Mechanism and regulation of selenoprotein synthesis". Annual Review of Nutrition. 23 (1): 17–40. doi:10.1146/annurev.nutr.23.011702.073318. PMID 12524431.
48. Krzycki JA (December 2005). "The direct genetic encoding of pyrrolysine". Current Opinion in Microbiology. 8 (6): 706–12. doi:10.1016/j.mib.2005.10.009. PMID 16256420.
49. Théobald-Dietrich A, Giegé R, Rudinger-Thirion J (2005). "Evidence for the existence in mRNAs of a hairpin element responsible for ribosome dependent pyrrolysine insertion into proteins". Biochimie. 87 (9–10): 813–7. doi:10.1016/j.biochi.2005.03.006. PMID 16164991.
50. Vermeer C (March 1990). "Gamma-carboxyglutamate-containing proteins and the vitamin K-dependent carboxylase". The Biochemical Journal. 266 (3): 625–36. doi:10.1042/bj2660625. PMC 1131186 . PMID 2183788.
51. Bhattacharjee A, Bansal M (March 2005). "Collagen structure: the Madras triple helix and the current scenario". IUBMB Life. 57 (3): 161–72. doi:10.1080/15216540500090710. PMID 16036578.
52. Park MH (February 2006). "The post-translational synthesis of a polyamine-derived amino acid, hypusine, in the eukaryotic translation initiation factor 5A (eIF5A)". Journal of Biochemistry. 139 (2): 161–9. doi:10.1093/jb/mvj034. PMC 2494880 . PMID 16452303.
53. Blenis J, Resh MD (December 1993). "Subcellular localization specified by protein acylation and phosphorylation". Current Opinion in Cell Biology. 5 (6): 984–9. doi:10.1016/0955-0674(93)90081-Z. PMID 8129952.

54. Curis E, Nicolis I, Moinard C, Osowska S, Zerrouk N, Bénazeth S, Cynober L (November 2005). "Almost all about citrulline in mammals". Amino Acids. 29 (3): 177–205. doi:10.1007/s00726-005-0235-4. PMID 16082501.

55. Coxon KM, Chakauya E, Ottenhof HH, Whitney HM, Blundell TL, Abell C, Smith AG (August 2005). "Pantothenate biosynthesis in higher plants". Biochemical Society Transactions. 33 (Pt 4): 743–6. doi:10.1042/BST0330743. PMID 16042590.

56. Ketchem, Randal R.; Hu, Weidong; Cross, Timothy A. (1993). "High-Resolution Conformation of Gramicidin A in a Lipid Bilayer by Solid-State NMR". Science. 261 (5127): 1457–1460. doi:10.1126/science.7690158.

57. Khoury, George A.; Baliban, Richard C.; Floudas, Christodoulos A. (2011). "Proteome-Wide Post-Translational Modification Statistics: Frequency Analysis and Curation of the Swiss-Prot Database". Scientific Reports. 1 (90). doi:10.1038/srep00090. PMC 3201773 . PMID 22034591.

58. Kryukov GV, Castellano S, Novoselov SV, Lobanov AV, Zehtab O, Guigó R, Gladyshev VN (2003). "Characterization of mammalian selenoproteomes". Science. 300 (5624): 1439–1443. Bibcode:2003Sci...300.1439K. doi:10.1126/science.1083516. PMID 12775843.

59. Gromer, S., Urig, S., Becker, K. (2004) The Thioredoxin System – From Science to Clinic. Medicinal Research Reviews. 24(1):40–89.

60. Sakami W, Harrington H (1963). "Amino acid metabolism". Annual Review of Biochemistry. 32 (1): 355–98. doi:10.1146/annurev.bi.32.070163.002035. PMID 14144484.

61. Brosnan JT (April 2000). "Glutamate, at the interface between amino acid and carbohydrate metabolism". The Journal of Nutrition. 130 (4S Suppl): 988S–90S. PMID 10736367.

62. Young VR, Ajami AM (September 2001). "Glutamine: the emperor or his clothes?". The Journal of Nutrition. 131 (9 Suppl): 2449S–59S; discussion 2486S–7S. PMID 11533293.

63. Young VR (August 1994). "Adult amino acid requirements: the case for a major revision in current recommendations". The Journal of Nutrition. 124 (8 Suppl): 1517S–1523S. PMID 8064412.

64. Fürst P, Stehle P (June 2004). "What are the essential elements needed for the determination of amino acid requirements in humans?". The Journal of Nutrition. 134 (6 Suppl): 1558S–1565S. PMID 15173430.

65. Reeds PJ (July 2000). "Dispensable and indispensable amino acids for humans". The Journal of Nutrition. 130 (7): 1835S–40S. PMID 10867060.

66. Imura K, Okada A (January 1998). "Amino acid metabolism in pediatric patients". Nutrition. 14 (1): 143–8. doi:10.1016/S0899-9007(97)00230-X. PMID 9437700.

67. Lourenço R, Camilo ME (2002). "Taurine: a conditionally essential amino acid in humans? An overview in health and disease". Nutrición Hospitalaria. 17 (6): 262–70. PMID 12514918.

68. Holtcamp, W. (2012). "The emerging science of BMAA: do cyanobacteria contribute to neurodegenerative disease?". Environmental Health Perspectives. 120 (3): a110–a116. doi:10.1289/ehp.120-a110. PMC 3295368 . PMID 22382274.

69. Cox, PA, Davis, DA, Mash, DC, Metcalf, JS, Banack, SA. (2015). "Dietary exposure to an environmental toxin triggers neurofibrillary tangles and amyloid deposits in the brain". Proceedings of the Royal Society B. 283 (1823): 20152397. doi:10.1098/rspb.2015.2397. PMC 4795023 . PMID 26791617.

70. Brook MS, Wilkinson DJ, Phillips BE, Perez-Schindler J, Philp A, Smith K, Atherton PJ (January 2016). "Skeletal muscle homeostasis and plasticity in youth and ageing: impact of nutrition and exercise". Acta Physiologica. 216 (1): 15–41. doi:10.1111/apha.12532. PMC 4843955 . PMID 26010896.

71. Lipton JO, Sahin M (October 2014). "The neurology of mTOR". Neuron. 84 (2): 275–291. doi:10.1016/j.neuron.2014.09.034. PMC 4223653 . PMID 25374355.

72. Figure 2: The mTOR Signaling Pathway

73. Phillips SM (May 2014). "A brief review of critical processes in exercise-induced muscular hypertrophy". Sports Med. 44 Suppl 1: S71–S77. doi:10.1007/s40279-014-0152-3. PMC 4008813 . PMID 24791918.

74. Broadley KJ (March 2010). "The vascular effects of trace amines and amphetamines". Pharmacol. Ther. 125 (3): 363–375. doi:10.1016/j.pharmthera.2009.11.005. PMID 19948186.

75. Lindemann L, Hoener MC (May 2005). "A renaissance in trace amines inspired by a novel GPCR family". Trends Pharmacol. Sci. 26 (5): 274–281. doi:10.1016/j.tips.2005.03.007. PMID 15860375.

76. Wang X, Li J, Dong G, Yue J (February 2014). "The endogenous substrates of brain CYP2D". Eur. J. Pharmacol. 724: 211–218. doi:10.1016/j.ejphar.2013.12.025. PMID 24374199. The highest level of brain CYP2D activity was found in the substantia nigra ... The in vitro and in vivo studies have shown the contribution of the alternative CYP2D-mediated dopamine synthesis to the concentration of this neurotransmitter although the classic biosynthetic route to dopamine from tyrosine is active. ... Tyramine levels are especially high in the basal ganglia and limbic system, which are thought to be related to individual behavior and emotion (Yu et al., 2003c). ... Rat CYP2D isoforms (2D2/2D4/2D18) are less efficient than human CYP2D6 for the generation of dopamine from p-tyramine. The Km values of the CYP2D isoforms are as follows: CYP2D6 (87–121 μm) ≈ CYP2D2 ≈ CYP2D18 > CYP2D4 (256 μm) for m-tyramine and CYP2D4 (433 μm) > CYP2D2 ≈ CYP2D6 > CYP2D18 (688 μm) for p-tyramine

77. Savelieva KV, Zhao S, Pogorelov VM, Rajan I, Yang Q, Cullinan E, Lanthorn TH (2008). Bartolomucci A, ed. "Genetic disruption of both tryptophan hydroxylase genes dramatically reduces serotonin and affects behavior in models sensitive to antidepressants". PLoS ONE. 3 (10): e3301. Bibcode:2008PLoSO...3.3301S. doi:10.1371/journal.pone.0003301. PMC 2565062 . PMID 18923670.

78. Shemin D, Rittenberg D (1 December 1946). "The biological utilization of glycine for the synthesis of the protoporphyrin of hemoglobin". Journal of Biological Chemistry. 166 (2): 621–5. PMID 20276176.

79. Tejero J, Biswas A, Wang ZQ, Page RC, Haque MM, Hemann C, Zweier JL, Misra S, Stuehr DJ (November 2008). "Stabilization and characterization of a heme-oxy reaction intermediate in inducible nitric-oxide synthase". The Journal of Biological Chemistry. 283 (48): 33498–507. doi:10.1074/jbc.M806122200. PMC 2586280 . PMID 18815130.

80. Rodríguez-Caso C, Montañez R, Cascante M, Sánchez-Jiménez F, Medina MA (August 2006). "Mathematical modeling of polyamine metabolism in mammals". The Journal of Biological Chemistry. 281 (31): 21799–812. doi:10.1074/jbc.M602756200. PMID 16709566.

81. Stryer, Lubert; Berg, Jeremy Mark; Tymoczko, John L. (2002). Biochemistry. San Francisco: W.H. Freeman. pp. 693–8. ISBN 0-7167-4684-0.

82. Hylin, John W. (1969). "Toxic peptides and amino acids in foods and feeds". Journal of Agricultural and Food Chemistry. 17 (3): 492–6. doi:10.1021/jf60163a003.

83. Turner BL, Harborne JB (1967). "Distribution of canavanine in the plant kingdom". Phytochemistry. 6 (6): 863–66. doi:10.1016/S0031-9422(00)86033-1.

84. Ekanayake S, Skog K, Asp NG (May 2007). "Canavanine content in sword beans (Canavalia gladiata): analysis and effect of processing". Food and Chemical Toxicology. 45 (5): 797–803. doi:10.1016/j.fct.2006.10.030. PMID 17187914.

85. Rosenthal GA (2001). "L-Canavanine: a higher plant insecticidal allelochemical". Amino Acids. 21 (3): 319–30. doi:10.1007/s007260170017. PMID 11764412.

86. Hammond AC (May 1995). "Leucaena toxicosis and its control in ruminants". Journal of Animal Science. 73 (5): 1487–92. PMID 7665380.

87. Leuchtenberger W, Huthmacher K, Drauz K (November 2005). "Biotechnological production of amino acids and derivatives: current status and prospects". Applied Microbiology and Biotechnology. 69 (1): 1–8. doi:10.1007/s00253-005-0155-y. PMID 16195792.

88. Ashmead, H. DeWayne (1993). The Role of Amino Acid Chelates in Animal Nutrition. Westwood: Noyes Publications.

89. Garattini S (April 2000). "Glutamic acid, twenty years later". The Journal of Nutrition. 130 (4S Suppl): 901S–9S. PMID 10736350.

90. Stegink LD (July 1987). "The aspartame story: a model for the clinical testing of a food additive". The American Journal of Clinical Nutrition. 46 (1 Suppl): 204–15. PMID 3300262.

91. Albion Laboratories, Inc. "Albion Ferrochel Website". Retrieved 12 July 2011.

92. Ashmead, H. DeWayne (1986). Foliar Feeding of Plants with Amino Acid Chelates. Park Ridge: Noyes Publications.

93. Turner EH, Loftis JM, Blackwell AD (March 2006). "Serotonin a la carte: supplementation with the serotonin precursor 5-hydroxytryptophan". Pharmacology

& Therapeutics. 109 (3): 325–38. doi:10.1016/j.pharmthera.2005.06.004. PMID 16023217.

94. Kostrzewa RM, Nowak P, Kostrzewa JP, Kostrzewa RA, Brus R (March 2005). “Peculiarities of L: -DOPA treatment of Parkinson’s disease”. Amino Acids. 28 (2): 157–64. doi:10.1007/s00726-005-0162-4. PMID 15750845.
95. Heby O, Persson L, Rentala M (August 2007). “Targeting the polyamine biosynthetic enzymes: a promising approach to therapy of African sleeping sickness, Chagas’ disease, and leishmaniasis”. Amino Acids. 33 (2): 359–66. doi:10.1007/s00726-007-0537-9. PMID 17610127.
96. Cruz-Vera LR, Magos-Castro MA, Zamora-Romo E, Guarneros G (2004). “Ribosome stalling and peptidyl-tRNA drop-off during translational delay at AGA codons”. Nucleic Acids Research. 32. 18 (15): 4462–8. doi:10.1093/nar/gkh784. PMC 516057 . PMID 15317870.
97. Coghlan, Andy (225 October 2012). “Molecules ‹too dangerous for nature’ kill cancer cells”. New Scientist. Check date values in: |date=(help)
98. “Lethal DNA tags could keep innocent people out of jail”. New Scientist. 2 May 2013.
99. Hanessian S (1993). “Reflections on the total synthesis of natural products: Art, craft, logic, and the chiron approach”. Pure and Applied Chemistry. 65 (6): 1189–204. doi:10.1351/pac199365061189.
100. Blaser, Hans Ulrich (1992). “The chiral pool as a source of enantioselective catalysts and auxiliaries”. Chemical Reviews. 92 (5): 935–52. doi:10.1021/cr00013a009.
101. Sanda F, Endo T (1999). “Syntheses and functions of polymers based on amino acids”. Macromolecular Chemistry and Physics. 200 (12): 2651–61. doi:10.1002/(SICI)1521-3935(19991201)200:12<2651::AID-MACP2651>3.0.CO;2-P.
102. Gross RA, Kalra B (2002). “Biodegradable Polymers for the Environment”. Science. 297 (5582): 803–807. Bibcode:2002Sci...297..803G. doi:10.1126/science.297.5582.803. PMID 12161646.
103. Low, K. C.; Wheeler, A. P.; Koskan, L. P. (1996). Commercial poly(aspartic acid) and Its Uses. Advances in Chemistry Series. 248. Washington, D.C.: American Chemical Society.
104. Thombre SM, Sarwade BD (2005). “Synthesis and Biodegradability of Polyaspartic Acid: A Critical Review”. Journal of Macromolecular Science, Part A. 42 (9): 1299–1315. doi:10.1080/10601320500189604.
105. Bourke SL, Kohn J (2003). “Polymers derived from the amino acid l-tyrosine: polycarbonates, polyarylates and copolymers with poly(ethylene glycol)”. Advanced Drug Delivery Reviews. 55 (4): 447–466. doi:10.1016/S0169-409X(03)00038-3. PMID 12706045.
106. Elmore, Donald Trevor; Barrett, G. C. (1998). Amino acids and peptides. Cambridge, UK: Cambridge University Press. pp. 48–60. ISBN 0-521-46827-2.

107. Konara S, Gagnona K, Clearfield A, Thompson C, Hartle J, Ericson C, Nelson C (2010). "Structural determination and characterization of copper and zinc bis-glycinates with X-ray crystallography and mass spectrometry". Journal of Coordination Chemistry. 63 (19): 3335–47. doi:10.1080/00958972.2010.514336.

108. Gutteridge A, Thornton JM (November 2005). "Understanding nature's catalytic toolkit". Trends in Biochemical Sciences. 30 (11): 622–9. doi:10.1016/j.tibs.2005.09.006. PMID 16214343.

109. McMurry, John (1996). Organic chemistry. Pacific Grove, CA, USA: Brooks/Cole. p. 1064. ISBN 0-534-23832-7.

110. Strecker, Adolph (1850). "Ueber die künstliche Bildung der Milchsäure und einen neuen, dem Glycocoll homologen Körper". Justus Liebigs Annalen der Chemie. 75 (1): 27–45. doi:10.1002/jlac.18500750103.

111. Strecker, Adolph (1854). "Ueber einen neuen aus Aldehyd – Ammoniak und Blausäure entstehenden Körper". Justus Liebigs Annalen der Chemie. 91 (3): 349–51. doi:10.1002/jlac.18540910309.

112. Masumoto S, Usuda H, Suzuki M, Kanai M, Shibasaki M (May 2003). "Catalytic enantioselective Strecker reaction of ketoimines". Journal of the American Chemical Society. 125 (19): 5634–5. doi:10.1021/ja034980. PMID 12733893.

113. Davis FA, Reddy RE, Portonovo PS (1994). "Asymmetric strecker synthesis using enantiopure sulfinimines: A convenient synthesis of α-amino acids". Tetrahedron Letters. 35 (50): 9351–9354. doi:10.1016/S0040-4039(00)78540-6.

114. Ishitani H, Komiyama S, Hasegawa Y, Kobayashi S (2000). "Catalytic Asymmetric Strecker Synthesis. Preparation of Enantiomerically Pure α-Amino Acid Derivatives from Aldimines and Tributyltin Cyanide or Achiral Aldehydes, Amines, and Hydrogen Cyanide Using a Chiral Zirconium Catalyst". Journal of the American Chemical Society. 122 (5): 762–6. doi:10.1021/ja9935207.

115. Huang J, Corey EJ (2004). "A New Chiral Catalyst for the Enantioselective Strecker Synthesis of α-Amino Acids". Orgic Letters. 62 (6): 5027–9. doi:10.1021/ol047698w. PMID 15606127.

116. Duthaler, Rudolf O. (1994). "Recent developments in the stereoselective synthesis of α-aminoacids". Tetrahedron. 50 (6): 1539–1650. doi:10.1016/S0040-4020(01)80840-1.

117. Ibba M, Söll D (May 2001). "The renaissance of aminoacyl-tRNA synthesis". EMBO Reports. 2 (5): 382–7. doi:10.1093/embo-reports/kve095. PMC 1083889 . PMID 11375928.

118. Lengyel P, Söll D (June 1969). "Mechanism of protein biosynthesis". Bacteriological Reviews. 33 (2): 264–301. PMC 378322 . PMID 4896351.

119. Wu G, Fang YZ, Yang S, Lupton JR, Turner ND (March 2004). "Glutathione metabolism and its implications for health". The Journal of Nutrition. 134 (3): 489–92. PMID 14988435.

120. Meister A (November 1988). "Glutathione metabolism and its selective modification". The Journal of Biological Chemistry. 263 (33): 17205–8. PMID 3053703.

121. Carpino, Louis A. (1992). "1-Hydroxy-7-azabenzotriazole. An efficient peptide coupling additive". Journal of the American Chemical Society. 115 (10): 4397–8. doi:10.1021/ja00063a082.

122. Marasco D, Perretta G, Sabatella M, Ruvo M (October 2008). "Past and future perspectives of synthetic peptide libraries". Current Protein & Peptide Science. 9 (5): 447–67. doi:10.2174/138920308785915209. PMID 18855697.

123. Jones, Russell Celyn; Buchanan, Bob B.; Gruissem, Wilhelm (2000). Biochemistry & molecular biology of plants. Rockville, Md: American Society of Plant Physiologists. pp. 371–2. ISBN 0-943088-39-9.

124. Brosnan JT, Brosnan ME (June 2006). "The sulfur-containing amino acids: an overview". The Journal of Nutrition. 136 (6 Suppl): 1636S–1640S. PMID 16702333.

125. Kivirikko KI, Pihlajaniemi T (1998). "Collagen hydroxylases and the protein disulfide isomerase subunit of prolyl 4-hydroxylases". Advances in Enzymology and Related Areas of Molecular Biology. Advances in Enzymology – and Related Areas of Molecular Biology. 72: 325–98. doi:10.1002/9780470123188.ch9. ISBN 9780470123188. PMID 9559057.

126. Whitmore L, Wallace BA (May 2004). "Analysis of peptaibol sequence composition: implications for in vivo synthesis and channel formation". European Biophysics Journal. 33 (3): 233–7. doi:10.1007/s00249-003-0348-1. PMID 14534753.

127. Alexander L, Grierson D (October 2002). "Ethylene biosynthesis and action in tomato: a model for climacteric fruit ripening". Journal of Experimental Botany. 53 (377): 2039–55. doi:10.1093/jxb/erf072. PMID 12324528.

128. Stipanuk, M. H. (2006). Biochemical, physiological, & molecular aspects of human nutrition (2 ed.): Saunders Elsevier.

129. Stryer, Lubert; Berg, Jeremy Mark; Tymoczko, John L. (2002). Biochemistry. San Francisco: W.H. Freeman. pp. 639–49. ISBN 0-7167-4684-0.

130. Urry DW (2004). "The change in Gibbs free energy for hydrophobic association: Derivation and evaluation by means of inverse temperature transitions". Chemical Physics Letters. 399 (1–3): 177–83. doi:10.1016/S0009-2614(04)01565-9.

131. Magee T, Seabra MC (April 2005). "Fatty acylation and prenylation of proteins: what's hot in fat". Current Opinion in Cell Biology. 17 (2): 190–6. doi:10.1016/j.ceb.2005.02.003. PMID 15780596.

132. Pilobello KT, Mahal LK (June 2007). "Deciphering the glycocode: the complexity and analytical challenge of glycomics". Current Opinion in Chemical Biology. 11 (3): 300–5. doi:10.1016/j.cbpa.2007.05.002. PMID 17500024.

133. Smotrys JE, Linder ME (2004). "Palmitoylation of intracellular signaling proteins: regulation and function". Annual Review of Biochemistry. 73 (1): 559–87. doi:10.1146/annurev.biochem.73.011303.073954. PMID 15189153.

134. Hausman, Robert E.; Cooper, Geoffrey M. (2004). The cell: a molecular approach. Washington, D.C: ASM Press. p. 51. ISBN 0-87893-214-3.

135. Kyte J, Doolittle RF (May 1982). "A simple method for displaying the hydropathic character of a protein". Journal of Molecular Biology. 157 (1): 105–32. doi:10.1016/0022-2836(82)90515-0. PMID 7108955.

136. Freifelder, D. (1983). Physical Biochemistry (2nd ed.). W. H. Freeman and Company. ISBN 0-7167-1315-2.[page needed]

137. Kozlowski, Lukasz P. (2016-10-26). "Proteome-pI: proteome isoelectric point database". Nucleic Acids Research: gkw978. doi:10.1093/nar/gkw978. PMID 27789699.

138. Aasland, Rein; Abrams, Charles; Ampe, Christophe; Ball, Linda J.; Bedford, Mark T.; Cesareni, Gianni; Gimona, Mario; Hurley, James H.; Jarchau, Thomas (20 February 2002). "Normalization of nomenclature for peptide motifs as ligands of modular protein domains". FEBS Letters. 513 (1): 141–144. doi:10.1016/S0014-5793(01)03295-1.

139. "A One-Letter Notation for Amino Acid Sequences". European Journal of Biochemistry. 5 (2): 151–153. 1 July 1968. doi:10.1111/j.1432-1033.1968.tb00350.x.

140. Suchanek M, Radzikowska A, Thiele C (April 2005). "Photo-leucine and photo-methionine allow identification of protein-protein interactions in living cells". Nature Methods. 2 (4): 261–7. doi:10.1038/nmeth752. PMID 15782218.

CHAPTER 3

DNA Structure and Replication

Chapter Contents

- **Introduction**
- **DNA Structures**
- **DNA Polymerase**
- **Function**
- **Variation Across Species**

DNA replication is a process whereby a single DNA molecule produces two identical copies. Each side of the DNA's double-helix structure extends in the opposite direction, allowing it to split down the middle. Once split, each side serves as a template for a new strand. In the case of prokaryotic cells, the process occurs in the cytoplasm, whereas in eukaryotic cells, the replication occurs in the nucleus.

INTRODUCTION

In molecular biology, DNA replication is the biological process of producing two identical replicas of DNA from one original DNA molecule. This process occurs in all living organisms and is the basis for biological inheritance. DNA is made up of a double helix of two complementary strands. During replication, these strands are separated. Each strand of the original DNA molecule then serves as a template for the production of its counterpart, a process referred to as semiconservative replication. Cellular proofreading and error-checking mechanisms ensure near perfect fidelity for DNA replication.

In a cell, DNA replication begins at specific locations, or origins of replication, in the genome. Unwinding of DNA at the origin and synthesis of new strands results in replication forks growing bi-directionally from the origin. A number of proteins are associated with the replication fork to help in the initiation and continuation of DNA synthesis. Most prominently, DNA polymerase synthesizes the new strands by adding nucleotides that complement each (template) strand. DNA replication occurs during the S-stage of interphase.

DNA replication can also be performed *in vitro* (artificially, outside a cell). DNA polymerases isolated from cells and artificial DNA primers can be used to initiate DNA synthesis at known sequences in a template DNA molecule. The polymerase chain reaction (PCR), a common laboratory technique, cyclically applies such artificial synthesis to amplify a specific target DNA fragment from a pool of DNA.

DNA STRUCTURES

DNA usually exists as a double-stranded structure, with both strands coiled together to form the characteristic double-helix. Each single strand of DNA is a chain of four types of nucleotides. Nucleotides in DNA contain a deoxyribose sugar, a phosphate, and a nucleobase. The four types of nucleotide correspond to the four nucleobases adenine, cytosine, guanine, and thymine, commonly abbreviated as A,C, G and T. Adenine and guanine are purine bases, while cytosine and thymine are pyrimidines. These nucleotides form phosphodiester bonds, creating the phosphate-deoxyribose backbone of the DNA double helix with the nuclei bases pointing inward (i.e., toward the opposing strand). Nucleotides (bases) are matched between strands through hydrogen bonds to form base pairs. Adenine pairs with thymine (two hydrogen bonds), and guanine pairs with cytosine (stronger: three hydrogen bonds).

DNA strands have a directionality, and the different ends of a single strand are called the "3' (three-prime) end" and the "5' (five-prime) end". By convention, if the base sequence of a single strand of DNA is given, the left end of the sequence is the 5' end, while the right end of the sequence is the 3' end. The strands of the double helix are anti-parallel with one being 5' to 3', and the opposite strand 3' to 5'. These terms refer to the carbon atom in deoxyribose to which the next phosphate in the chain attaches. Directionality has consequences in DNA synthesis, because DNA polymerase can synthesize DNA in only one direction by adding nucleotides to the 3' end of a DNA strand.

The pairing of complementary bases in DNA (through hydrogen bonding) means that the information contained within each strand is redundant. Phosphodiester (intra-strand) bonds are stronger than hydrogen (inter-strand) bonds. This allows the strands to be separated from one another. The nucleotides on a single strand can therefore be used to reconstruct nucleotides on a newly synthesized partner strand.

DNA POLYMERASE

In molecular biology, DNA polymerases are enzymes that synthesize DNA molecules from deoxyribonucleotides, the building blocks of DNA. These enzymes are essential to DNA replication and usually work in pairs to create two identical DNA strands from a single original DNA molecule. During this process, DNA polymerase "reads" the existing DNA strands to create two new strands that match the existing ones.

These enzymes catalyze the following chemical reaction deoxynucleoside triphosphate + $DNA_n \rightleftharpoons$ diphosphate + DNA_{n+1}

Catalyses DNA-template-directed extension of the 3'- end of a DNA strand by one nucleotide at a time.

Every time a cell divides, DNA polymerases are required to help duplicate the cell's DNA, so that a copy of the original DNA molecule can be passed to each daughter cell. In this way, genetic information is passed down from generation to generation.

Before replication can take place, an enzyme called helicase unwinds the DNA molecule from its tightly woven form. This opens up or "unzips" the double-stranded DNA to give two single strands of DNA that can be used as templates for replication.

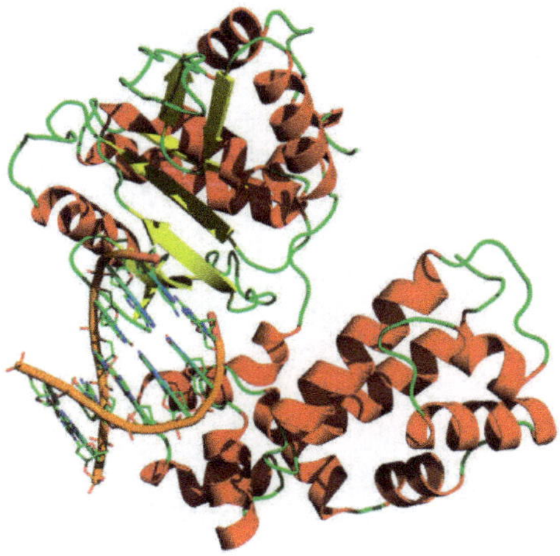

3D structure of the DNA-binding helix-turn-helix motifs in human DNA polymerase beta (based on PDB file 7ICG)

In 1956, Arthur Kornberg and colleagues discovered DNA polymerase I (Pol I), in *Escherichia coli*. They described the DNA replication process by which DNA polymerase copies the base sequence of a template DNA strand. Kornberg was later awarded the Nobel Prize in Physiology or Medicine in 1959 for this work. DNA polymerase II was also discovered by Kornberg and Malcolm E. Gefter in 1970 while further elucidating the role of Pol I in *E. coli* DNA replication.

FUNCTION

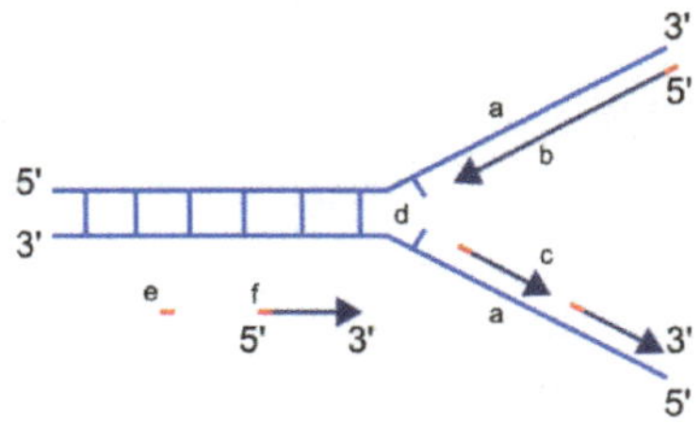

DNA polymerase moves along the old strand in the 3'-5' direction, creating a new strand having a 5'-3' direction.

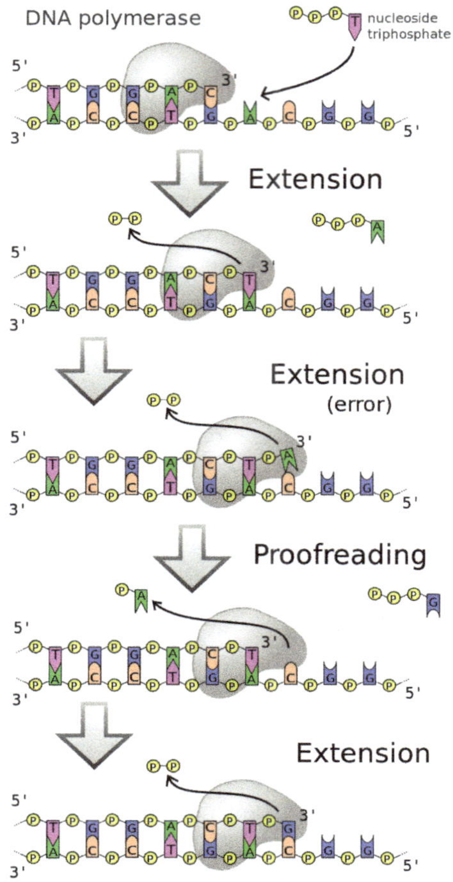

DNA polymerase with proofreading ability

The main function of DNA polymerase is to synthesize DNA from deoxyribonucleotides, the building blocks of DNA. The DNA copies are created by the pairing of nucleotides to bases present on each strand of the original DNA molecule. This pairing always occurs in specific combinations, with cytosine along with guanine, and thymine along with adenine, forming two separate pairs, respectively. By contrast, RNA polymerases synthesize RNA from ribonucleotides from either RNA or DNA.

When synthesizing new DNA, DNA polymerase can add free nucleotides only to the 3' end of the newly forming strand. This results in elongation of the newly forming strand in a 5'-3' direction. No known DNA polymerase is able to begin a new chain (*de novo*); it can only add a nucleotide onto a pre-existing 3'-OH group, and therefore needs a primer at which it can add the first nucleotide. Primers consist of RNA or DNA bases (or both). In DNA replication, the first two bases are always RNA, and are synthesized by another enzyme called primase. Enzymes, helicase and topoisomerase II, are required to unwind DNA from a double-strand structure to a single-strand structure to facilitate replication of each strand consistent with the semiconservative model of DNA replication.

It is important to note that the directionality of the newly forming strand (the daughter strand) is opposite to the direction in which DNA polymerase moves along the template strand. Since DNA polymerase requires a free 3' OH group for initiation of synthesis, it can synthesize in only one direction by extending the 3' end of the preexisting nucleotide chain. Hence, DNA polymerase moves along the template strand in a 3'-5' direction, and the daughter strand is formed in a 5'-3' direction. This difference enables the resultant double-strand DNA formed to be composed of two DNA strands that are antiparallel to each other.

The function of DNA polymerase is not quite perfect, with the enzyme making about one mistake for every billion base pairs copied. Error correction is a property of some, but not all DNA polymerases. This process corrects mistakes in newly synthesized DNA. When an incorrect base pair is recognized, DNA polymerase moves backwards by one base pair of DNA. The 3'-5' exonuclease activity of the enzyme allows the incorrect base pair to be excised (this activity is known as *proofreading*). Following base excision, the polymerase can re-insert the correct base and replication can continue forwards. This preserves the integrity of the original DNA strand that is passed onto the daughter cells.

Fidelity is very important in DNA replication. Mismatches in DNA base pairing can potentially result in dysfunctional proteins and could lead to cancer. Many DNA polymerase contain an exonuclease domain, which acts in detecting base pair mismatches and further performs in the removal of the incorrect nucleotide to be replaced by the correct one. The shape and the interactions accommodating the Watson and Crick base pair are what primarily contribute to the detection or error. Hydrogen bonds play a key role in base pair binding and interaction. The loss of an interaction, which occurs at a mismatch, is said to trigger a shift in the balance, for the binding of the template-primer, from the polymerase, to the exonuclease domain. In addition, an incorporation of a wrong nucleotide causes a retard in DNA polymerization. This delay gives time for the DNA to be switched from the polymerase site to the exonuclease site. Different conformational changes and loss of interaction occur at different mismatches. In a purine:pyrimidine mismatch there is a displacement of the pyrimidine towards the major groove and the purine towards the minor groove. Relative to the shape of DNA polymerase's binding pocket, steric clashes occur between the purine and residues in the minor groove, and important van der Waals and electrostatic interactions are lost by the pyrimidine. Pyrimidine:pyrimidine and purine:purine mismatches present less notable changes since the bases are displaced towards the major groove, and less steric hindrance is experienced. However, although the different mismatches result in different steric properties, DNA polymerase is still able to detect and differentiate them so uniformly and maintain fidelity in DNA replication.

Structure

The known DNA polymerases have highly conserved structure, which means that their overall catalytic subunits vary very little from species to species, independent of their domain structures. Conserved structures usually indicate important, irreplaceable functions of the cell, the maintenance of which provides evolutionary advantages. The shape can be described as resembling a right hand with thumb, finger, and palm domains. The palm domain appears to function in catalyzing the transfer of phosphoryl groups in the phosphoryl transfer reaction. DNA is bound to the palm when the enzyme is active. This reaction is believed to be catalyzed by a two-metal-ion mechanism. The finger domain functions to bind the nucleoside triphosphates with the template base. The thumb domain plays a potential role in the processivity, translocation, and positioning of the DNA.

Processivity

DNA polymerase's rapid catalysis is due to its processive nature. Processivity is a characteristic of enzymes that function on polymeric substrates. In the case of DNA polymerase, the degree of processivity refers to the average number of nucleotides added each time the enzyme binds a template. The average DNA polymerase requires about one second locating and binding a primer/template junction. Once it is bound, a nonprocessive DNA polymerase adds nucleotides at a rate of one nucleotide per second. Processive DNA polymerases, however, add multiple nucleotides per second, drastically increasing the rate of DNA synthesis. The degree of processivity is directly proportional to the rate of DNA synthesis. The rate of DNA synthesis in a living cell was first determined as the rate of phage T4 DNA elongation in phage infected *E. coli*. During the period of exponential DNA increase at 37 °C, the rate was 749 nucleotides per second.

DNA polymerase's ability to slide along the DNA template allows increased processivity. There is a dramatic increase in processivity at the replication fork. This increase is facilitated by the DNA polymerase's association with proteins known as the sliding DNA clamp. The clamps are multiple protein subunits associated in the shape of a ring. Using the hydrolysis of ATP, a class of proteins known as the sliding clamp loading proteins open up the ring structure of the sliding DNA clamps allowing binding to and release from the DNA strand. Protein-protein interaction with the clamp prevents DNA polymerase from diffusing from the DNA template, thereby ensuring that the enzyme binds the same primer/template junction and continues replication. DNA polymerase changes conformation, increasing affinity to the clamp when associated with it and decreasing affinity when it completes the replication of a stretch of DNA to allow release from the clamp.

VARIATION ACROSS SPECIES

Based on sequence homology, DNA polymerases can be further subdivided into seven different families: A, B, C, D, X, Y, and RT.

Some viruses also encode special DNA polymerases, such as Hepatitis B virus DNA polymerase. These may selectively replicate viral DNA through a variety of mechanisms. Retroviruses encode an unusual DNA polymerase called reverse transcriptase, which is an RNA-dependent DNA polymerase (RdDp). It polymerizes DNA from a template of RNA.

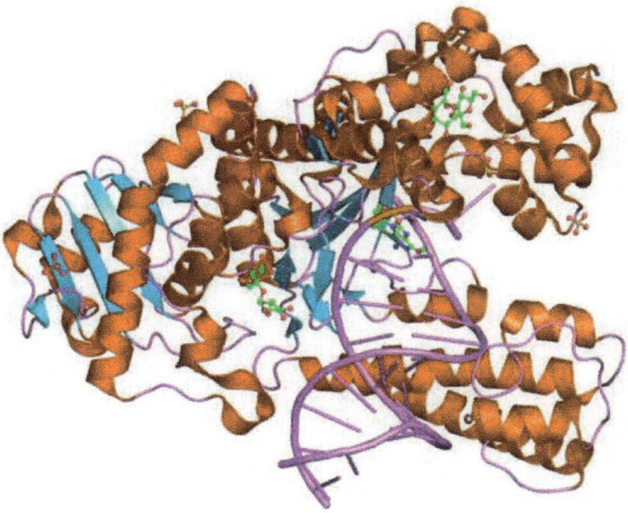

c:o6-methyl-guanine pair in the polymerase-2 basepair position

Family	Types of DNA polymerase	Species	Examples
A	Replicative and Repair Polymerases	Eukaryotic and Prokaryotic	T7 DNA polymerase, Pol I, and DNA Polymerase γ
B	Replicative and Repair Polymerases	Eukaryotic and Prokaryotic	Pol II, Pol B, Pol ζ, Pol α, δ, and ε
C	Replicative Polymerases	Prokaryotic	Pol III
D	Replicative Polymerases	Euryarchaeota	Not well-characterized
X	Replicative and Repair Polymerases	Eukaryotic	Pol β, Pol σ, Pol λ, Pol μ, and Terminal deoxynucleotidyl transferase
Y	Replicative and Repair Polymerases	Eukaryotic and Prokaryotic	Pol ι (iota), Pol κ (kappa), Pol η (eta),Pol IV, and Pol V
RT	Replicative and Repair Polymerases	Viruses, Retroviruses, and Eukaryotic	Telomerase, Hepatitis B virus

Prokaryotic DNA Polymerases

Pol I

Prokaryotic family A polymerases include the DNA polymerase I (Pol I) enzyme, which is encoded by the *polA* gene and ubiquitous among prokaryotes. This repair polymerase is involved in excision repair with both 3'-5' and 5'-3' exonuclease activity and processing of Okazaki fragments generated during lagging strand synthesis. Pol I is the most abundant polymerase, accounting for >95% of polymerase activity in *E. coli*; yet cells lacking Pol I have been found suggesting Pol I activity can be replaced by the other four polymerases. Pol I adds ~15-20 nucleotides per second, thus showing poor processivity. Instead, Pol I starts adding nucleotides at the RNA primer:template junction known as the origin of replication (ori). Approximately 400 bp downstream from the origin, the Pol III holoenzyme is assembled and takes over replication at a highly processive speed and nature.

Pol II

DNA polymerase II, a family B polymerase, is a polB gene product also known as DinA. Pol II has 3'-5' exonuclease activity and participates in DNA repair, replication restart to bypass lesions, and its cell presence can jump from ~30-50 copies per cell to ~200-300 during SOS induction. Pol II is also thought to be a backup to Pol III as it can interact with holoenzyme proteins and assume a high level of processivity. The main role of Pol II is thought to be the ability to direct polymerase activity at the replication fork and helped stalled Pol III bypass terminal mismatches.

Pol III

DNA polymerase III holoenzyme is the primary enzyme involved in DNA replication in *E. coli* and belongs to family C polymerases. It consists of three assemblies: the pol III core, the beta sliding clamp processivity factor, and the clamp-loading complex. The core consists of three subunits: α, the polymerase activity hub, ε, exonucleolytic proofreader, and θ, which may act as a stabilizer for ε. The holoenzyme contains two cores, one for each strand, the lagging and leading. The beta sliding clamp processivity factor is also present in duplicate, one for each core, to create a clamp that encloses DNA allowing for high processivity. The third assembly is a seven-subunit ($\tau 2\gamma\delta\delta'\chi\psi$) clamp loader complex. Recent research has classified Family C polymerases as a subcategory of Family X with no eukaryotic equivalents.

Pol IV

In *E. coli*, DNA polymerase IV (Pol 4) is an error-prone DNA polymerase involved in non-targeted mutagenesis. Pol IV is a Family Y polymerase expressed by the dinB gene that is switched on via SOS induction caused by stalled polymerases at the replication fork. During SOS induction, Pol IV production is increased tenfold and one of the functions during this time is to interfere with Pol III holoenzyme processivity. This creates a checkpoint, stops replication, and allows time to repair DNA lesions via the appropriate repair pathway. Another function of Pol IV is to perform translesion synthesis at the stalled replication fork like, for example, bypassing N2-deoxyguanine adducts at a faster rate than transversing undamaged DNA. Cells lacking dinB gene have a higher rate of mutagenesis caused by DNA damaging agents.

Pol V

DNA polymerase V (Pol V) is a Y-family DNA polymerase that is involved in SOS response and translesion synthesis DNA repair mechanisms.Transcription of Pol V via the umuDC genes is highly regulated to produce only Pol V when damaged DNA is present in the cell generating an SOS response. Stalled polymerases causes RecA to bind to the ssDNA, which causes the LexA protein to autodigest. LexA then loses its ability to repress the transcription of the umuDC operon. The same RecA-ssDNA nucleoprotein posttranslationally modifies the UmuD protein into UmuD' protein. UmuD and UmuD' form a heterodimer that interacts with UmuC, which in turn activates umuC's polymerase catalytic activity on damaged DNA.

Eukaryotic DNA Polymerase

Polymerases β, λ, σ and μ (Beta, Lambda, Sigma, and Mu)

Family X polymerases contain the well-known eukaryotic polymerase pol β (beta), as well as other eukaryotic polymerases such as Pol σ (sigma), Pol λ (lambda), Pol μ (mu), and Terminal deoxynucleotidyl transferase (TdT). Family X polymerases are found mainly in vertebrates, and a few are found in plants and fungi. These polymerases have highly conserved regions that include two helix-hairpin-helix motifs that are imperative in the DNA-polymerase interactions. One motif is located in the 8 kDa domain that interacts with downstream DNA and one motif is located in the thumb domain that interacts with the primer strand. Pol β, encoded by POLB gene, is required for short-patch base excision repair, a DNA repair pathway that is essential for repairing alkylated or oxidized bases as well as abasic sites. Pol λ and Pol μ, encoded by the POLL and POLM genes respectively, are involved in non-homologous end-joining, a mechanism for rejoining DNA double-strand breaks due to hydrogen peroxide and ionizing radiation, respectively. TdT is expressed only in lymphoid tissue, and adds "n nucleotides" to double-strand breaks formed during V(D)J recombination to promote immunological diversity.

Polymerases α, δ and ε (Alpha, Delta, and Epsilon)

Pol α (alpha), Pol δ (delta), and Pol ε (epsilon) are members of Family B Polymerases and are the main polymerases involved with nuclear DNA replication. Pol α complex (pol α-DNA primase complex) consists of four subunits: the catalytic subunit POLA1, the regulatory subunit POLA2, and the small and the large primase subunits PRIM1 and PRIM2 respectively. Once primase has created the RNA primer, Pol α starts replication elongating the primer with ~20 nucleotides. Due to its high processivity, Pol δ takes over the leading and lagging strand synthesis from Pol α. Pol δ is expressed by genes POLD1, creating the catalytic subunit, POLD2, POLD3, and POLD4 creating the other subunits that interact with Proliferating Cell Nuclear Antigen (PCNA), which is a DNA clamp that allows Pol δ to possess processivity. Pol ε is encoded by the POLE1, the catalytic subunit, POLE2, and POLE3 gene. It has been reported that the function of Pol ε is to extend the leading strand during replication, while Pol δ primarily replicates the lagging strand; however, recent evidence suggested that Pol δ might have a role in replicating the leading strand of DNA as well. Pol ε's C-terminus region is thought to be essential to cell vitality as well. The C-terminus region is thought to provide a checkpoint before entering anaphase, provide stability to the holoenzyme, and add proteins to the holoenzyme necessary for initiation of replication.

Polymerases H, I and K (Eta, Iota, and Kappa)

Pol η (eta), Pol ι (iota), and Pol κ (kappa), are Family Y DNA polymerases involved in the DNA repair by translesion synthesis and encoded by genes POLH, POLI, and POLK respectively. Members of Family Y have five common motifs to aid in binding the substrate and primer terminus and they all include the typical right hand thumb, palm and finger domains with added domains like little finger (LF), polymerase-associated domain (PAD), or wrist. The active site, however, differs between family members due to the different lesions being repaired. Polymerases in Family Y are low-fidelity polymerases, but have been proven to do more good than harm as mutations that affect the polymerase can cause various diseases, such as skin cancer and Xeroderma Pigmentosum Variant (XPS). The importance of these polymerases is evidenced by the fact that gene encoding DNA polymerase η is referred as XPV, because loss of this gene results in the disease Xeroderma Pigmentosum Variant. Pol η is particularly important for allowing accurate translesion synthesis of DNA damage resulting from ultraviolet radiation. The functionality of Pol κ is not completely understood, but researchers have found two probable functions. Pol κ is thought to act as an extender or an inserter of a specific base at certain DNA lesions. All three translesion synthesis polymerases, along with Rev1, are recruited to damaged lesions via stalled replicative DNA polymerases. There are two pathways of damage repair leading researchers to conclude that the chosen pathway depends on which strand contains the damage, the leading or lagging strand.

Polymerases Rev1 and ζ (Zeta)

Pol ζ another B family polymerase, is made of two subunits Rev3, the catalytic subunit, and Rev7, which increases the catalytic function of the polymerase, and is involved in translesion synthesis. Pol ζ lacks 3’ to 5’ exonuclease activity, is unique in that it can extend primers with terminal mismatches. Rev1 has three regions of interest in the BRCT domain, ubiquitin-binding domain, and C-terminal domain and has dCMP transferase ability, which adds deoxycytidine opposite lesions that would stall replicative polymerases Pol δ and Pol ε. These stalled polymerases activate ubiquitin complexes that in turn disassociate replication polymerases and recruit Pol ζ and Rev1. Together Pol ζ and Rev1 add deoxycytidine and Pol ζ extends past the lesion. Through a yet undetermined process, Pol ζ disassociates and replication polymerases reassociate and continue replication. Pol ζ and Rev1 are not required for replication, but loss of REV3 gene in budding yeast can cause increased sensitivity to DNA-damaging agents due to collapse of replication forks where replication polymerases have stalled.

Telomerase

Telomerase is a ribonucleoprotein recruited to replicate ends of linear chromosomes because normal DNA polymerase cannot replicate the ends, or telomere. The single-strand 3’ overhang of the double-strand chromosome with the sequence 5’-TTAGGG-3’ recruits telomerase. Telomerase acts like other DNA polymerases by extending the 3' end, but, unlike other DNA polymerases, telomerase does not require a template. The TERT subunit, an example of a reverse transcriptase, uses the RNA subunit to form the primer–template junction that allows telomerase to extend the 3' end of chromosome ends. The gradual decrease in size of telomeres as the result of many replications over a lifetime are thought to be associated with the effects of aging.

Polymerases Γ and Θ (Gamma and Theta)

Pol γ (gamma) and Pol θ (theta) are Family A polymerases. Pol γ, encoded by the POLG gene, is the only mtDNA polymerase and therefore replicates, repairs, and has proofreading 3'-5' exonuclease and 5' dRP lyase activities. Any mutation that leads to limited or non-functioning Pol γ has a significant effect on mtDNA and is the most common cause of autosomal inherited mitochondrial disorders. Pol γ contains a C-terminus polymerase domain and an N-terminus 3'-5' exonuclease domain that are connected via the linker region, which binds the accessory subunit. The accessory subunit binds DNA and is required for processivity of Pol γ. Point mutation A467T in the linker region is responsible for more than one-third of all Pol γ-associated mitochondrial disorders. While many homologs of Pol θ, encoded by the POLQ gene, are found in eukaryotes, its function is not clearly understood. The sequence of amino acids in the C-terminus is what classifies Pol θ as Family A polymerase, although the error rate for Pol θ is more closely related to Family Y polymerases. Pol θ extends mismatched primer termini and can bypass abasic sites by adding a nucleotide. It also has Deoxyribophosphodiesterase (dRPase) activity in the polymerase domain and can show ATPase activity in close proximity to ssDNA.

Polymerase N (Nu)

Reverse Transcriptase

Retroviruses encode an unusual DNA polymerase called reverse transcriptase, which is an RNA-dependent DNA polymerase (RdDp) that synthesizes DNA from a template of RNA. The reverse transcriptase family contain both DNA polymerase functionality and RNase H functionality, which degrades RNA base-paired to DNA. An example of a retrovirus is HIV.

Replication Process

DNA replication, like all biological polymerization processes, proceeds in three enzymatically catalyzed and coordinated steps: initiation, elongation and termination.

Initiation

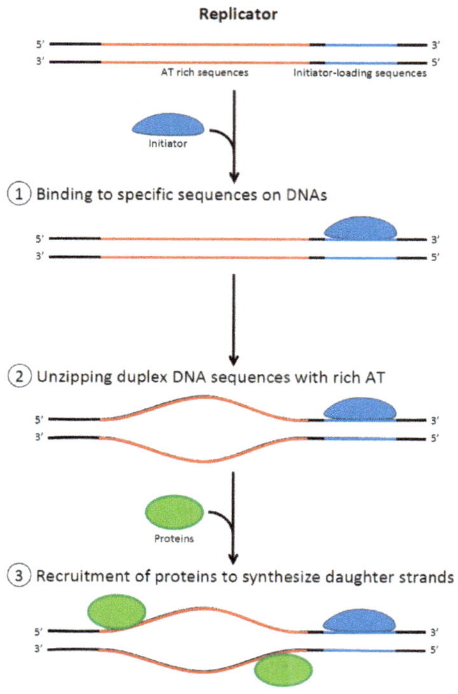

Role of initiators for initiation of DNA replication.

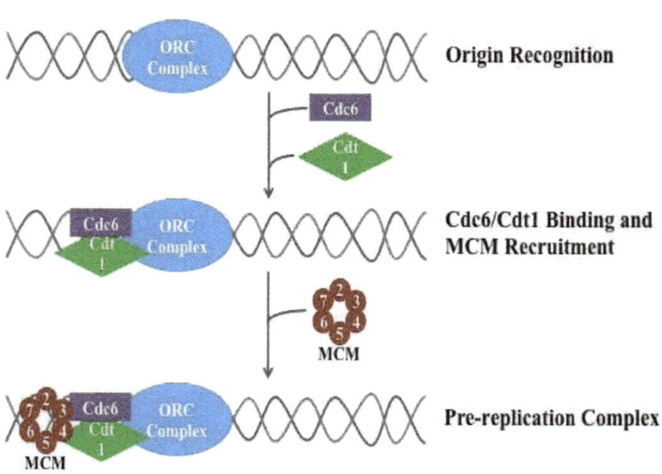

Formation of pre-replication complex.

For a cell to divide, it must first replicate its DNA. This process is initiated at particular points in the DNA, known as "origins", which are targeted by initiator proteins. In *E.*

coli this protein is DnaA; in yeast, this is the origin recognition complex. Sequences used by initiator proteins tend to be "AT-rich" (rich in adenine and thymine bases), because A-T base pairs have two hydrogen bonds (rather than the three formed in a C-G pair) and thus are easier to strand separate. Once the origin has been located, these initiators recruit other proteins and form the pre-replication complex, which unzips the double-stranded DNA.

Elongation

DNA polymerase has 5'-3' activity. All known DNA replication systems require a free 3' hydroxyl group before synthesis can be initiated (note: the DNA template is read in 3' to 5' direction whereas a new strand is synthesized in the 5' to 3' direction—this is often confused). Four distinct mechanisms for DNA synthesis are recognized:

- All cellular life forms and many DNA viruses, phages and plasmids use a primase to synthesize a short RNA primer with a free 3' OH group which is subsequently elongated by a DNA polymerase.
- The retroelements (including retroviruses) employ a transfer RNA that primes DNA replication by providing a free 3′ OH that is used for elongation by the reverse transcriptase.
- In the adenoviruses and the φ29 family of bacteriophages, the 3' OH group is provided by the side chain of an amino acid of the genome attached protein (the terminal protein) to which nucleotides are added by the DNA polymerase to form a new strand.
- In the single stranded DNA viruses — a group that includes the circoviruses, the geminiviruses, the parvoviruses and others — and also the many phages and plasmids that use the rolling circle replication (RCR) mechanism, the RCR endonuclease creates a nick in the genome strand (single stranded viruses) or one of the DNA strands (plasmids). The 5′ end of the nicked strand is transferred to a tyrosine residue on the nuclease and the free 3′ OH group is then used by the DNA polymerase to synthesize the new strand.

The first is the best known of these mechanisms and is used by the cellular organisms. In this mechanism, once the two strands are separated, primase adds RNA primers to the template strands. The leading strand receives one RNA primer while the lagging strand receives several. The leading strand is continuously extended from the primer by a DNA polymerase with high processivity, while the lagging strand is extended discontinuously from each primer forming Okazaki fragments. RNase removes the primer RNA fragments, and a low processivity DNA polymerase distinct from the replicative polymerase enters to fill the gaps. When this is complete, a single nick on the leading strand and several nicks on the lagging strand can be found. Ligase works to fill these nicks in, thus completing the newly replicated DNA molecule.

The primase used in this process differs significantly between bacteria and archaea/eukaryotes. Bacteria use a primase belonging to the DnaG protein superfamily which contains a catalytic domain of the TOPRIM fold type. The TOPRIM fold contains an α/β core with four conserved strands in a Rossmann-like topology. This structure is also found in the catalytic domains of topoisomerase Ia, topoisomerase II, the OLD-family nucleases and DNA repair proteins related to the RecR protein.

The primase used by archaea and eukaryotes, in contrast, contains a highly derived version of the RNA recognition motif (RRM). This primase is structurally similar to many viral RNA-dependent RNA polymerases, reverse transcriptases, cyclic nucleotide generating cyclases and DNA polymerases of the A/B/Y families that are involved in DNA replication and repair. In eukaryotic replication, the primase forms a complex with Pol α.

Multiple DNA polymerases take on different roles in the DNA replication process. In *E. coli*, DNA Pol III is the polymerase enzyme primarily responsible for DNA replication. It assembles into a replication complex at the replication fork that exhibits extremely high processivity, remaining intact for the entire replication cycle. In contrast, DNA Pol I is the enzyme responsible for replacing RNA primers with DNA. DNA Pol I has a 5' to 3' exonuclease activity in addition to its polymerase activity, and uses its exonuclease activity to degrade the RNA primers ahead of it as it extends the DNA strand behind it, in a process called nick translation. Pol I is much less processive than Pol III because its primary function in DNA replication is to create many short DNA regions rather than a few very long regions.

In eukaryotes, the low-processivity enzyme, Pol α, helps to initiate replication because it forms a complex with primase. In eukaryotes, leading strand synthesis is thought to be conducted by Pol ε; however, this view has recently been challenged, suggesting a role for Pol δ. Primer removal is completed Pol δ while repair of DNA during replication is completed by Pol ε.

As DNA synthesis continues, the original DNA strands continue to unwind on each side of the bubble, forming a replication fork with two prongs. In bacteria, which have a single origin of replication on their circular chromosome, this process creates a "theta structure" (resembling the Greek letter theta: θ). In contrast, eukaryotes have longer linear chromosomes and initiate replication at multiple origins within these.>

Replication Fork

The replication fork is a structure that forms within the nucleus during DNA replication. It is created by helicases, which break the hydrogen bonds holding the two DNA strands together. The resulting structure has two branching "prongs", each one made up of a single strand of DNA. These two strands serve as the template for the leading and lagging strands, which will be created as DNA polymerase matches complementary nucleotides to the templates; the templates may be properly referred to as the leading strand template and the lagging strand template.

DNA is always synthesized in the 5' to 3' direction. Since the leading and lagging strand templates are oriented in opposite directions at the replication fork, a major issue is how to achieve synthesis of nascent (new) lagging strand DNA, whose direction of synthesis is opposite to the direction of the growing replication fork.

Leading Strand

The leading strand is the strand of nascent DNA which is being synthesized in the same direction as the growing replication fork. A polymerase "reads" the leading strand *template* and adds complementary nucleotides to the nascent leading strand on a continuous basis.

Lagging Strand

The lagging strand is the strand of nascent DNA whose direction of synthesis is opposite to the direction of the growing replication fork. Because of its orientation, replication of the lagging strand is more complicated as compared to that of the leading strand. As a consequence, the DNA polymerase on this strand is seen to "lag behind" the other strand.

The lagging strand is synthesized in short, separated segments. On the lagging strand *template*, a primase "reads" the template DNA and initiates synthesis of a short complementary RNA primer. A DNA polymerase extends the primed segments, forming Okazaki fragments. The RNA primers are then removed and replaced with DNA, and the fragments of DNA are joined together by DNA ligase.

Dynamics at the Replication Fork

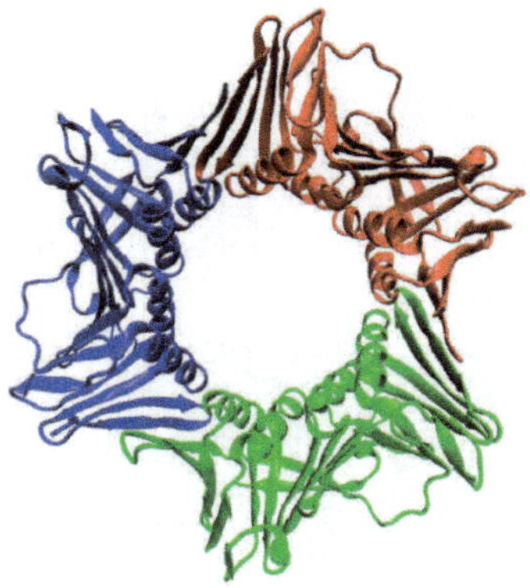

The assembled human DNA clamp, a trimer of the protein PCNA.

As helicase unwinds DNA at the replication fork, the DNA ahead is forced to rotate. This process results in a build-up of twists in the DNA ahead. This build-up forms a torsional resistance that would eventually halt the progress of the replication fork. Topoisomerases are enzymes that temporarily break the strands of DNA, relieving the tension caused by unwinding the two strands of the DNA helix; topoisomerases (including DNA gyrase) achieve this by adding negative supercoils to the DNA helix.

Bare single-stranded DNA tends to fold back on itself forming secondary structures; these structures can interfere with the movement of DNA polymerase. To prevent this, single-strand binding proteins bind to the DNA until a second strand is synthesized, preventing secondary structure formation.

Clamp proteins form a sliding clamp around DNA, helping the DNA polymerase maintain contact with its template, thereby assisting with processivity. The inner face of the clamp enables DNA to be threaded through it. Once the polymerase reaches the end of the template or detects double-stranded DNA, the sliding clamp undergoes a conformational change that releases the DNA polymerase. Clamp-loading proteins are used to initially load the clamp, recognizing the junction between template and RNA primers.[:274-5]

DNA Replication Proteins

At the replication fork, many replication enzymes assemble on the DNA into a complex molecular machine called the replisome. The following is a list of major DNA replication enzymes that participate in the replisome:

Enzyme	Function in DNA replication
DNA Helicase	Also known as helix destabilizing enzyme. Helicase separates the two strands of DNA at the Replication Fork behind the topoisomerase.
DNA Polymerase	The enzyme responsible for catalyzing the addition of nucleotide substrates to DNA in the 5' to 3' direction during DNA replication. Also performs proof-reading and error correction. There exist many different types of DNA Polymerase, (such as the DNA Polymerase III)each of which perform different functions in different types of cells.
DNA clamp	A protein which prevents elongating DNA polymerases from dissociating from the DNA parent strand.
Single-Strand Binding (SSB) Proteins	Bind to ssDNA and prevent the DNA double helix from re-annealing after DNA helicase unwinds it, thus maintaining the strand separation, and facilitating the synthesis of the nascent strand.
Topoisomerase	Relaxes the DNA from its super-coiled nature.
DNA Gyrase	Relieves strain of unwinding by DNA helicase; this is a specific type of topoisomerase
DNA Ligase	Re-anneals the semi-conservative strands and joins Okazaki Fragments of the lagging strand.
Primase	Provides a starting point of RNA (or DNA) for DNA polymerase to begin synthesis of the new DNA strand.
Telomerase	Lengthens telomeric DNA by adding repetitive nucleotide sequences to the ends of **eukaryotic chromosomes**. This allows germ cells and stem cells to avoid the Hayflick limit on cell division.

Replication Machinery

Replication machineries consist of factors involved in DNA replication and appearing on template ssDNAs. Replication machineries include primosotors are replication enzymes; DNA polymerase, DNA helicases, DNA clamps and DNA topoisomerases, and replication proteins; e.g. single-stranded DNA binding proteins (SSB). In the replication machineries these components coordinate. In most of the bacteria, all of the factors involved in DNA replication are located on replication forks and the complexes stay on the forks during DNA replication. These replication machineries are called replisomes or DNA replicase systems. These terms are generic terms for proteins located on replication forks. In eukaryotic and some bacterial cells the replisomes are not formed.

Since replication machineries do not move relatively to template DNAs such as factories, they are called a replication factory. In an alternative figure, DNA factories are similar to projectors and DNAs are like as cinematic films passing constantly into the projectors. In the replication factory model, after both DNA helicases for leading stands and lagging strands are loaded on the template DNAs, the helicases run along the DNAs into each other. The helicases remain associated for the remainder of replication process. Peter Meister et al. observed directly replication sites in budding yeast by monitoring green fluorescent protein(GFP)-tagged DNA polymerases α. They detected DNA replication of pairs of the tagged loci spaced apart symmetrically from a replication origin and found that the distance between the pairs decreased markedly by time. This finding suggests that the mechanism of DNA replication goes with DNA factories. That is, couples of replication factories are loaded on replication origins and the factories associated with each other. Also, template DNAs move into the factories, which bring extrusion of the template ssDNAs and nascent DNAs. Meister's finding is the first direct evidence of replication factory model. Subsequent research has shown that DNA helicases form dimers in many eukaryotic cells and bacterial replication machineries stay in single intranuclear location during DNA synthesis.

The replication factories perform disentanglement of sister chromatids. The disentanglement is essential for distributing the chromatids into daughter cells after DNA replication. Because sister chromatids after DNA replication hold each other by **Cohesin** rings, there is the only chance for the disentanglement in DNA replication. Fixing of replication machineries as replication factories can improve the success rate of DNA replication. If replication forks move freely in chromosomes, catenation of nuclei is aggravated and impedes mitotic segregation.

Termination

Eukaryotes initiate DNA replication at multiple points in the chromosome, so replication forks meet and terminate at many points in the chromosome; these are not known to be regulated in any particular way. Because eukaryotes have linear chromosomes, DNA replication is unable to reach the very end of the chromosomes, but ends at the telomere region of repetitive DNA close to the ends. This shortens the telomere of the daughter DNA strand. Shortening of the telomeres is a normal process in somatic cells. As a result, cells can only divide a certain number of times before the DNA loss prevents further division. (This is known as the Hayflick limit.) Within the germ cell line, which passes DNA to the next generation, telomerase extends the repetitive sequences of the telomere region to prevent degradation. Telomerase can become mistakenly active in somatic cells, sometimes leading to cancer formation. Increased telomerase activity is one of the hallmarks of cancer.

Termination requires that the progress of the DNA replication fork must stop or be blocked. Termination at a specific locus, when it occurs, involves the interaction between two components: (1) a termination site sequence in the DNA, and (2) a protein which binds to this sequence to physically stop DNA replication. In various bacterial species, this is named the DNA replication terminus site-binding protein, or Ter protein.

Because bacteria have circular chromosomes, termination of replication occurs when the two replication forks meet each other on the opposite end of the parental chromosome. *E. coli* regulates this process through the use of termination sequences that, when bound by the Tus protein, enable only one direction of replication fork to pass through. As a result, the replication forks are constrained to always meet within the termination region of the chromosome.

Regulation

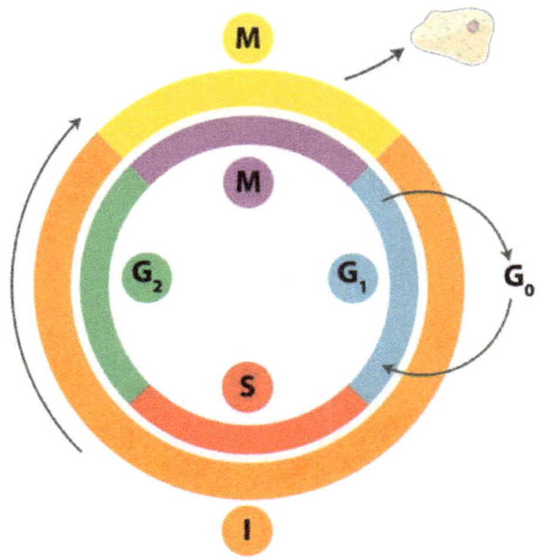

The cell cycle of eukaryotic cells.

Eukaryotes

Within eukaryotes, DNA replication is controlled within the context of the cell cycle. As the cell grows and divides, it progresses through stages in the cell cycle; DNA replication takes place during the S phase (synthesis phase). The progress of the eukaryotic cell through the cycle is controlled by cell cycle checkpoints. Progression through checkpoints is controlled through complex interactions between various proteins, including cyclins and cyclin-dependent kinases. Unlike bacteria, eukaryotic DNA replicates in the confines of the nucleus.

The G1/S checkpoint (or restriction checkpoint) regulates whether eukaryotic cells enter the process of DNA replication and subsequent division. Cells that do not proceed through this checkpoint remain in the G0 stage and do not replicate their DNA.

Replication of chloroplast and mitochondrial genomes occurs independently of the cell cycle, through the process of D-loop replication.

Replication Focus

In vertebrate cells, replication sites concentrate into positions called **replication foci**. Replication sites can be detected by immunostaining daughter strands and replication enzymes and monitoring GFP-tagged replication factors. By these methods it is found that replication foci of varying size and positions appear in S phase of cell division and their number per nucleus is far smaller than the number of genomic replication forks.

P. Heun et al.(2001) tracked GFP-tagged replication foci in budding yeast cells and revealed that replication origins move constantly in G1 and S phase and the dynamics decreased significantly in S phase. Traditionally, replication sites were fixed on spatial structure of chromosomes by nuclear matrix or lamins. The Heun's results denied the traditional concepts, budding yeasts don't have lamins, and support that replication origins self-assemble and form replication foci.

By firing of replication origins, controlled spatially and temporally, the formation of replication foci is regulated. D. A. Jackson et al.(1998) revealed that neighboring origins fire simultaneously in mammalian cells. Spatial juxtaposition of replication sites brings clustering of replication forks. The clustering do rescue of stalled replication forks and favors normal progress of replication forks. Progress of replication forks is inhibited by many factors; collision with proteins or with complexes binding strongly on DNA, deficiency of dNTPs, nicks on template DNAs and so on. If replication forks stall and the remaining sequences from the stalled forks are not replicated, the daughter strands have nick obtained un-replicated sites. The un-replicated sites on one parent's strand hold the other strand together but not daughter strands. Therefore, the resulting sister chromatids cannot separate from each other and cannot divide into 2 daughter cells. When neighboring origins fire and a fork from one origin is stalled, fork from other origin access on an opposite direction of the stalled fork and duplicate the un-replicated sites. As other mechanism of the rescue there is application of dormant replication origins that excess origins don't fire in normal DNA replication.

Bacteria

Most bacteria do not go through a well-defined cell cycle but instead continuously copy their DNA; during rapid growth, this can result in the concurrent occurrence of multiple rounds of replication. In *E. coli*, the best-characterized bacteria, DNA replication is regulated through several mechanisms, including: the hemimethylation and sequestering of the origin sequence, the ratio of adenosine triphosphate (ATP) to adenosine diphosphate (ADP), and the levels of protein DnaA. All these control the binding of initiator proteins to the origin sequences.

Because *E. coli* methylates GATC DNA sequences, DNA synthesis results in hemimethylated sequences. This hemimethylated DNA is recognized by the protein SeqA, which binds and sequesters the origin sequence; in addition, DnaA (required for initiation of replication) binds less well to hemimethylated DNA. As a result, newly replicated origins are prevented from immediately initiating another round of DNA replication.

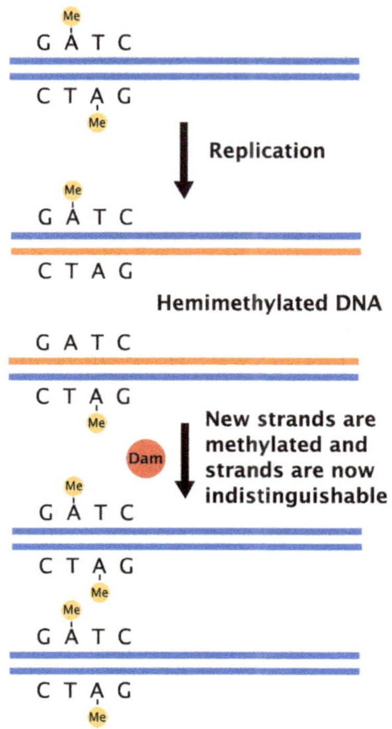

Dam methylates adenine of GATC sites after replication.

ATP builds up when the cell is in a rich medium, triggering DNA replication once the cell has reached a specific size. ATP competes with ADP to bind to DnaA, and the DnaA-ATP complex is able to initiate replication. A certain number of DnaA proteins are also required for DNA replication — each time the origin is copied, the number of binding sites for DnaA doubles, requiring the synthesis of more DnaA to enable another initiation of replication.

Polymerase Chain Reaction

Researchers commonly replicate DNA *in vitro* using the polymerase chain reaction (PCR). PCR uses a pair of primers to span a target region in template DNA, and then polymerizes partner strands in each direction from these primers using a thermostable DNA polymerase. Repeating this process through multiple cycles amplifies the targeted DNA region. At the start of each cycle, the mixture of template and primers is heated, separating the newly synthesized molecule and template. Then, as the mixture cools, both of these become templates for annealing of new primers, and the polymerase extends from these. As a result, the number of copies of the target region doubles each round, increasing exponentially.

Recombinant DNA is a genetic technology that allows a scientist to combine genes from one organism with that of another creating DNA that would never naturally occur.

Recombinant DNA (rDNA) molecules are DNA molecules formed by laboratory methods of genetic recombination (such as molecular cloning) to bring together genetic material from multiple sources, creating sequences that would not otherwise be found in the genome. Recombinant DNA is possible because DNA molecules from all organisms share the same chemical structure. They differ only in the nucleotide sequence within that identical overall structure.

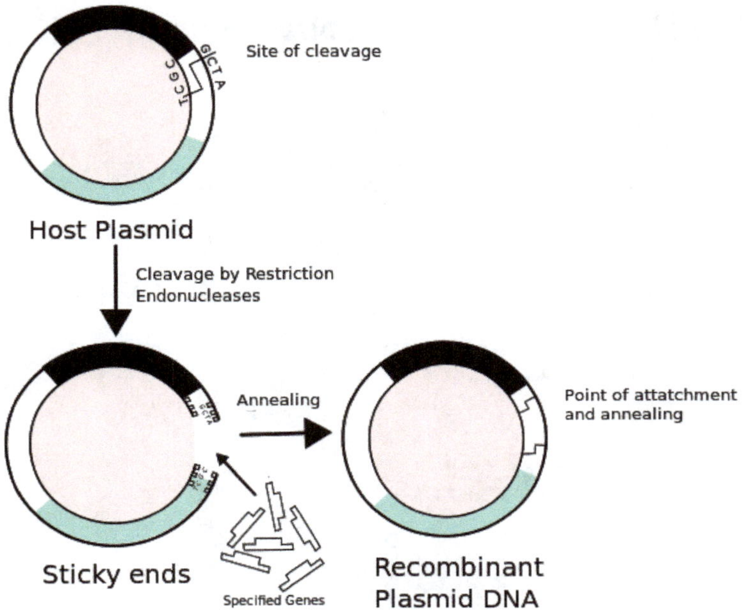

Construction of recombinant DNA, in which a foreign DNA fragment is inserted into a plasmid vector. In this example, the gene indicated by the white color is inactivated upon insertion of the foreign DNA fragment.

Recombinant DNA is the general name for a piece of DNA that has been created by the combination of at least two strands. Recombinant DNA molecules are sometimes called chimeric DNA, because they can be made of material from two different species, like the mythical chimera. R-DNA technology uses palindromic sequences and leads to the production of sticky and blunt ends.

The DNA sequences used in the construction of recombinant DNA molecules can originate from any species. For example, plant DNA may be joined to bacterial DNA, or human DNA may be joined with fungal DNA. In addition, DNA sequences that do not occur anywhere in nature may be created by the chemical synthesis of DNA, and incorporated into recombinant molecules. Using recombinant DNA technology and synthetic DNA, literally any DNA sequence may be created and introduced into any of a very wide range of living organisms.

Proteins that can result from the expression of recombinant DNA within living cells are termed recombinant proteins. When recombinant DNA encoding a protein is introduced into a host organism, the recombinant protein is not necessarily produced. Expression of foreign proteins requires the use of specialized expression vectors and often necessitates significant restructuring by foreign coding sequences.

Recombinant DNA differs from genetic recombination in that the former results from artificial methods in the test tube, while the latter is a normal biological process that results in the remixing of existing DNA sequences in essentially all organisms.

CREATING RECOMBINANT DNA

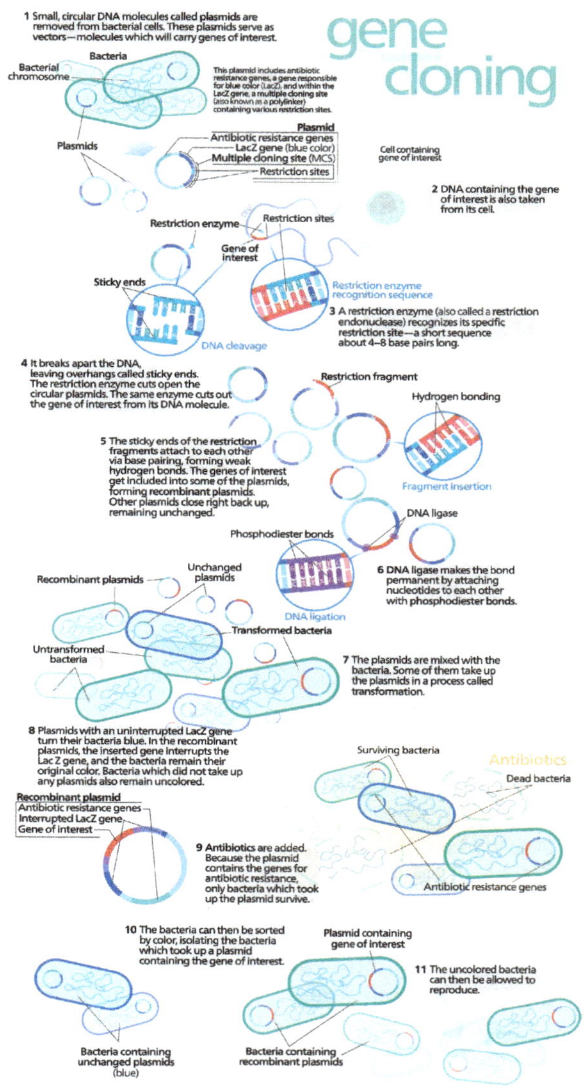

Molecular cloning is the laboratory process used to create recombinant DNA. It is one of two widely used methods, along with polymerase chain reaction (PCR) used to direct the replication of any specific DNA sequence chosen by the experimentalist. There are two fundamental differences between the methods. One is that molecular cloning involves replication of the DNA within a living cell, while PCR replicates DNA in the test tube, free of living cells. The other difference is that cloning involves cutting and pasting DNA sequences, while PCR only copies/amplifies an existing sequence.

Formation of recombinant DNA requires a cloning vector, a DNA molecule that replicates within a living cell. Vectors are generally derived from plasmids or viruses, and represent relatively small segments of DNA that contain necessary genetic signals for replication, as well as additional elements for convenience in inserting foreign DNA, identifying cells that contain recombinant DNA, and, where appropriate, expressing the foreign DNA. The choice of vector for molecular cloning depends on the choice of host organism, the size of the DNA to be cloned, and whether and how the foreign DNA is to be expressed. The DNA segments can be combined by using a variety of methods, such as restriction enzyme/ligase cloning or Gibson assembly.

In standard cloning protocols, the cloning of any DNA fragment essentially involves seven steps: (1) Choice of host organism and cloning vector, (2) Preparation of vector DNA, (3) Preparation of DNA to be cloned, (4) Creation of recombinant DNA, (5) Introduction of recombinant DNA into the host organism, (6) Selection of organisms containing recombinant DNA, and (7) Screening for clones with desired DNA inserts and biological properties. *These steps are described in some detail in a related article (molecular cloning).*

Expression of Recombinant DNA

Following transplantation into the host organism, the foreign DNA contained within the recombinant DNA construct may or may not be expressed. That is, the DNA may simply be replicated without expression, or it may be transcribed and translated at a recombinant protein is produced. Generally speaking, expression of a foreign gene requires restructuring the gene to include sequences that are required for producing an mRNA molecule that can be used by the host's translational apparatus (e.g. promoter, translational initiation signal, and transcriptional terminator). Specific changes to the host organism may be made to improve expression of the ectopic gene. In addition, changes may be needed to the coding sequences as well, to optimize translation, make the protein soluble, direct the recombinant protein to the proper cellular or extracellular location, and stabilize the protein from degradation.

Properties of Organisms Containing Recombinant DNA

In most cases, organisms containing recombinant DNA have apparently normal phenotypes. That is, their appearance, behavior and metabolism are usually unchanged, and the only way to demonstrate the presence of recombinant sequences is to examine the DNA itself, typically using a polymerase chain reaction (PCR) test. Significant exceptions exist, and are discussed below.

If the rDNA sequences encode a gene that is expressed, then the presence of RNA and/or protein products of the recombinant gene can be detected, typically using RT-PCR or western hybridization methods. Gross phenotypic changes are not the norm, unless the recombinant gene has been chosen and modified so as to generate biological activity in the host organism. Additional phenotypes that are encountered include toxicity to the host organism induced by the recombinant gene product, especially if it is over-expressed or expressed within inappropriate cells or tissues.

In some cases, recombinant DNA can have deleterious effects even if it is not expressed. One mechanism by which this happens is insertional inactivation, in which the rDNA becomes inserted into a host cell's gene. In some cases, researchers use this phenomenon to "knock out" genes to determine their biological function and importance. Another mechanism by which rDNA insertion into chromosomal DNA can affect gene expression is by inappropriate activation of previously unexpressed host cell genes. This can happen, for example, when a recombinant DNA fragment containing an active promoter becomes located next to a previously silent host cell gene, or when a host cell gene that functions to restrain gene expression undergoes insertional inactivation by recombinant DNA.

APPLICATIONS OF RECOMBINANT DNA TECHNOLOGY

A group of GloFish fluorescent fish

Recombinant DNA is widely used in biotechnology, medicine and research. Today, recombinant proteins and other products that result from the use of DNA technology are found in essentially every western pharmacy, doctor›s or veterinarian›s office, medical testing laboratory, and biological research laboratory. In addition, organisms that have been manipulated using recombinant DNA technology, as well as products derived from those organisms, have found their way into many farms, supermarkets, home medicine cabinets, and even pet shops, such as those that sell GloFish and other genetically modified animals.

The most common application of recombinant DNA is in basic research, in which the technology is important to most current work in the biological and biomedical sciences. Recombinant DNA is used to identify, map and sequence genes, and to determine their function. rDNA probes are employed in analyzing gene expression within individual cells, and throughout the tissues of whole organisms. Recombinant proteins are widely used as reagents in laboratory experiments and to generate antibody probes for examining protein synthesis within cells and organisms.

Many additional practical applications of recombinant DNA are found in industry, food production, human and veterinary medicine, agriculture, and bioengineering. Some specific examples are identified below.

Recombinant Chymosin

Found in rennet, chymosin is an enzyme required to manufacture cheese. It was the first genetically engineered food additive used commercially. Traditionally, processors obtained chymosin from rennet, a preparation derived from the fourth stomach of milk-fed calves. Scientists engineered a non-pathogenic strain (K-12) of *E. coli* bacteria for large-scale laboratory production of the enzyme. This microbiologically produced recombinant enzyme, identical structurally to the calf derived enzyme, costs less and is produced in abundant quantities. Today about 60% of U.S. hard cheese is made with genetically engineered chymosin. In 1990, FDA granted chymosin «generally recognized as safe» (GRAS) status based on data showing that the enzyme was safe.

Recombinant Human Insulin

Almost completely replaced insulin obtained from animal sources (e.g. pigs and cattle) for the treatment of insulin-dependent diabetes. A variety of different recombinant insulin preparations are in widespread use. Recombinant insulin is synthesized by inserting the human insulin gene into *E. coli*, or yeast (saccharomyces cerevisiae) which then produces insulin for human use.

Recombinant Human Growth Hormone (HGH, Somatotropin)

Administered to patients whose pituitary glands generate insufficient quantities to support normal growth and development. Before recombinant HGH became available, HGH for therapeutic use was obtained from pituitary glands of cadavers. This unsafe practice led to some patients developing Creutzfeldt–Jakob disease. Recombinant HGH eliminated this problem, and is now used therapeutically. It has also been misused as a performance-enhancing drug by athletes and others. DrugBank entry

Recombinant Blood Clotting Factor VIII

A blood-clotting protein that is administered to patients with forms of the bleeding disorder hemophilia, who are unable to produce factor VIII in quantities sufficient to support normal blood coagulation. Before the development of recombinant factor VIII, the protein was obtained by processing large quantities of human blood from multiple donors, which carried a very high risk of transmission of blood borne infectious diseases, for example HIV and hepatitis B. DrugBank entry

Recombinant Hepatitis B Vaccine

Hepatitis B infection is controlled through the use of a recombinant hepatitis B vaccine, which contains a form of the hepatitis B virus surface antigen that is produced in yeast cells. The development of the recombinant subunit vaccine was an important and necessary development because hepatitis B virus, unlike other common viruses such as polio virus, cannot be grown in vitro. Vaccine information from Hepatitis B Foundation

Diagnosis of Infection With HIV

Each of the three widely used methods for diagnosing HIV infection has been developed using recombinant DNA. The antibody test (ELISA or western blot) uses a recombinant HIV protein to test for the presence of antibodies that the body has produced in response to an HIV infection. The DNA test looks for the presence of HIV genetic material using reverse transcription polymerase chain reaction (RT-PCR). Development of the RT-PCR test was made possible by the molecular cloning and sequence analysis of HIV genomes. HIV testing page from US Centers for Disease Control (CDC)

Golden rice

A recombinant variety of rice that has been engineered to express the enzymes responsible for β-carotene biosynthesis. This variety of rice holds substantial promise for reducing the incidence of vitamin A deficiency in the world's population. Golden rice is not currently in use, pending the resolution of regulatory and intellectual property issues.

Herbicide-resistant crops

Commercial varieties of important agricultural crops (including soy, maize/corn, sorghum, canola, alfalfa and cotton) have been developed that incorporate a recombinant gene that results in resistance to the herbicide glyphosate (trade name *Roundup*), and simplifies weed control by glyphosate application. These crops are in common commercial use in several countries.

Insect-resistant crops

Bacillus thuringeiensis is a bacterium that naturally produces a protein (Bt toxin) with insecticidal properties. The bacterium has been applied to crops as an insect-control strategy for many years, and this practice has been widely adopted in agriculture and gardening. Recently, plants have been developed that express a recombinant form of the bacterial protein, which may effectively control some insect predators. Environmental issues associated with the use of these transgenic crops have not been fully resolved.

History of Recombinant DNA

The idea of recombinant DNA was first proposed by Peter Lobban, a graduate student of Prof. Dale Kaiser in the Biochemistry Department at Stanford University Medical School. The first publications describing the successful production and intracellular replication of recombinant DNA appeared in 1972 and 1973. Stanford University applied for a US patent on recombinant DNA in 1974, listing the inventors as Stanley N. Cohen and Herbert W. Boyer; this patent was awarded in 1980. The first licensed drug generated using recombinant DNA technology was human insulin, developed by Genentech and Licensed by Eli Lilly and Company.

Controversy

Scientists associated with the initial development of recombinant DNA methods recognized that the potential existed for organisms containing recombinant DNA to have undesirable or dangerous properties. At the 1975 Asilomar Conference on Recombinant DNA, these concerns were discussed and a voluntary moratorium on recombinant DNA research was initiated for experiments that were considered particularly risky. This moratorium was widely observed until the National Institutes of Health (USA) developed and issued formal guidelines for rDNA work. Today, recombinant DNA molecules and recombinant proteins are usually not regarded as dangerous. However, concerns remain about some organisms that express recombinant DNA, particularly when they leave the laboratory and are introduced into the environment or food chain. These concerns are discussed in the articles on genetically modified organisms and genetically modified food controversies.

REFERENCES

1. Campbell, Neil A. & Reece, Jane B.. (2002). Biology (6th ed.). San Francisco: Addison Wesley. pp. 375–401. ISBN 0-201-75054-6.
2. Peter Walter; Alberts, Bruce; Johnson, Alexander S.; Lewis, Julian; Raff, Martin C.; Roberts, Keith (2008). Molecular Biology of the Cell (5th edition, Extended version). New York: Garland Science. ISBN 0-8153-4111-3.. Fourth edition is available online through the NCBI Bookshelf: link
3. Berg, Jeremy Mark; Tymoczko, John L.; Stryer, Lubert (2010). Biochemistry, 7th ed. (Biochemistry (Berg)). W.H. Freeman & Company. ISBN 1-4292-2936-5. Fifth edition available online through the NCBI Bookshelf: link
4. Watson, James D. (2007). Recombinant DNA: Genes and Genomes: A Short Course. San Francisco: W.H. Freeman. ISBN 0-7167-2866-4.

5. Russell, David W.; Sambrook, Joseph (2001). Molecular cloning: a laboratory manual. Cold Spring Harbor, N.Y: Cold Spring Harbor Laboratory. ISBN 0-87969-576-5.
6. Hannig, G.; Makrides, S. (1998). "Strategies for optimizing heterologous protein expression in Escherichia coli". Trends in Biotechnology. 16 (2): 54–60. doi:10.1016/S0167-7799(97)01155-4. PMID 9487731.
7. Brondyk, W. H. (2009). "Chapter 11 Selecting an Appropriate Method for Expressing a Recombinant Protein". Methods in enzymology. Methods in Enzymology. 463: 131–147. doi:10.1016/S0076-6879(09)63011-1. ISBN 9780123745361. PMID 19892171.
8. Brown, Terry (2006). Gene Cloning and DNA Analysis: an Introduction. Cambridge, MA: Blackwell Pub. ISBN 1-4051-1121-6.
9. Ye, X.; Al-Babili, S.; Klöti, A.; Zhang, J.; Lucca, P.; Beyer, P.; Potrykus, I. (2000). "Engineering the provitamin A (beta-carotene) biosynthetic pathway into (carotenoid-free) rice endosperm". Science. 287 (5451): 303–305. doi:10.1126/science.287.5451.303. PMID 10634784.
10. Koller, B. H.; Smithies, O. (1992). "Altering Genes in Animals by Gene Targeting". Annual Review of Immunology. 10: 705–730. doi:10.1146/annurev.iy.10.040192.003421. PMID 1591000.
11. Donna U. Vogt and Mickey Parish. (1999) Food Biotechnology in the United States: Science, Regulation, and Issues
12. Gualandi-Signorini, A.; Giorgi, G. (2001). "Insulin formulations--a review". European review for medical and pharmacological sciences. 5 (3): 73–83. PMID 12004916.
13. DrugBank: Insulin Regular (DB00030)
14. Von Fange, T.; McDiarmid, T.; MacKler, L.; Zolotor, A. (2008). "Clinical inquiries: Can recombinant growth hormone effectively treat idiopathic short stature?". The Journal of family practice. 57 (9): 611–612. PMID 18786336.
15. Fernandez, M.; Hosey, R. (2009). "Performance-enhancing drugs snare nonathletes, too". The Journal of family practice. 58 (1): 16–23. PMID 19141266.
16. Manco-Johnson, M. J. (2010). "Advances in the Care and Treatment of Children with Hemophilia". Advances in Pediatrics. 57 (1): 287–294. doi:10.1016/j.yapd.2010.08.007. PMID 21056743.
17. Paine, J. A.; Shipton, C. A.; Chaggar, S.; Howells, R. M.; Kennedy, M. J.; Vernon, G.; Wright, S. Y.; Hinchliffe, E.; Adams, J. L.; Silverstone, A. L.; Drake, R. (2005). "Improving the nutritional value of Golden Rice through increased pro-vitamin a content". Nature Biotechnology. 23 (4): 482–487. doi:10.1038/nbt1082. PMID 15793573.
18. Deccan Herald, " Foreign group roots for 'golden rice' in India", March 18, 2015 http://www.deccanherald.com/content/466247/foreign-group-roots-golden-rice.html
19. Funke, T.; Han, H.; Healy-Fried, M.; Fischer, M.; Schönbrunn, E. (2006).

"Molecular basis for the herbicide resistance of Roundup Ready crops". Proceedings of the National Academy of Sciences. 103 (35): 13010–13015. doi:10.1073/pnas.0603638103. PMC 1559744 . PMID 16916934.

20. Mendelsohn, M.; Kough, J.; Vaituzis, Z.; Matthews, K. (2003). "Are Bt crops safe?". Nature Biotechnology. 21 (9): 1003–1009. doi:10.1038/nbt0903-1003. PMID 12949561.
21. Lear, J. (1978). Recombinant DNA: The Untold Story. New York: Crown Publishers. p. 43.
22. Jackson, D.; Symons, R.; Berg, P. (1972). "Biochemical method for inserting new genetic information into DNA of Simian Virus 40: Circular SV40 DNA molecules containing lambda phage genes and the galactose operon of Escherichia coli". Proceedings of the National Academy of Sciences of the United States of America. 69 (10): 2904–2909. doi:10.1073/pnas.69.10.2904. PMC 389671 . PMID 4342968.
23. Mertz, J. E.; Davis, R. W. (1972). "Cleavage of DNA by R 1 restriction endonuclease generates cohesive ends". Proceedings of the National Academy of Sciences of the United States of America. 69 (11): 3370–4. doi:10.1073/pnas.69.11.3370. PMC 389773 . PMID 4343968.
24. Lobban, P.; Kaiser, A. (1973). "Enzymatic end-to end joining of DNA molecules". Journal of Molecular Biology. 78 (3): 453–471. doi:10.1016/0022-2836(73)90468-3. PMID 4754844.
25. Cohen, S.; Chang, A.; Boyer, H.; Helling, R. (1973). "Construction of biologically functional bacterial plasmids in vitro". Proceedings of the National Academy of Sciences of the United States of America. 70 (11): 3240–3244. doi:10.1073/pnas.70.11.3240. PMC 427208 . PMID 4594039.
26. Hughes, S. (2001). "Making dollars out of DNA. The first major patent in biotechnology and the commercialization of molecular biology, 1974-1980". Isis; an international review devoted to the history of science and its cultural influences. 92 (3): 541–575. doi:10.1086/385281. PMID 11810894.
27. Johnson, I. S. (1983). "Human insulin from recombinant DNA technology". Science. 219 (4585): 632–637. doi:10.1126/science.6337396. PMID 6337396.
28. Stryer L, Berg JM, Tymoczko JL (2002). Biochemistry (5th ed.). San Francisco: W.H. Freeman. ISBN 0-7167-4955-6.
29. Schomburg I, Chang A, Placzek S, Söhngen C, Rother M, Lang M, Munaretto C, Ulas S, Stelzer M, Grote A, Scheer M, Schomburg D (January 2013). "BRENDA in 2013: integrated reactions, kinetic data, enzyme function data, improved disease classification: new options and contents in BRENDA". Nucleic Acids Research. 41 (Database issue): D764–72. doi:10.1093/nar/gks1049. PMC 3531171 . PMID 23203881.
30. Radzicka A, Wolfenden R (January 1995). "A proficient enzyme". Science. 267 (5194): 90–931. Bibcode:1995Sci...267...90R. doi:10.1126/science.7809611. PMID 7809611.
31. Callahan BP, Miller BG (December 2007). "OMP decarboxylase—An enigma

persists". Bioorganic Chemistry. 35 (6): 465–9. doi:10.1016/j.bioorg.2007.07.004. PMID 17889251.

32. de Réaumur RA (1752). "Observations sur la digestion des oiseaux". Histoire de l'academie royale des sciences. 1752: 266, 461.
33. Williams HS (1904). A History of Science: in Five Volumes. Volume IV: Modern Development of the Chemical and Biological Sciences. Harper and Brothers.
34. Payen A, Persoz JF (1833). "Mémoire sur la diastase, les principaux produits de ses réactions et leurs applications aux arts industriels" [Memoir on diastase, the principal products of its reactions, and their applications to the industrial arts]. Annales de chimie et de physique. 2nd (in French). 53: 73–92.
35. Manchester KL (December 1995). "Louis Pasteur (1822–1895)–chance and the prepared mind". Trends in Biotechnology. 13 (12): 511–5. doi:10.1016/S0167-7799(00)89014-9. PMID 8595136.
36. Kühne coined the word "enzyme" in: Kühne W (1877). "Über das Verhalten verschiedener organisirter und sog. ungeformter Fermente" [On the behavior of various organized and so-called unformed ferments]. Verhandlungen des naturhistorisch-medicinischen Vereins zu Heidelberg. new series (in German). 1 (3): 190–193. The relevant passage occurs on page 190: "Um Missverständnissen vorzubeugen und lästige Umschreibungen zu vermeiden schlägt Vortragender vor, die ungeformten oder nicht organisirten Fermente, deren Wirkung ohne Anwesenheit von Organismen und ausserhalb derselben erfolgen kann, als Enzyme zu bezeichnen." (Translation: In order to obviate misunderstandings and avoid cumbersome periphrases, [the author, a university lecturer] suggests designating as "enzymes" the unformed or not organized ferments, whose action can occur without the presence of organisms and outside of the same.)
37. Holmes FL (2003). "Enzymes". In Heilbron JL. The Oxford Companion to the History of Modern Science. Oxford: Oxford University Press. p. 270.
38. "Eduard Buchner". Nobel Laureate Biography. Nobelprize.org. Retrieved 23 February 2015.
39. "Eduard Buchner – Nobel Lecture: Cell-Free Fermentation". Nobelprize.org. 1907. Retrieved 23 February 2015.
40. The naming of enzymes by adding the suffix "-ase" to the substrate on which the enzyme acts, has been traced to French scientist Émile Duclaux (1840–1904), who intended to honor the discoverers of diastase – the first enzyme to be isolated – by introducing this practice in his book Duclaux E (1899). Traité de microbiologie: Diastases, toxines et venins [Microbiology Treatise: diastases , toxins and venoms] (in French). Paris, France: Masson and Co. See Chapter 1, especially page 9.
41. Willstätter R (1927). "Faraday lecture. Problems and methods in enzyme research". Journal of the Chemical Society (Resumed): 1359. doi:10.1039/JR9270001359. quoted in Blow D (April 2000). "So do we understand how enzymes work?" (pdf). Structure (London, England : 1993). 8 (4): R77–R81. doi:10.1016/S0969-2126(00)00125-8. PMID 10801479.

42. “Nobel Prizes and Laureates: The Nobel Prize in Chemistry 1946”. Nobelprize. org. Retrieved 23 February 2015.

43. Blake CC, Koenig DF, Mair GA, North AC, Phillips DC, Sarma VR (May 1965). “Structure of hen egg-white lysozyme. A three-dimensional Fourier synthesis at 2 Ångström resolution”. Nature. 206 (4986): 757–61. Bibcode:1965Natur.206..757B. doi:10.1038/206757a0. PMID 5891407.

44. Johnson LN, Petsko GA (1999). “David Phillips and the origin of structural enzymology”. Trends Biochem. Sci. 24 (7): 287–9. doi:10.1016/S0968-0004(99)01423-1. PMID 10390620.

45. Nomenclature Committee. “Classification and Nomenclature of Enzymes by the Reactions they Catalyse”. International Union of Biochemistry and Molecular Biology (NC-IUBMB). School of Biological and Chemical Sciences, Queen Mary, University of London.

46. Nomenclature Committee. “EC 2.7.1.1”. International Union of Biochemistry and Molecular Biology (NC-IUBMB). School of Biological and Chemical Sciences, Queen Mary, University of London.

47. Anfinsen CB (July 1973). “Principles that govern the folding of protein chains”. Science. 181 (4096): 223–30. Bibcode:1973Sci...181..223A. doi:10.1126/science.181.4096.223. PMID 4124164.

48. Imperfect DNA replication results in mutations. Berg JM, Tymoczko JL, Stryer L, Clarke ND (2002). “Chapter 27: DNA Replication, Recombination, and Repair”. Biochemistry. W.H. Freeman and Company. ISBN 0-7167-3051-0.

49. Alberts B, Johnson A, Lewis J, Raff M, Roberts K, Walter P (2002). “5DNA Replication, Repair, and Recombination”. Molecular Biology of the Cell. Garland Science. ISBN 0-8153-3218-1.

50. Berg JM, Tymoczko JL, Stryer L, Clarke ND (2002). “Chapter 27, Section 4: DNA Replication of Both Strands Proceeds Rapidly from Specific Start Sites”. Biochemistry. W.H. Freeman and Company. ISBN 0-7167-3051-0.

51. Alberts, B., et al., Molecular Biology of the Cell, Garland Science, 4th ed., 2002, pp. 238–240 ISBN 0-8153-3218-1

52. Allison, Lizabeth A. Fundamental Molecular Biology. Blackwell Publishing. 2007. p.112 ISBN 978-1-4051-0379-4

53. Berg JM, Tymoczko JL, Stryer L, Clarke ND (2002). Biochemistry. W.H. Freeman and Company. ISBN 0-7167-3051-0. Chapter 27, Section 2: DNA Polymerases Require a Template and a Primer

54. McCulloch, Scott D; Kunkel, Thomas A (January 2008). “The fidelity of DNA synthesis by eukaryotic replicative and translesion synthesis polymerases”. Cell Research. 18 (1): 148–161. doi:10.1038/cr.2008.4. PMC 3639319 . PMID 18166979.

55. McCarthy, David; Minner, Charles; Bernstein, Harris; Bernstein, Carol (October 1976). “DNA elongation rates and growing point distributions of wild-type phage T4 and a DNA-delay amber mutant”. Journal of Molecular Biology. 106 (4):

963–981. doi:10.1016/0022-2836(76)90346-6. PMID 789903.

56. Drake JW (1970) The Molecular Basis of Mutation. Holden-Day, San Francisco ISBN 0816224501 ISBN 978-0816224500
57. Alberts B, Johnson A, Lewis J, Raff M, Roberts K, Walter P (2002). Molecular Biology of the Cell. Garland Science. ISBN 0-8153-3218-1.Chapter 5: DNA Replication Mechanisms
58. Weigel C, Schmidt A, Rückert B, Lurz R, Messer W (November 1997). "DnaA protein binding to individual DnaA boxes in the Escherichia coli replication origin, oriC". The EMBO Journal. 16 (21): 6574–83. doi:10.1093/emboj/16.21.6574. PMC 1170261 . PMID 9351837.
59. Lodish H, Berk A, Zipursky LS, Matsudaira P, Baltimore D, Darnell J (2000). Molecular Cell Biology. W. H. Freeman and Company. ISBN 0-7167-3136-3.12.1. General Features of Chromosomal Replication: Three Common Features of Replication Origins
60. Aravind, L.; Leipe, D. D.; Koonin, E. V. (1998). "Toprim—a conserved catalytic domain in type IA and II topoisomerases, DnaG-type primases, OLD family nucleases and RecR proteins". Nucleic Acids Research. 26 (18): 4205–4213. doi:10.1093/nar/26.18.4205. PMC 147817 . PMID 9722641.
61. Frick, David; Richardson, Charles (July 2001). "DNA Primases". Annual Review of Biochemistry. 70: 39–80. doi:10.1146/annurev.biochem.70.1.39. PMID 11395402.
62. Barry, Elizabeth R.; Bell, Stephen D. (8 December 2006). "DNA Replication in the Archaea". Microbiology and Molecular Biology Reviews. 70 (4): 876–887. doi:10.1128/MMBR.00029-06. PMC 1698513 . PMID 17158702.
63. Stillman, Bruce (July 2015). "Reconsidering DNA Polymerases at the Replication Fork in Eukaryotes". Molecular Cell. 59 (2): 139–141. doi:10.1016/j.molcel.2015.07.004. PMC 4636199 . PMID 26186286.
64. Distinguishing the pathways of primer removal during Eukaryotic Okazaki fragment maturation Contributor Author Rossi, Marie Louise. Date Accessioned: 2009-02-23T17:05:09Z. Date Available: 2009-02-23T17:05:09Z. Date Issued: 2009-02-23T17:05:09Z. Identifier Uri: http://hdl.handle.net/1802/6537. Description: Dr. Robert A. Bambara, Faculty Advisor. Thesis (PhD) – School of Medicine and Dentistry, University of Rochester. UR only until January 2010. UR only until January 2010.
65. Huberman, Joel A.; Riggs, Arthur D. (March 1968). "On the mechanism of DNA replication in mammalian chromosomes". Journal of Molecular Biology. 32 (2): 327–341. doi:10.1016/0022-2836(68)90013-2. PMID 5689363.
66. Alberts B, Johnson A, Lewis J, Raff M, Roberts K, Walter P (2002). Molecular Biology of the Cell. Garland Science. ISBN 0-8153-3218-1.DNA Replication Mechanisms: DNA Topoisomerases Prevent DNA Tangling During Replication
67. Reece, Richard J.; Maxwell, Anthony; Wang, James C. (26 September 2008). "DNA Gyrase: Structure and Function". Critical Reviews in Biochemistry and Molecular Biology. 26 (3-4): 335–375. doi:10.3109/10409239109114072. PMID 1657531.

68. Alberts B, Johnson A, Lewis J, Raff M, Roberts K, Walter P (2002). Molecular Biology of the Cell. Garland Science. ISBN 0-8153-3218-1.DNA Replication Mechanisms: Special Proteins Help to Open Up the DNA Double Helix in Front of the Replication Fork
69. Griffiths A.J.F.; Wessler S.R.; Lewontin R.C.; Carroll S.B. (2008). Introduction to Genetic Analysis. W. H. Freeman and Company. ISBN 0-7167-6887-9.[Chapter 7: DNA: Structure and Replication. pg 283–290]
70. "Will the Hayflick limit keep us from living forever?". Howstuffworks. Retrieved January 20, 2015.
71. James D. Watson et al. (2008), "Molecular Biology of the gene", Pearson Education: 237
72. Peter Meister, Angela Taddeil, Susan M. Gasser(June 2006), "In and out of the Replication Factory", Cell 125 (7): 1233–1235
73. TA Brown (2002). Genomes. BIOS Scientific Publishers. ISBN 1-85996-228-9.13.2.3. Termination of replication
74. Alberts B, Johnson A, Lewis J, Raff M, Roberts K, Walter P (2002). Molecular Biology of the Cell. Garland Science. ISBN 0-8153-3218-1.Intracellular Control of Cell-Cycle Events: S-Phase Cyclin-Cdk Complexes (S-Cdks) Initiate DNA Replication Once Per Cycle
75. Brown, TA (2002). "13". Genomes (2nd ed.). Oxford: Wiley-Liss.
76. Tobiason DM, Seifert HS (2006). "The Obligate Human Pathogen, Neisseria gonorrhoeae, Is Polyploid". PLoS Biology. 4 (6): e185. doi:10.1371/journal.pbio.0040185. PMC 1470461 . PMID 16719561.
77. Slater, Steven; Wold, Sture; Lu, Min; Boye, Erik; Skarstad, Kirsten; Kleckner, Nancy (September 1995). "E. coli SeqA protein binds oriC in two different methyl-modulated reactions appropriate to its roles in DNA replication initiation and origin sequestration". Cell. 82 (6): 927–936. doi:10.1016/0092-8674(95)90272-4. PMID 7553853.
78. Saiki, Randall; Gelfand, David H.; Stoffel, Susanne; Scharf, Stephen J.; Higuchi, Russell; Horn, Glenn T.; Mullis, Kary B.; Erlich, Henry A. (29 January 1988). "Primer-directed enzymatic amplification of DNA with a thermostable DNA polymerase". Science. 239 (4839): 487–491. doi:10.1126/science.2448875. PMID 2448875. Retrieved 7 April 2016.
79. Bollum, F.J. (1960). "Calf thymus polymerase". J. Biol. Chem. 235: 2399–2403. PMID 13802334.
80. Falaschi, A.; Kornberg, A. (1966). "Biochemical studies of bacterial sporulation. II. Deoxy-ribonucleic acid polymerase in spores of Bacillus subtilis". J. Biol. Chem. 241 (7): 1478–1482. PMID 4957767.
81. Lehman, I.R.; Bessman, M.J.; Simms, E.S.; Kornberg, A. (1958). "Enzymatic synthesis of deoxyribonucleic acid. I. Preparation of substrates and partial purification of an enzyme from Escherichia coli". J. Biol. Chem. 233 (1): 163–170. PMID 13563462.

82. Richardson, C.C.; Schildkraut, C.L.; Aposhian, H.V.; Kornberg, A. (1964). "Enzymatic synthesis of deoxyribonucleic acid. XIV. Further purification and properties of deoxyribonucleic acid polymerase of Escherichia coli". J. Biol. Chem. 239: 222–232. PMID 14114848.

83. Schachman, H.K.; Adler, J.; Radding, C.M.; Lehman, I.R.; Kornberg, A. (1960). "Enzymatic synthesis of deoxyribonucleic acid. VII. Synthesis of a polymer of deoxyadenylate and deoxythymidylate". J. Biol. Chem. 235: 3242–3249. PMID 13747134.

84. Zimmerman, B.K. (1966). "Purification and properties of deoxyribonucleic acid polymerase from Micrococcus lysodeikticus". J. Biol. Chem. 241 (9): 2035–2041. PMID 5946628.

85. "The Nobel Prize in Physiology or Medicine 1959". Nobel Foundation. Retrieved December 1, 2012.

86. Tessman I, Kennedy MA (February 1994). "DNA polymerase II of Escherichia coli in the bypass of abasic sites in vivo". Genetics. 136 (2): 439–48. PMC 1205799 . PMID 7908652.

87. Garrett, Grisham (2013). Biochemistry. Mary Finch.

88. Hunter, Brown; Anand, Kennard (1986). "Structure of an adenine-cytosine base pair in DNA and its implications for mismatch repair". Nature.

89. Swan, Johnson, Prakash, Prakash, and Aggarwal (September 2009). "Structural basis of high-fidelity DNA synthesis by yeast DNA polymerase δ". Nature Structural & Molecular Biology.

90. Steitz TA; common mechanisms (June 1999). "DNA polymerases: structural diversity". J. Biol. Chem. 274 (25): 17395–8. doi:10.1074/jbc.274.25.17395. PMID 10364165.

91. Losick R, Watson JD, Baker TA, Bell S, Gann A, Levine MW (2008). Molecular biology of the gene (6th ed.). San Francisco: Pearson/Benjamin Cummings. ISBN 0-8053-9592-X.

92. McCarthy D, Minner C, Bernstein H, Bernstein C (October 1976). "DNA elongation rates growing point distributions of wild-type phage T4 and a DNA-delay amber mutant". J. Mol. Biol. 106 (4): 963–81. doi:10.1016/0022-2836(76)90346-6. PMID 789903.

93. Boehm, Elizabeth M.; Powers, Kyle T.; Kondratick, Christine M.; Spies, Maria; Houtman, Jon C. D.; Washington, M. Todd (2016-04-15). "The Proliferating Cell Nuclear Antigen (PCNA)-interacting Protein (PIP) Motif of DNA Polymerase η Mediates Its Interaction with the C-terminal Domain of Rev1". The Journal of Biological Chemistry. 291 (16): 8735–8744. doi:10.1074/jbc.M115.697938. ISSN 1083-351X. PMC 4861442 . PMID 26903512.

94. Maga G, Hubscher U, Spadari S, Villani G (2010). DNA Polymerases: Discovery, Characterization Functions in Cellular DNA Transactions. World Scientific Publishing Company. ISBN 981-4299-16-2.

95. Camps, Manel; Loeb, Lawrence A. (28 October 2014). "When Pol I Goes into

High Gear: Processive DNA Synthesis by Pol I in the Cell". Cell Cycle. 3 (2): 114–116. doi:10.4161/cc.3.2.651.

96. Banach-Orlowska, Magdalena; Fijalkowska, Iwona J.; Schaaper, Roel M.; Jonczyk, Piotr (5 September 2005). "DNA polymerase II as a fidelity factor in chromosomal DNA synthesis in Escherichia coli". Molecular Microbiology. 58 (1): 61–70. doi:10.1111/j.1365-2958.2005.04805.x. PMID 16164549.
97. Olson MW, Dallmann HG, McHenry CS (December 1995). "DnaX complex of Escherichia coli DNA polymerase III holoenzyme. The chi psi complex functions by increasing the affinity of tau gamma for delta.delta' to a physiologically relevant range". J. Biol. Chem. 270 (49): 29570–7. doi:10.1074/jbc.270.49.29570. PMID 7494000.
98. "DNA Polymerase Families". News-medical.net. 2014-05-06. Retrieved 2014-06-28.
99. Goodman MF (2002). "Error-prone repair DNA polymerases in prokaryotes eukaryotes". Annu. Rev. Biochem. 71: 17–50. doi:10.1146/annurev.biochem.71.083101.124707. PMID 12045089.
100. Mori T (2012). "Escherichia coli DinB inhibits replication fork progression without significantly inducing the SOS response". Genes Genet Syst. 87 (2): 75–87. doi:10.1266/ggs.87.75. PMID 22820381.
101. Jarosz DF (2007). "Proficient accurate bypass of persistent DNA lesions by DinB DNA polymerases.". Cell Cycle. 6 (7): 817–22. doi:10.4161/cc.6.7.4065. PMID 17377496.
102. Patel M, Jiang Q, Woodgate R, Cox MM, Goodman MF (June 2010). "A new model for SOS-induced mutagenesis: how RecA protein activates DNA polymerase V". Crit. Rev. Biochem. Mol. Biol. 45 (3): 171–84. doi:10.3109/10409238.2010.480968. PMC 2874081 . PMID 20441441.
103. Sutton MD, Walker GC (July 2001). "Managing DNA polymerases: coordinating DNA replication, DNA repair DNA recombination". Proc. Natl. Acad. Sci. U.S.A. 98 (15): 8342–9. doi:10.1073/pnas.111036998. PMC 37441 . PMID 11459973.
104. Yamtich J, Sweasy JB (May 2010). "DNA polymerase family X: function, structure cellular roles". Biochim. Biophys. Acta. 1804 (5): 1136–50. doi:10.1016/j.bbapap.2009.07.008. PMC 2846199 . PMID 19631767.
105. Chansky, Michael Lieberman, Allan Marks, Alisa Peet ; illustrations by Matthew (2012). Marks' basic medical biochemistry : a clinical approach (4th ed.). Philadelphia: Wolter Kluwer Health/Lippincott Williams & Wilkins. p. chapter13. ISBN 160831572X.
106. Chung DW, Zhang JA, Tan CK, Davie EW, So AG, Downey KM (December 1991). "Primary structure of the catalytic subunit of human DNA polymerase delta chromosomal location of the gene". Proc. Natl. Acad. Sci. U.S.A. 88 (24): 11197–201. doi:10.1073/pnas.88.24.11197. PMC 53101 . PMID 1722322.
107. Johnson RE, Klassen R, Prakash L, Prakash S (July 16, 2015). "A Major Role of DNA Polymerase δ in Replication of Both the Leading Lagging DNA

Strands". Mol. Cell. 59 (2): 163–175. doi:10.1016/j.molcel.2015.05.038. PMC 4517859 . PMID 26145172.

108. Edwards S, Li CM, Levy DL, Brown J, Snow PM, Campbell JL (April 2003). "Saccharomyces cerevisiae DNA polymerase epsilon polymerase sigma interact physically and functionally, suggesting a role for polymerase epsilon in sister chromatid cohesion". Mol. Cell. Biol. 23 (8): 2733–48. doi:10.1128/mcb.23.8.2733-2748.2003. PMC 152548 . PMID 12665575.

109. Ohmori H, Hanafusa T, Ohashi E, Vaziri C (2009). "Separate roles of structured unstructured regions of Y-family DNA polymerases". Adv Protein Chem Struct Biol. Advances in Protein Chemistry and Structural Biology. 78: 99–146. doi:10.1016/S1876-1623(08)78004-0. ISBN 9780123748270. PMC 3103052 . PMID 20663485.

110. Gan GN, Wittschieben JP, Wittschieben B, Wood RD (January 2008). "DNA polymerase zeta (pol zeta) in higher eukaryotes". Cell Res. 18 (1): 174–83. doi:10.1038/cr.2007.117. PMID 18157155.

111. Zhang L, Chan SS, Wolff DJ (July 2011). "Mitochondrial disorders of DNA polymerase γ dysfunction: from anatomic to molecular pathology diagnosis". Arch. Pathol. Lab. Med. 135 (7): 925–34. doi:10.1043/2010-0356-RAR.1 (inactive 2016-09-12). PMC 3158670 . PMID 21732785.

112. Stumpf JD, Copeland WC (January 2011). "Mitochondrial DNA replication disease: insights from DNA polymerase γ mutations". Cell. Mol. Life Sci. 68 (2): 219–33. doi:10.1007/s00018-010-0530-4. PMC 3046768 . PMID 20927567.

113. Hogg M, Sauer-Eriksson AE, Johansson E (March 2012). "Promiscuous DNA synthesis by human DNA polymerase θ". Nucleic Acids Res. 40 (6): 2611–22. doi:10.1093/nar/gkr1102. PMC 3315306 . PMID 22135286.

CHAPTER 4

RNA Synthesis and Procesing

Chapter Contents

- **Introduction**
- **Transfer RNA**
- **Nucleic Acid Structure**

Ribonucleic acid (RNA) is a chain of nucleotides present in the cells of all life. This chain has a number of important functions for living organisms, ranging from the regulation of gene expression to assistance with copying genes. Severo Ochoa, Robert Holley, and Carl Woese all played critical roles in discovering RNA and understanding how it worked, and more research is constantly being performed.

INTRODUCTION

Ribonucleic acid (RNA) is a polymeric molecule essential in various biological roles in coding, decoding, regulation, and expression of genes. RNA and DNA are nucleic acids, and, along with proteins and carbohydrates, constitute the four major macromolecules essential for all known forms of life. Like DNA, RNA is assembled as a chain of nucleotides, but unlike DNA it is more often found in nature as a single-strand folded onto itself, rather than a paired double-strand. Cellular organisms use messenger RNA (mRNA) to convey genetic information (using the letters G, U, A, and C to denote the nitrogenous bases guanine, uracil, adenine, and cytosine) that directs synthesis of specific proteins. Many viruses encode their genetic information using an RNA genome.

Some RNA molecules play an active role within cells by catalyzing biological reactions, controlling gene expression, or sensing and communicating responses to cellular signals. One of these active processes is protein synthesis, a universal function where RNA molecules direct the assembly of proteins on ribosomes. This process uses transfer RNA (tRNA) molecules to deliver amino acids to the ribosome, where ribosomal RNA (rRNA) then links amino acids together to form proteins.

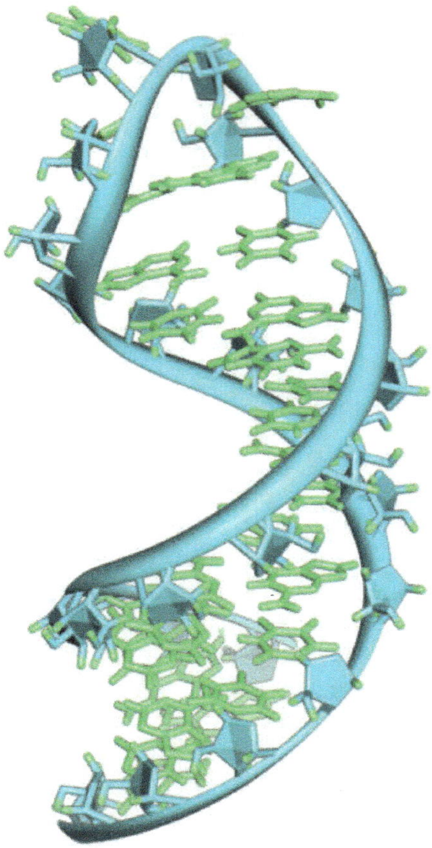

A hairpin loop from a pre-mRNA. Highlighted are the nucleobases (green) and the ribose-phosphate backbone (blue). Note that this is a single strand of RNA that folds back upon itself.

TRANSFER RNA

A transfer RNA (abbreviated tRNA and formerly referred to as sRNA, for soluble RNA) is an adaptor molecule composed of RNA, typically 76 to 90 nucleotides in length, that serves as the physical link between the mRNA and the amino acid sequence of proteins. It does this by carrying an amino acid to the protein synthetic machinery of a cell (ribosome) as directed by a three-nucleotide sequence (codon) in a messenger RNA (mRNA). As such, tRNAs are a necessary component of translation, the biological synthesis of new proteins in accordance with the genetic code.

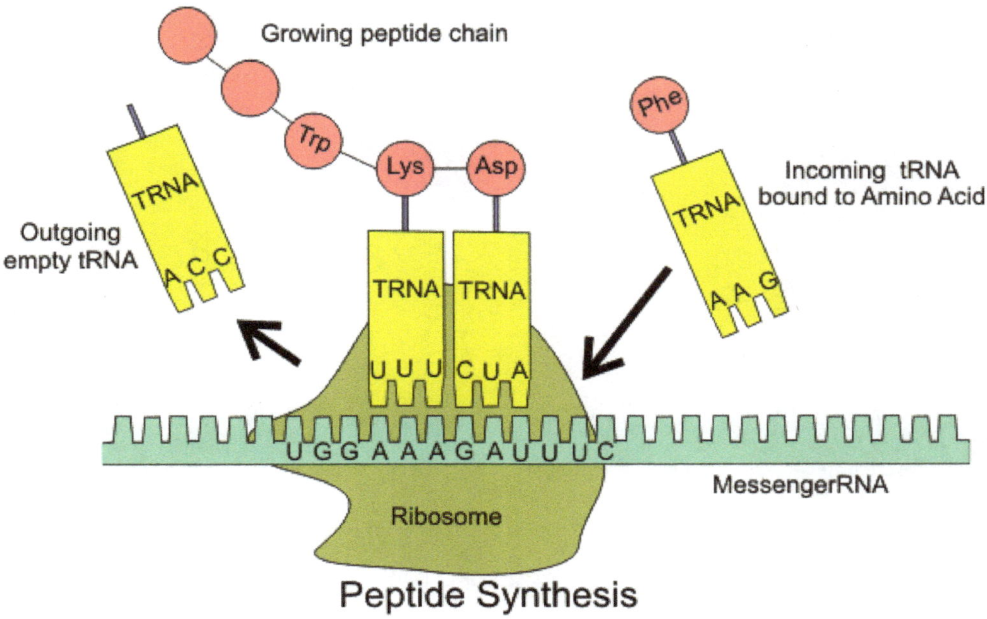

The interaction of tRNA and mRNA in protein synthesis.

While the specific nucleotide sequence of an mRNA specifies which amino acids are incorporated into the protein product of the gene from which the mRNA is transcribed, the role of ***tRNA*** is to specify which sequence from the genetic code corresponds to which amino acid. The mRNA encodes a protein as a series of contiguous codons, each of which is recognized by a particular tRNA. One end of the tRNA matches the genetic code in a three-nucleotide sequence called the anticodon. The anticodon forms three base pairs with a codon in mRNA during protein biosynthesis. On the other end of the tRNA is a covalent attachment to the amino acid that corresponds to the anticodon sequence. Each type of tRNA molecule can be attached to only one type of amino acid, so each organism has many types of tRNA. Because the genetic code contains multiple codons that specify the same amino acid, there are several tRNA molecules bearing different anticodons which carry the same amino acid.

The covalent attachment to the tRNA 3' end is catalyzed by enzymes called aminoacyl tRNA synthetases. During protein synthesis, tRNAs with attached amino acids are delivered to the ribosome by proteins called elongation factors, which aid in association of the tRNA with the ribosome, synthesis of the new polypeptide and translocation (movement) of the ribosome along the mRNA. If the tRNA›s anticodon matches the mRNA, another tRNA already bound to the ribosome transfers the growing polypeptide chain from its 3' end to the amino acid attached to the 3' end of the newly delivered tRNA, a reaction catalyzed by the ribosome. A large number of the individual nucleotides in a tRNA molecule may be chemically modified, often by methylation or deamidation. These unusual bases sometimes affect the tRNA›s interaction with ribosomes and sometimes occur in the anticodon to alter base-pairing properties.

Structure

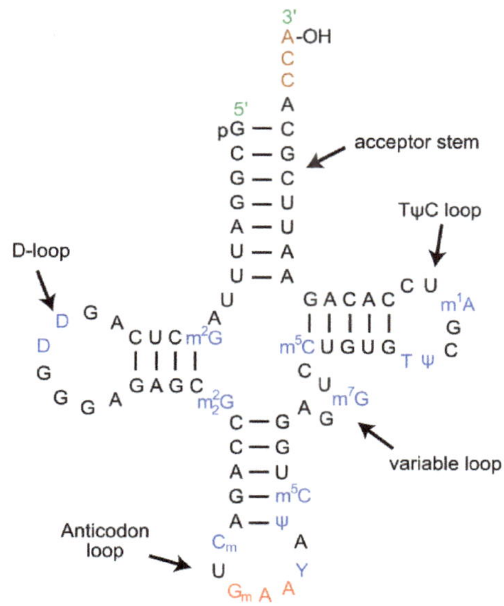

Secondary *cloverleaf structure* of tRNA^{Phe} from yeast.

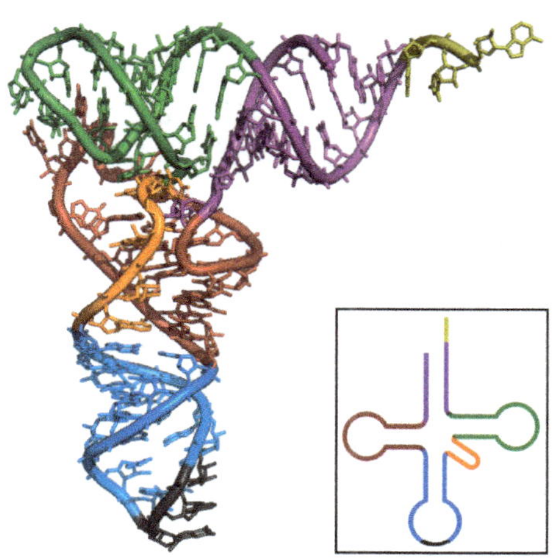

Tertiary structure of tRNA. *CCA tail* in yellow, *Acceptor stem* in purple, *Variable loop* in orange, *D arm* in red, *Anticodon arm* in blue with *Anticodon* in black, *T arm* in green.

The structure of tRNA can be decomposed into its primary structure, its secondary structure (usually visualized as the *cloverleaf structure*), and its tertiary structure (all tRNAs have a similar L-shaped 3D structure that allows them to fit into the P and A sites of the ribosome). The cloverleaf structure becomes the 3D L-shaped structure through coaxial stacking of the helices, which is a common RNA tertiary structure motif.

The lengths of each arm, as well as the loop ‹diameter›, in a tRNA molecule vary from species to species.

The tRNA structure consists of the following:

- A 5›-terminal phosphate group.
- The acceptor stem is a 7- to 9-base pair (bp) stem made by the base pairing of the 5›-terminal nucleotide with the 3›-terminal nucleotide (which contains the CCA 3›-terminal group used to attach the amino acid). The acceptor stem may contain non-Watson-Crick base pairs.
- The CCA tail is a cytosine-cytosine-adenine sequence at the 3' end of the tRNA molecule. The amino acid loaded onto the tRNA by aminoacyl tRNA synthetases, to form aminoacyl-tRNA, is covalently bonded to the 3›-hydroxyl group on the CCA tail. This sequence is important for the recognition of tRNA by enzymes and critical in translation. In prokaryotes, the CCA sequence is transcribed in some tRNA sequences. In most prokaryotic tRNAs and eukaryotic tRNAs, the CCA sequence is added during processing and therefore does not appear in the tRNA gene.
- The D arm is a 4- to 6-bp stem ending in a loop that often contains dihydrouridine.
- The anticodon arm is a 5-bp stem whose loop contains the anticodon. The tRNA 5›-to-3› primary structure contains the anticodon but in reverse order, since 3›-to-5› directionality is required to read the mRNA from 5›-to-3›.
- The T arm is a 4- to 5- bp stem containing the sequence TΨC where Ψ is pseudouridine, a modified uridine.
- Bases that have been modified, especially by methylation (e.g. tRNA (guanine-N7-)-methyltransferase), occur in several positions throughout the tRNA. The first anticodon base, or wobble-position, is sometimes modified to inosine (derived from adenine), pseudouridine or lysidine (derived from cytosine).

Anticodon

An anticodon is a unit made up of three nucleotides that correspond to the three bases of the codon on the mRNA. Each tRNA contains a distinct anticodon triplet sequence that can base-pair to one or more codons for an amino acid. Some anticodons can pair with more than one codon due to a phenomenon known as wobble base pairing. Frequently, the first nucleotide of the anticodon is one not found on mRNA: inosine, which can hydrogen bond to more than one base in the corresponding codon position.[:29.3.9] In the genetic code, it is common for a single amino acid to be specified by all four third-position possibilities, or at least by both pyrimidines and purines; for example, the amino acid glycine is coded for by the codon sequences GGU, GGC, GGA, and GGG. Other modified nucleotides may also appear at the first anticodon position - sometimes known as the "wobble position" - resulting in subtle changes to the genetic code, as for example in mitochondria.

To provide a one-to-one correspondence between tRNA molecules and codons that specify amino acids, 61 types of tRNA molecules would be required per cell. However, many cells contain fewer than 61 types of tRNAs because the wobble base is capable of binding to several, though not necessarily all, of the codons that specify a particular amino acid. A minimum of 31 tRNA are required to translate, unambiguously, all 61 sense codons of the standard genetic code.

Aminoacylation

Aminoacylation is the process of adding an aminoacyl group to a compound. It covalently links an amino acid to the CCA 3' end of a tRNA molecule.

Each tRNA is aminoacylated (or *charged*) with a specific amino acid by an aminoacyl tRNA synthetase. There is normally a single aminoacyl tRNA synthetase for each amino acid, despite the fact that there can be more than one tRNA, and more than one anticodon, for an amino acid. Recognition of the appropriate tRNA by the synthetases is not mediated solely by the anticodon, and the acceptor stem often plays a prominent role.

Reaction:

- amino acid + ATP → aminoacyl-AMP + PPi
- aminoacyl-AMP + tRNA → aminoacyl-tRNA + AMP

Certain organisms can have one or more aminoacyl tRNA synthetases missing. This leads to charging of the tRNA by a chemically related amino acid. An enzyme or enzymes modify the charged amino acid to the final one. For example, *Helicobacter pylori* has glutaminyl tRNA synthetase missing. Thus, glutamate tRNA synthetase charges tRNA-glutamine(tRNA-Gln) with glutamate. An amidotransferase then converts the acid side chain of the glutamate to the amide, forming the correctly charged gln-tRNA-Gln.

Binding to Ribosome

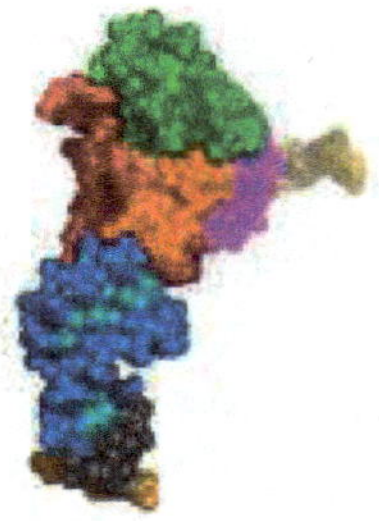

The range of conformations adopted by tRNA as it transits the A/T through P/E sites on the ribosome. The Protein Data Bank (PDB) codes for the structural models used as end points of the animation are given. Both tRNAs are modeled as phenylalanine-specific tRNA from *Escherichia coli*, with the A/T tRNA as a homology model of the deposited coordinates. Color coding as shown for tRNA tertiary structure. Adapted from.

The ribosome has three binding sites for tRNA molecules that span the space between the two ribosomal subunits: the A (aminoacyl), P (peptidyl), and E (exit) sites. In addition, the ribosome has two other sites for tRNA binding that are used during mRNA decoding or during the initiation of protein synthesis. These are the T site (named elongation factor Tu) and I site (initiation). By convention, the tRNA binding sites are denoted with the site on the small ribosomal subunit listed first and the site on the large ribosomal subunit listed second. For example, the A site is often written A/A, the P site, P/P, and the E site, E/E. The binding proteins like L27, L2, L14, L15, L16 at the A- and P- sites have been determined by affinity labeling by A.P. Czernilofsky et al. (*Proc. Natl. Acad. Sci, USA*, pp 230–234, 1974).

Once translation initiation is complete, the first aminoacyl tRNA is located in the P/P site, ready for the elongation cycle described below. During translation elongation, tRNA first binds to the ribosome as part of a complex with elongation factor Tu (EF-Tu) or its eukaryotic (eEF-1) or archaeal counterpart. This initial tRNA binding site is called the A/T site. In the A/T site, the A-site half resides in the small ribosomal subunit where the mRNA decoding site is located. The mRNA decoding site is where the mRNA codon is read out during translation. The T-site half resides mainly on the large ribosomal subunit where EF-Tu or eEF-1 interacts with the ribosome. Once mRNA decoding is complete, the aminoacyl-tRNA is bound in the A/A site and is ready for the next peptide bond to be formed to its attached amino acid. The peptidyl-tRNA, which transfers the growing polypeptide to the aminoacyl-tRNA bound in the A/A site, is bound in the P/P site. Once the peptide bond is formed, the tRNA in the P/P site is deacylated, or has a free 3' end, and the tRNA in the A/A site carries the growing polypeptide chain. To allow for the next elongation cycle, the tRNAs then move through hybrid A/P and P/E binding sites, before completing the cycle and residing in the P/P and E/E sites. Once the A/A and P/P tRNAs have moved to the P/P and E/E sites, the mRNA has also moved over by one codon and the A/T site is vacant, ready for the next round of mRNA decoding. The tRNA bound in the E/E site then leaves the ribosome.

The P/I site is actually the first to bind to aminoacyl tRNA, which is delivered by an initiation factor called IF2 in bacteria. However, the existence of the P/I site in eukaryotic or archaeal ribosomes has not yet been confirmed. The P-site protein L27 has been determined by affinity labeling by E. Collatz and A.P. Czernilofsky (*FEBS Lett.*, Vol. 63, pp 283–286, 1976).

tRNA Genes

Organisms vary in the number of tRNA genes in their genome. The nematode worm *C. elegans*, a commonly used model organism in genetics studies, has 29,647 genes in its nuclear genome, of which 620 code for tRNA. The budding yeast *Saccharomyces cerevisiae* has 275 tRNA genes in its genome.

In the human genome, which, according to January 2013 estimates, has about 20,848 protein coding genes in total, there are 497 nuclear genes encoding cytoplasmic tRNA molecules, and 324 tRNA-derived pseudogenes—tRNA genes thought to be no longer functional (although pseudo tRNAs have been shown to be involved in antibiotic resistance in bacteria). Regions in nuclear chromosomes, very similar in sequence to mitochondrial tRNA genes, have also been identified (tRNA-lookalikes). These tRNA-lookalikes are also considered part of the nuclear mitochondrial DNA (genes transferred from the mitochondria to the nucleus).

As with all eukaryotes, there are 22 mitochondrial tRNA genes in humans. Mutations in some of these genes have been associated with severe diseases like the MELAS syndrome.

Cytoplasmic tRNA genes can be grouped into 49 families according to their anticodon features. These genes are found on all chromosomes, except 22 and Y chromosome. High clustering on 6p is observed (140 tRNA genes), as well on 1 chromosome.

The HGNC, in collaboration with the Genomic tRNA Database (GtRNAdb) and experts in the field, has approved unique names for human genes that encode tRNAs.

Evolution

Genomic tRNA content is a differentiating feature of genomes among biological domains of life: Archaea present the simplest situation in terms of genomic tRNA content with a uniform number of gene copies, Bacteria have an intermediate situation and Eukarya present the most complex situation. Eukarya present not only more tRNA gene content than the other two kingdoms but also a high variation in gene copy number among different isoacceptors, and this complexity seem to be due to duplications of tRNA genes and changes in anticodon specificity.

Evolution of the tRNA gene copy number across different species has been linked to the appearance of specific tRNA modification enzymes (uridine methyltransferases in Bacteria, and adenosine deaminases in Eukarya), which increase the decoding capacity of a given tRNA. As an example, tRNAAla encodes four different tRNA isoacceptors (AGC, UGC, GGC and CGC). In Eukarya, AGC isoacceptors are extremely enriched in gene copy number in comparison to the rest of isoacceptors, and this has been correlated with its A-to-I modification of its wobble base. This same trend has been shown for most amino acids of eukaryal species. Indeed, the effect of these two tRNA modifications is also seen in codon usage bias. Highly expressed genes seem to be enriched in codons that are exclusively using codons that will be decoded by these modified tRNAs, which suggests a possible role of these codons—and consequently of these tRNA modifications—in translation efficiency.

tRNA-Derived Fragments

tRNA-derived fragments (or tRFs) are short molecules that emerge after cleavage of the mature tRNAs or the precursor transcript. Both cytoplasmic and mitochondrial tRNAs can produce fragments. There are at least four structural types of tRFs believed to originate from mature tRNAs, including the relatively long tRNA halves and short 5'-tRFs, 3'-tRFs and i-tRFs. The precursor tRNA can be cleaved to produce molecules from the 5' leader or 3' trail sequences. Cleavage enzymes include Angiogenin, Dicer, RNase Z and RNase P. Especially in the case of Angiogenin, the tRFs have a characteristically unusual cyclic phosphate at their 3' end and a hydroxyl group at the 5' end.

tRFs have multiple dependencies and roles. They exhibit significant changes between sexes, among races and disease status. Functionally, they can be loaded on Ago and act through RNAi pathways, participate in the formation of stress granules, displace mRNAs from RNA-binding proteins or inhibit translation. At the system or the organismal level, the four types of tRFs have a diverse spectrum of activities. Functionally, tRFs are associated with viral infection, cancer, cell proliferation and also with epigenetic transgenerational regulation of metabolism.

tRFs are not restricted to humans but have been shown to exist in multiple organisms.

Two online tools are available for those wishing to learn more about tRFs: the framework for the interactive exploration of mitochondrial and nuclear tRNA fragments (MINTbase) and the relational database of Transfer RNA related Fragments(tRFdb),. MINTbase also provides a naming scheme for the naming of tRFs called tRF-license plates that is genome independent.

tRNA Biogenesis

In eukaryotic cells, tRNAs are transcribed by RNA polymerase III as pre-tRNAs in the nucleus. RNA polymerase III recognizes two highly conserved downstream promoter sequences: the 5› intragenic control region (5›-ICR, D-control region, or A box), and the 3›-ICR (T-control region or B box) inside tRNA genes. The first promoter begins at +8 of mature tRNAs and the second promoter is located 30-60 nucleotides downstream of the first promoter. The transcription terminates after a stretch of four or more thymidines.

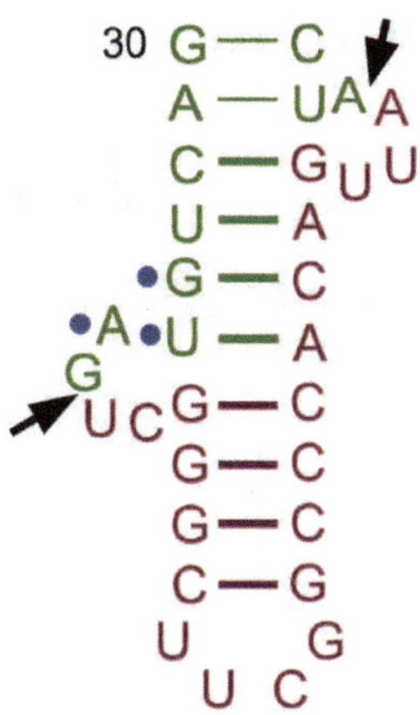

Bulge-helix-bulge motive of tRNA intron

Pre-tRNAs undergo extensive modifications inside the nucleus. Some pre-tRNAs contain introns that are spliced, or cut, to form the functional tRNA molecule; in bacteria these self-splice, whereas in eukaryotes and archaea they are removed by tRNA-splicing endonucleases. Eukaryotic pre-tRNA contains bulge-helix-bulge (BHB) structure motif that is important for recognition and precise splicing of tRNA intron by endonucleases. This motif position and structure are evolutionary conserved. However, some organisms, such as unicellular algae have a non-canonical position of BHB-motif as well as 5›- and 3›-ends of the spliced intron sequence. The 5› sequence is removed by RNase P, whereas the 3› end is removed by the tRNase Z enzyme. A notable exception is in the archaeon

Nanoarchaeum equitans, which does not possess an RNase P enzyme and has a promoter placed such that transcription starts at the 5' end of the mature tRNA. The non-templated 3› CCA tail is added by a nucleotidyl transferase.Before tRNAs are exported into the cytoplasm by Los1/Xpo-t, tRNAs are aminoacylated. The order of the processing events is not conserved. For example, in yeast, the splicing is not carried out in the nucleus but at the cytoplasmic side of mitochondrial membranes.

The existence of tRNA was first hypothesized by Francis Crick, based on the assumption that there must exist an adapter molecule capable of mediating the translation of the RNA alphabet into the protein alphabet. Significant research on structure was conducted in the early 1960s by Alex Rich and Don Caspar, two researchers in Boston, the Jacques Fresco group in Princeton University and a United Kingdom group at King›s College London. In 1965, Robert W. Holley of Cornell University reported the primary structure and suggested three secondary structures. tRNA was first crystallized in Madison, Wisconsin, by Robert M. Bock. The cloverleaf structure was ascertained by several other studies in the following years and was finally confirmed using X-ray crystallography studies in 1974. Two independent groups, Kim Sung-Hou working under Alexander Rich and a British group headed by Aaron Klug, published the same crystallography findings within a year.

NUCLEIC ACID STRUCTURE

Nucleic acid structure refers to the structure of nucleic acids such as DNA and RNA. Chemically speaking, DNA and RNA are very similar. Nucleic acid structure is often divided into four different levels: primary, secondary, tertiary and quaternary.

Primary Structure

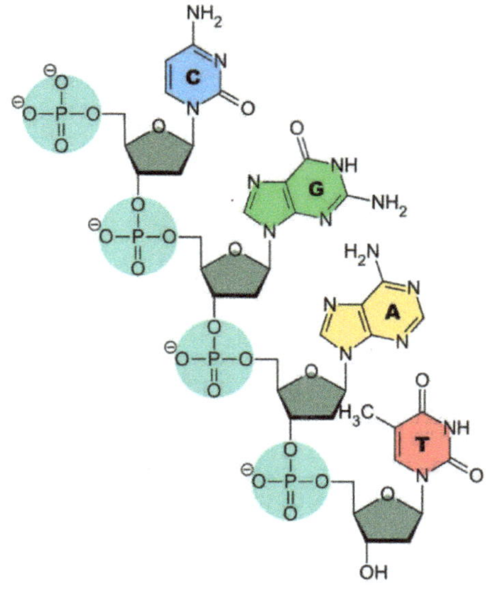

Chemical structure of DNA

Primary structure consists of a linear sequence of nucleotides that are linked together by phosphodiester bonds. It is this linear sequence of nucleotides that make up the Primary structure of DNA or RNA. Nucleotides consist of 3 components:

- Nitrogenous base
    - Adenine
    - Guanine
    - Cytosine
    - Thymine (present in DNA only)
    - Uracil (present in RNA only)
- 5-carbon sugar which is called deoxyribose (found in DNA) and ribose (found in RNA).
- One or more phosphate groups.

The nitrogen bases adenine and guanine are purine in structure and form a glycosidic bond between their 9 nitrogen and the 1' -OH group of the deoxyribose. Cytosine, thymine and uracil are pyrimidines, hence the glycosidic bonds forms between their 1 nitrogen and the 1› -OH of the deoxyribose. For both the purine and pyrimidine bases, the phosphate group forms a bond with the deoxyribose sugar through an ester bond between one of its negatively charged oxygen groups and the 5› -OH of the sugar. The polarity in DNA and RNA is derived from the oxygen and nitrogen atoms in the backbone. Nucleic acids are formed when nucleotides come together through phosphodiester linkages between the 5› and 3› carbon atoms.A Nucleic acid sequence is the order of nucleotides within a DNA (GACT) or RNA (GACU) molecule that is determined by a series of letters. Sequences are presented from the 5' to 3' end and determine the covalent structure of the entire molecule. Sequences can be complementary to another sequence in that the base on each position is complementary as well as in the reverse order. An example of a complementary sequence to AGCT is TCGA. DNA is double-stranded containing both a sense strand and an antisense strand. Therefore, the complementary sequence will be to the sense strand.

Nucleic acid design can be used to create nucleic acid complexes with complicated secondary structures such as this four-arm junction. These four strands associate into this structure because it maximizes the number of correct base pairs, with A›s matched to T›s and C›s matched to G›s. Image from Mao, 2004.

Complexes with Alkali Metal Ions.

There are three potential metal binding groups on nucleic acids: phosphate, sugar and base moieties. Solid-state structure of complexes with alkali metal ions have been reviewed.

Secondary Structure

Secondary structure is the set of interactions between bases, i.e., which parts of strands are bound to each other. In DNA double helix, the two strands of DNA are held together by hydrogen bonds. The nucleotides on one strand base pairs with the nucleotide on the other strand. The secondary structure is responsible for the shape that the nucleic acid assumes. The bases in the DNA are classified as Purines and Pyrimidines. The purines are Adenine and Guanine. Purines consist of a double ring structure, a six membered and a five membered ring containing nitrogen. The pyrimidines are Cytosine and Thymine. It has a single ringed structure, a six membered ring containing nitrogen. A purine base always pairs with a pyrimidine base (Guanosine (G) pairs with Cytosine(C) and Adenine(A) pairs with Thymine (T) or Uracil (U)). DNA's secondary structure is predominantly determined by base-pairing of the two polynucleotide strands wrapped around each other to form a double helix. There is also a major groove and a minor groove on the double helix.

Tertiary Structure

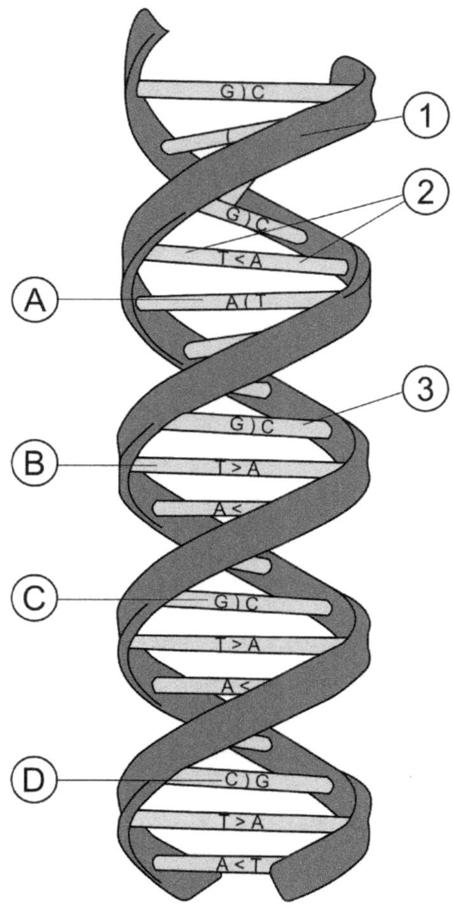

DNA structure and bases

The secondary structure of RNA consists of a single polynucleotide. Base pairing in RNA occurs when RNA folds between complementarity regions. Both single- and double-stranded regions are often found in RNA molecules. The antiparallel strands form a helical shape. The four basic elements in the secondary structure of RNA are helices, loops, bulges, and junctions. Stem-loop or hairpin loop is the most common element of RNA secondary structure. Stem-loop is formed when the RNA chains fold back on themselves to form a double helical tract called the stem, the unpaired nucleotides forms single stranded region called the loop. Secondary structure of RNA can be predicted by experimental data on the secondary structure elements, helices, loops and bulges. Bulges and internal loops are formed by separation of the double helical tract on either one strand (bulge) or on both strands (internal loops) by unpaired nucleotides. A Tetraloop is a four-base pairs hairpin RNA structure. There are three common families of tetraloop in ribosomal RNA: UNCG, GNRA, and CUUG (N is one of the four nucleotides and R is a purine).UNCG is the most stable tetraloop. Pseudoknot is a RNA secondary structure first identified in turnip yellow mosaic virus. Pseudoknots are formed when nucleotides from the hairpin loop pairs with a single stranded region outside of the hairpin to form a helical segment. H-type fold pseudoknots are best characterized. In H-type fold, nucleotides in the hairpin loop pairs with the bases outside the hairpin stem forming second stem and loop. This causes formation of pseudoknots with two stems and two loops. Pseudoknots are functional elements in RNA structure having diverse function and found in most classes of RNA. DotKnot-PW method is used for comparative pseudoknots prediction .The main points in the DotKnot-PW method is scoring the similarities found in stems, secondary elements and H-type pseudoknots.

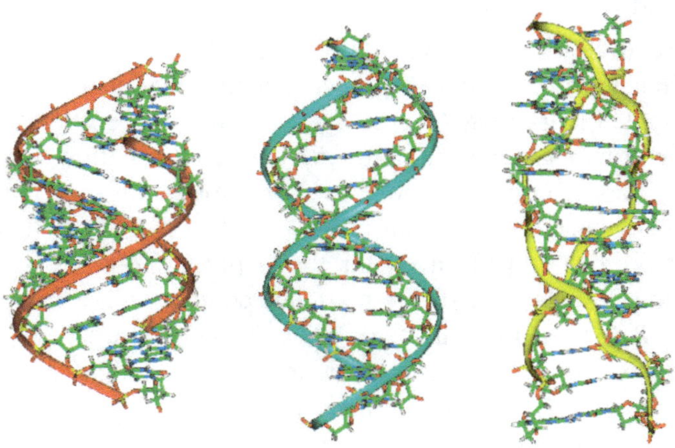

A-B-Z-DNA Side View

Tertiary structure refers to the locations of the atoms in three-dimensional space, taking into consideration geometrical and steric constraints. It is a higher order than the secondary structure, in which large-scale folding in a linear polymer occurs and the entire chain is folded into a specific 3-dimensional shape. There are 4 areas in which the structural forms of DNA can differ.

- Handedness - right or left
- Length of the helix turn
- Number of base pairs per turn
- Difference in size between the major and minor grooves

The tertiary arrangement of DNA›s double helix in space includes B-DNA, A-DNA and Z-DNA.

B-DNA is the most common form of DNA in vivo and is a more narrow, elongated helix than A-DNA. Its wide major groove makes it more accessible to proteins. On the other hand, it has a narrow minor groove. B-DNA›s favored conformations occur at high water concentrations; the hydration of the minor groove appears to favor B-DNA. B-DNA base pairs are nearly perpendicular to the helix axis. The sugar pucker which determines the shape of the a-helix, whether the helix will exist in the A-form or in the B-form, occurs at the C2›-endo.

A-DNA, is a form of the DNA duplex observed under dehydrating conditions. It is shorter and wider than B-DNA. RNA adopts this double helical form, and RNA-DNA duplexes are mostly A-form, but B-form RNA-DNA duplexes have been observed. In localized single strand dinucleotide contexts, RNA can also adopt the B-form without pairing to DNA. A-DNA has a deep, narrow major groove which does not make it easily accessible to proteins. On the other hand, its wide, shallow minor groove makes it accessible to proteins but with lower information content than the major groove. Its favored conformation is at low water concentrations. A-DNAs base pairs are tilted relative to the helix axis, and are displaced from the axis. The sugar pucker occurs at the C3›-endo and in RNA 2›-OH inhibits C2›-endo conformation. Long considered little more than a laboratory artifice, A-DNA is now known to have several biological functions.

Z-DNA is a relatively rare left-handed double-helix. Given the proper sequence and superhelical tension, it can be formed in vivo but its function is unclear. It has a more narrow, more elongated helix than A or B. Z-DNA›s major groove is not really a groove, and it has a narrow minor groove. The most favored conformation occurs when there are high salt concentrations. There are some base substitutions but they require an alternating purine-pyrimidine sequence. The N2-amino of G H-bonds to 5› PO, which explains the slow exchange of protons and the need for the G purine. Z-DNA base pairs are nearly perpendicular to the helix axis. Z-DNA does not contain single base-pairs but rather a GpC repeat with P-P distances varying for GpC and CpG. On the GpC stack there is good base overlap, whereas on the CpG stack there is less overlap. Z-DNA›s zigzag backbone is due to the C sugar conformation compensating for G glycosidic bond conformation. The conformation of G is syn, C2›-endo and for C it is anti, C3›-endo.

A linear DNA molecule having free ends can rotate, to adjust to changes of various dynamic processes in the cell, by changing how many times the two chains of its double helix twist around each other. Some DNA molecules are circular and are topologically constrained. More recently circular RNA was descirbed as well to be a natural pervasive class of nucleic acids, expressed in many organisms (see circRNA).

A covalently closed, circular DNA (also known as cccDNA) is topologically constrained as the number of times the chains coiled around one other cannot change. This cccDNA can be supercoiled, which is the tertiary structure of DNA. Supercoiling is characterized by the linking number, twist and writhe. The linking number (Lk) for circular DNA is defined as the number of times one strand would have to pass through the other strand to completely separate the two strands. The linking number for circular DNA can only be changed by breaking of a covalent bond in one of the two strands. Always an integer, the linking number of a cccDNA is the sum of two components: twists (Tw) and writhes (Wr).

$$Lk = Tw + Wr$$

Twists are the number of times the two strands of DNA are twisted around each other. Writhes are number of times the DNA helix crosses over itself. DNA in cells is negatively supercoiled and has the tendency to unwind. Hence the separation of strands is easier in negatively supercoiled DNA than in relaxed DNA. The two components of supercoiled DNA are solenoid and plectonemic. The plectonemic supercoil is found in prokaryotes, while the solenoidal supercoiling is mostly seen in eukaryotes

Comparison with DNA

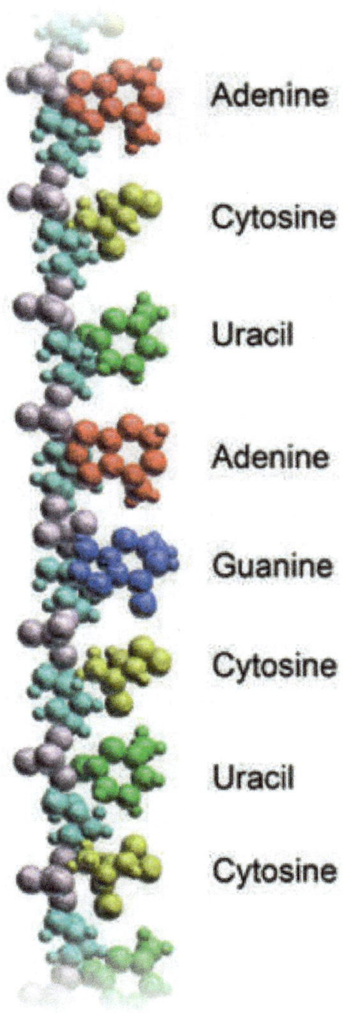

Bases in an RNA molecule.

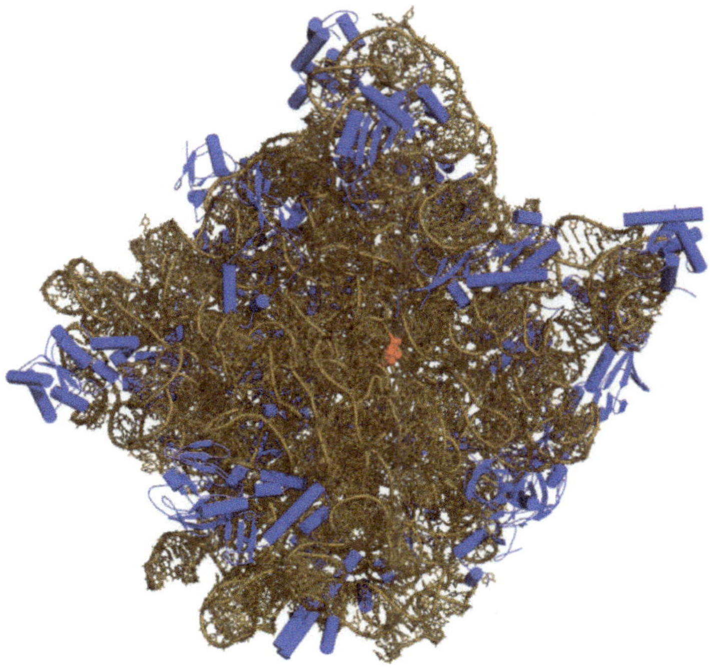

Three-dimensional representation of the 50S ribosomal subunit. Ribosomal RNA is in ochre, proteins in blue. The active site is a small segment of rRNA, indicated in red.

The chemical structure of RNA is very similar to that of DNA, but differs in three main ways:

- Unlike double-stranded DNA, RNA is a single-stranded molecule in many of its biological roles and has a much shorter chain of nucleotides.However, RNA can, by complementary base pairing, form intrastrand (i.e., single-strand) double helixes, as in tRNA.
- While DNA contains *deoxyribose*, RNA contains *ribose* (in deoxyribose there is no hydroxyl group attached to the pentose ring in the 2› position). These hydroxyl groups make RNA less stable than DNA because it is more prone to hydrolysis.
- The complementary base to adenine in DNA is thymine, whereas in RNA, it is uracil, which is an unmethylated form of thymine.

Like DNA, most biologically active RNAs, including mRNA, tRNA, rRNA, snRNAs, and other non-coding RNAs, contain self-complementary sequences that allow parts of the RNA to fold and pair with itself to form double helices. Analysis of these RNAs has revealed that they are highly structured. Unlike DNA, their structures do not consist of long double helices, but rather collections of short helices packed together into structures akin to proteins. In this fashion, RNAs can achieve chemical catalysis (like enzymes). For instance, determination of the structure of the ribosome—an enzyme that catalyzes peptide bond formation—revealed that its active site is composed entirely of RNA.

Structure

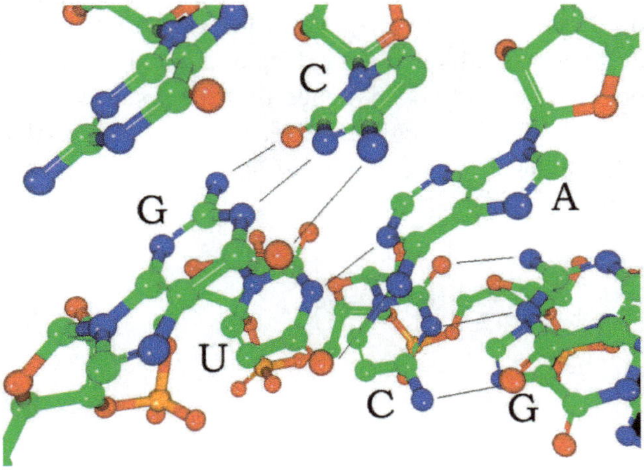

Watson-Crick base pairs in a siRNA (hydrogen atoms are not shown)

Each nucleotide in RNA contains a ribose sugar, with carbons numbered 1' through 5'. A base is attached to the 1' position, in general, adenine (A), cytosine (C), guanine (G), or uracil (U). Adenine and guanine are purines, cytosine and uracil are pyrimidines. A phosphate group is attached to the 3' position of one ribose and the 5' position of the next. The phosphate groups have a negative charge each, making RNA a charged molecule (polyanion). The bases form hydrogen bonds between cytosine and guanine, between adenine and uracil and between guanine and uracil. However, other interactions are possible, such as a group of adenine bases binding to each other in a bulge, or the GNRA tetraloop that has a guanine–adenine base-pair.

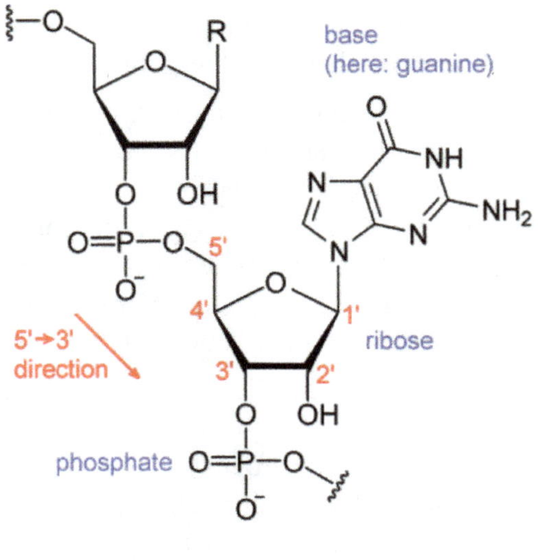

Chemical structure of RNA

An important structural feature of RNA that distinguishes it from DNA is the presence of a hydroxyl group at the 2' position of the ribose sugar. The presence of this functional group causes the helix to mostly adopt the A-form geometry, although in single strand dinucleotide contexts, RNA can rarely also adopt the B-form most commonly observed in DNA. The A-form geometry results in a very deep and narrow major groove and a shallow and wide minor groove. A second consequence of the presence of the 2›-hydroxyl group is that in conformationally flexible regions of an RNA molecule (that is, not involved in formation of a double helix), it can chemically attack the adjacent phosphodiester bond to cleave the backbone.

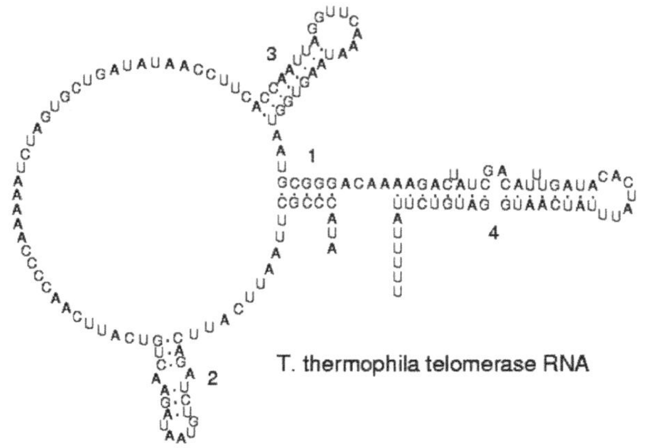

Secondary structure of a telomerase RNA.

RNA is transcribed with only four bases (adenine, cytosine, guanine and uracil), but these bases and attached sugars can be modified in numerous ways as the RNAs mature. Pseudouridine (Ψ), in which the linkage between uracil and ribose is changed from a C–N bond to a C–C bond, and ribothymidine (T) are found in various places (the most notable ones being in the TΨC loop of tRNA). Another notable modified base is hypoxanthine, a deaminated adenine base whose nucleoside is called inosine (I). Inosine plays a key role in the wobble hypothesis of the genetic code.

There are more than 100 other naturally occurring modified nucleosides, The greatest structural diversity of modifications can be found in tRNA, while pseudouridine and nucleosides with 2›-O-methylribose often present in rRNA are the most common. The specific roles of many of these modifications in RNA are not fully understood. However, it is notable that, in ribosomal RNA, many of the post-transcriptional modifications occur in highly functional regions, such as the peptidyl transferase center and the subunit interface, implying that they are important for normal function.

The functional form of single-stranded RNA molecules, just like proteins, frequently requires a specific tertiary structure. The scaffold for this structure is provided by secondary structural elements that are hydrogen bonds within the molecule. This leads to several recognizable "domains" of secondary structure like hairpin loops, bulges, and internal loops. Since RNA is charged, metal ions such as Mg^{2+} are needed to stabilise many secondary and tertiary structures.

The naturally occurring enantiomer of RNA is D-RNA composed of D-ribonucleotides. All chirality centers are located in the D-ribose. By the use of L-ribose or rather L-ribonucleotides, L-RNA can be synthesized. L-RNA is much more stable against degradation by RNase.

Like other structured biopolymers such as proteins, one can define topology of a folded RNA molecule. This is often done based on arrangement of intra-chain contacts within a folded RNA, termed as circuit topology.

Synthesis

Synthesis of RNA is usually catalyzed by an enzyme—RNA polymerase—using DNA as a template, a process known as transcription. Initiation of transcription begins with the binding of the enzyme to a promoter sequence in the DNA (usually found "upstream" of a gene). The DNA double helix is unwound by the helicase activity of the enzyme. The enzyme then progresses along the template strand in the 3' to 5' direction, synthesizing a complementary RNA molecule with elongation occurring in the 5' to 3' direction. The DNA sequence also dictates where termination of RNA synthesis will occur.

Primary transcript RNAs are often modified by enzymes after transcription. For example, a poly(A) tail and a 5› cap are added to eukaryotic pre-mRNA and introns are removed by the spliceosome.

There are also a number of RNA-dependent RNA polymerases that use RNA as their template for synthesis of a new strand of RNA. For instance, a number of RNA viruses (such as poliovirus) use this type of enzyme to replicate their genetic material. Also, RNA-dependent RNA polymerase is part of the RNA interference pathway in many organisms.

Types of RNA

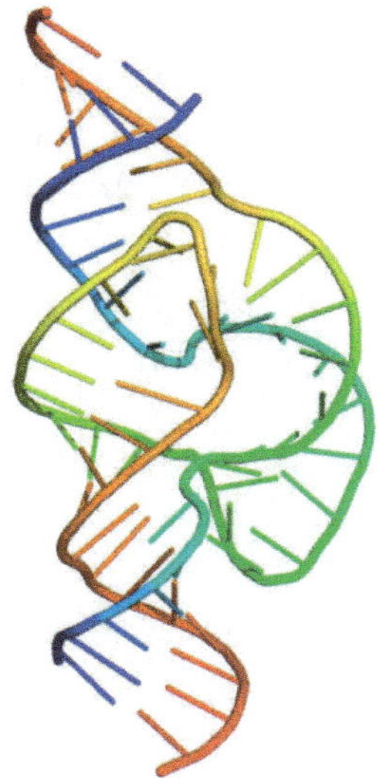

Structure of a hammerhead ribozyme, a ribozyme that cuts RNA

Messenger RNA (mRNA) is the RNA that carries information from DNA to the ribosome, the sites of protein synthesis (translation) in the cell. The coding sequence of the mRNA determines the amino acid sequence in the protein that is produced. However, many RNAs do not code for protein (about 97% of the transcriptional output is non-protein-coding in eukaryotes).

These so-called non-coding RNAs ("ncRNA») can be encoded by their own genes (RNA genes), but can also derive from mRNA introns. The most prominent examples of non-coding RNAs are transfer RNA (tRNA) and ribosomal RNA (rRNA), both of which are involved in the process of translation. There are also non-coding RNAs involved in gene regulation, RNA processing and other roles. Certain RNAs are able to catalyse chemical reactions such as cutting and ligating other RNA molecules, and the catalysis of peptide bond formation in the ribosome; these are known as ribozymes.

In Length

According to the length of RNA chain, RNA includes small RNA and long RNA. Usually, small RNAs are <200 nt in length, and long RNA are >200 nt. Long RNAs, also called large RNAs, mainly include long non-coding RNA (lncRNA) and mRNA. Small RNAs mainly include 5.8S ribosomal RNA (rRNA), 5S rRNA, transfer RNA (tRNA), microRNA (miRNA), small interfering RNA (siRNA), small nucleolar RNA (snoRNAs), Piwi-interacting RNA (piRNA), tRNA-derived small RNA (tsRNA) and small rDNA-derived RNA (srRNA).

In Translation

Messenger RNA (mRNA) carries information about a protein sequence to the ribosomes, the protein synthesis factories in the cell. It is coded so that every three nucleotides (a codon) corresponds to one amino acid. In eukaryotic cells, once precursor mRNA (pre-mRNA) has been transcribed from DNA, it is processed to mature mRNA. This removes its introns—non-coding sections of the pre-mRNA. The mRNA is then exported from the nucleus to the cytoplasm, where it is bound to ribosomes and translated into its corresponding protein form with the help of tRNA. In prokaryotic cells, which do not have nucleus and cytoplasm compartments, mRNA can bind to ribosomes while it is being transcribed from DNA. After a certain amount of time the message degrades into its component nucleotides with the assistance of ribonucleases.

Transfer RNA (tRNA) is a small RNA chain of about 80 nucleotides that transfers a specific amino acid to a growing polypeptide chain at the ribosomal site of protein synthesis during translation. It has sites for amino acid attachment and an anticodon region for codon recognition that binds to a specific sequence on the messenger RNA chain through hydrogen bonding.

Ribosomal RNA (rRNA) is the catalytic component of the ribosomes. Eukaryotic ribosomes contain four different rRNA molecules: 18S, 5.8S, 28S and 5S rRNA. Three of the rRNA molecules are synthesized in the nucleolus, and one is synthesized elsewhere. In the cytoplasm, ribosomal RNA and protein combine to form a nucleoprotein called a ribosome. The ribosome binds mRNA and carries out protein synthesis. Several ribosomes may be attached to a single mRNA at any time. Nearly all the RNA found in a typical eukaryotic cell is rRNA.

Transfer-messenger RNA (tmRNA) is found in many bacteria and plastids. It tags proteins encoded by mRNAs that lack stop codons for degradation and prevents the ribosome from stalling.

Regulatory RNAs

Several types of RNA can downregulate gene expression by being complementary to a part of an mRNA or a gene's DNA. MicroRNAs (miRNA; 21-22 nt) are found in eukaryotes and act through RNA interference (RNAi), where an effector complex of miRNA and enzymes can cleave complementary mRNA, block the mRNA from being translated, or accelerate its degradation.

While small interfering RNAs (siRNA; 20-25 nt) are often produced by breakdown of viral RNA, there are also endogenous sources of siRNAs. siRNAs act through RNA interference in a fashion similar to miRNAs. Some miRNAs and siRNAs can cause genes they target to be methylated, thereby decreasing or increasing transcription of those genes. Animals have Piwi-interacting RNAs (piRNA; 29-30 nt) that are active in germline cells and are thought to be a defense against transposons and play a role in gametogenesis.

Many prokaryotes have CRISPR RNAs, a regulatory system similar to RNA interference. Antisense RNAs are widespread; most downregulate a gene, but a few are activators of transcription. One way antisense RNA can act is by binding to an mRNA, forming double-stranded RNA that is enzymatically degraded. There are many long noncoding RNAs that regulate genes in eukaryotes, one such RNA is Xist, which coats one X chromosome in female mammals and inactivates it.

An mRNA may contain regulatory elements itself, such as riboswitches, in the 5› untranslated region or 3› untranslated region; these cis-regulatory elements regulate the activity of that mRNA. The untranslated regions can also contain elements that regulate other genes.

In RNA Processing

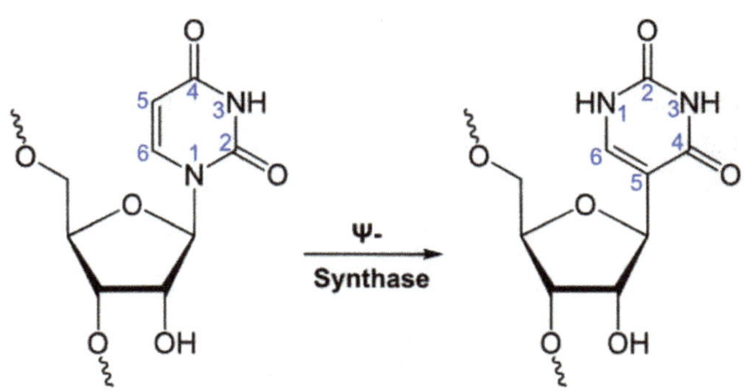

Uridine to pseudouridine is a common RNA modification.

Many RNAs are involved in modifying other RNAs. Introns are spliced out of pre-mRNA by spliceosomes, which contain several small nuclear RNAs (snRNA), or the introns can be ribozymes that are spliced by themselves. RNA can also be altered by having its nucleotides modified to nucleotides other than A, C, G and U. In eukaryotes, modifications of RNA nucleotides are in general directed by small nucleolar RNAs (snoRNA; 60-300 nt), found in the nucleolus and cajal bodies. snoRNAs associate with enzymes and guide them to a spot on an RNA by basepairing to that RNA. These enzymes then perform the nucleotide modification. rRNAs and tRNAs are extensively modified, but snRNAs and mRNAs can also be the target of base modification. RNA can also be methylated.

RNA Genomes

Like DNA, RNA can carry genetic information. RNA viruses have genomes composed of RNA that encodes a number of proteins. The viral genome is replicated by some of those proteins, while other proteins protect the genome as the virus particle moves to a new host cell. Viroids are another group of pathogens, but they consist only of RNA, do not encode any protein and are replicated by a host plant cell's polymerase.

In Reverse Transcription

Reverse transcribing viruses replicate their genomes by reverse transcribing DNA copies from their RNA; these DNA copies are then transcribed to new RNA. Retrotransposons also spread by copying DNA and RNA from one another, and telomerase contains an RNA that is used as template for building the ends of eukaryotic chromosomes.

Double-Stranded RNA

Double-stranded RNA (dsRNA) is RNA with two complementary strands, similar to the DNA found in all cells. dsRNA forms the genetic material of some viruses (double-stranded RNA viruses). Double-stranded RNA such as viral RNA or siRNA can trigger RNA interference in eukaryotes, as well as interferon response in vertebrates.

Circular RNA

Recently, it was shown that there is a single stranded covalently closed, *i.e.* circular form of RNA expressed throughout the animal and plant kingdom (see circRNA). circRNAs are thought to arise via a «back-splice» reaction where the spliceosome joins a downstream donor to an upstream acceptor splice site. So far the function of circRNAs is largely unknown, although for few examples a microRNA sponging activity has been demonstrated.

Key Discoveries in RNA Biology

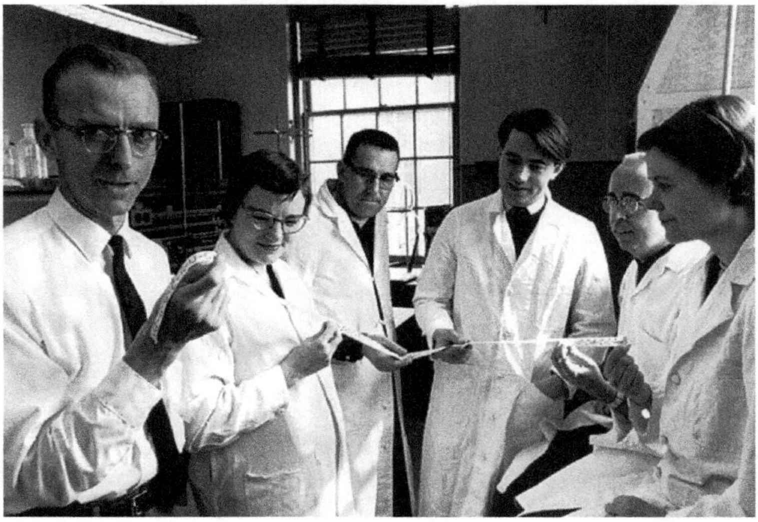

Robert W. Holley, left, poses with his research team.

Research on RNA has led to many important biological discoveries and numerous Nobel Prizes. Nucleic acids were discovered in 1868 by Friedrich Miescher, who called the material ‹nuclein› since it was found in the nucleus. It was later discovered that prokaryotic cells, which do not have a nucleus, also contain nucleic acids. The role of RNA in protein synthesis was suspected already in 1939. Severo Ochoa won the 1959 Nobel Prize in Medicine (shared with Arthur Kornberg) after he discovered an enzyme that can synthesize RNA in the laboratory. However, the enzyme discovered by Ochoa (polynucleotide phosphorylase) was later shown to be responsible for RNA degradation, not RNA synthesis. In 1956 Alex Rich and David Davies hybridized two separate strands of RNA to form the first crystal of RNA whose structure could be determined by X-ray crystallography.

The sequence of the 77 nucleotides of a yeast tRNA was found by Robert W. Holley in 1965, winning Holley the 1968 Nobel Prize in Medicine (shared with Har Gobind Khorana and Marshall Nirenberg). In 1967, Carl Woese hypothesized that RNA might be catalytic and suggested that the earliest forms of life (self-replicating molecules) could have relied on RNA both to carry genetic information and to catalyze biochemical reactions—an RNA world.

During the early 1970s, retroviruses and reverse transcriptase were discovered, showing for the first time that enzymes could copy RNA into DNA (the opposite of the usual route for transmission of genetic information). For this work, David Baltimore, Renato Dulbecco and Howard Temin were awarded a Nobel Prize in 1975. In 1976, Walter Fiers and his team determined the first complete nucleotide sequence of an RNA virus genome, that of bacteriophage MS2.

In 1977, introns and RNA splicing were discovered in both mammalian viruses and in cellular genes, resulting in a 1993 Nobel to Philip Sharp and Richard Roberts.

Catalytic RNA molecules (ribozymes) were discovered in the early 1980s, leading to a 1989 Nobel award to Thomas Cech and Sidney Altman. In 1990, it was found in *Petunia* that introduced genes can silence similar genes of the plant's own, now known to be a result of RNA interference.

At about the same time, 22 nt long RNAs, now called microRNAs, were found to have a role in the development of *C. elegans*. Studies on RNA interference gleaned a Nobel Prize for Andrew Fire and Craig Mello in 2006, and another Nobel was awarded for studies on the transcription of RNA to Roger Kornberg in the same year. The discovery of gene regulatory RNAs has led to attempts to develop drugs made of RNA, such as siRNA, to silence genes.

Evolution

In March 2015, complex DNA and RNA organic compounds of life, including uracil, cytosine and thymine, were reportedly formed in the laboratory under outer space conditions, using starter chemicals, such as pyrimidine, found in meteorites. Pyrimidine, like polycyclic aromatic hydrocarbons (PAHs), the most carbon-rich chemical found in the Universe, may have been formed in red giants or in interstellar dust and gas clouds, according to the scientists.

REFERENCES

1. Berg JM; Tymoczko JL; Stryer L (2002). Biochemistry (5th ed.). WH Freeman and Company. pp. 118–19, 781–808. ISBN 0-7167-4684-0. OCLC 179705944.
2. I. Tinoco & C. Bustamante (1999). "How RNA folds". J. Mol. Biol. 293 (2): 271–281. doi:10.1006/jmbi.1999.3001. PMID 10550208.
3. Higgs PG (2000). "RNA secondary structure: physical and computational aspects". Quarterly Reviews of Biophysics. 33 (3): 199–253. doi:10.1017/S0033583500003620. PMID 11191843.
4. Nissen P; Hansen J; Ban N; Moore PB; Steitz TA (2000). "The structural basis of ribosome activity in peptide bond synthesis". Science. 289 (5481): 920–30. Bibcode:2000Sci...289..920N. doi:10.1126/science.289.5481.920. PMID 10937990.
5. Lee JC; Gutell RR (2004). "Diversity of base-pair conformations and their occurrence in rRNA structure and RNA structural motifs". J. Mol. Biol. 344 (5): 1225–49. doi:10.1016/j.jmb.2004.09.072. PMID 15561141.
6. Barciszewski J; Frederic B; Clark C (1999). RNA biochemistry and biotechnology. Springer. pp. 73–87. ISBN 0-7923-5862-7. OCLC 52403776.
7. Salazar M; Fedoroff OY; Miller JM; Ribeiro NS; Reid BR (1992). "The DNA strand in DNAoRNA hybrid duplexes is neither B-form nor A-form in solution". Biochemistry. 32 (16): 4207–15. doi:10.1021/bi00067a007. PMID 7682844.
8. Sedova A; Banavali NK (2016). "RNA approaches the B-form in stacked single strand dinucleotide contexts". Biopolymers. 105 (2): 65–82. doi:10.1002/bip.22750. PMID 26443416.

9. Hermann T; Patel DJ (2000). "RNA bulges as architectural and recognition motifs". Structure. 8 (3): R47–R54. doi:10.1016/S0969-2126(00)00110-6. PMID 10745015.

10. Mikkola S; Stenman E; Nurmi K; Yousefi-Salakdeh E; Strömberg R; Lönnberg H (1999). "The mechanism of the metal ion promoted cleavage of RNA phosphodiester bonds involves a general acid catalysis by the metal aquo ion on the departure of the leaving group". Perkin transactions 2 (8): 1619–26. doi:10.1039/a903691a.

11. Jankowski JAZ; Polak JM (1996). Clinical gene analysis and manipulation: Tools, techniques and troubleshooting. Cambridge University Press. p. 14. ISBN 0-521-47896-0. OCLC 33838261.

12. Yu Q; Morrow CD (2001). "Identification of critical elements in the tRNA acceptor stem and TΨC loop necessary for human immunodeficiency virus type 1 infectivity". J Virol. 75 (10): 4902–6. doi:10.1128/JVI.75.10.4902-4906.2001. PMC 114245 . PMID 11312362.

13. Elliott MS; Trewyn RW (1983). "Inosine biosynthesis in transfer RNA by an enzymatic insertion of hypoxanthine". J. Biol. Chem. 259 (4): 2407–10. PMID 6365911.

14. Cantara, WA; Crain, PF; Rozenski, J; McCloskey, JA; Harris, KA; Zhang, X; Vendeix, FA; Fabris, D; Agris, PF (January 2011). "The RNA Modification Database, RNAMDB: 2011 update". Nucleic Acids Research. 39 (Database issue): D195–201. doi:10.1093/nar/gkq1028. PMC 3013656 . PMID 21071406.

15. Söll D; RajBhandary U (1995). TRNA: Structure, biosynthesis, and function. ASM Press. p. 165. ISBN 1-55581-073-X. OCLC 183036381.

16. Kiss T (2001). "Small nucleolar RNA-guided post-transcriptional modification of cellular RNAs". The EMBO Journal. 20 (14): 3617–22. doi:10.1093/emboj/20.14.3617. PMC 125535 . PMID 11447102.

17. King TH; Liu B; McCully RR; Fournier MJ (2002). "Ribosome structure and activity are altered in cells lacking snoRNPs that form pseudouridines in the peptidyl transferase center". Molecular Cell. 11 (2): 425–35. doi:10.1016/S1097-2765(03)00040-6. PMID 12620230.

18. Mathews DH; Disney MD; Childs JL; Schroeder SJ; Zuker M; Turner DH (2004). "Incorporating chemical modification constraints into a dynamic programming algorithm for prediction of RNA secondary structure". Proc. Natl. Acad. Sci. USA. 101 (19): 7287–92. Bibcode:2004PNAS..101.7287M. doi:10.1073/pnas.0401799101. PMC 409911 . PMID 15123812.

19. Tan ZJ; Chen SJ (2008). "Salt dependence of nucleic acid hairpin stability". Biophys. J. 95 (2): 738–52. Bibcode:2008BpJ....95..738T. doi:10.1529/biophysj.108.131524. PMC 2440479 . PMID 18424500.

20. Vater A; Klussmann S (January 2015). "Turning mirror-image oligonucleotides into drugs: the evolution of Spiegelmer therapeutics". Drug Discovery Today. 20 (1): 147–155. doi:10.1016/j.drudis.2014.09.004. PMID 25236655.

21. Nudler E; Gottesman ME (2002). "Transcription termination and anti-termination in E. coli". Genes to Cells. 7 (8): 755–68. doi:10.1046/j.1365-2443.2002.00563.x. PMID 12167155.
22. Jeffrey L Hansen; Alexander M Long; Steve C Schultz (1997). "Structure of the RNA-dependent RNA polymerase of poliovirus". Structure. 5 (8): 1109–22. doi:10.1016/S0969-2126(97)00261-X. PMID 9309225.
23. Ahlquist P (2002). "RNA-Dependent RNA Polymerases, Viruses, and RNA Silencing". Science. 296 (5571): 1270–73. Bibcode:2002Sci...296.1270A. doi:10.1126/science.1069132. PMID 12016304.
24. Cooper GC; Hausman RE (2004). The Cell: A Molecular Approach (3rd ed.). Sinauer. pp. 261–76, 297, 339–44. ISBN 0-87893-214-3. OCLC 174924833.
25. Mattick JS; Gagen MJ (1 September 2001). "The evolution of controlled multitasked gene networks: the role of introns and other noncoding RNAs in the development of complex organisms". Mol. Biol. Evol. 18 (9): 1611–30. doi:10.1093/oxfordjournals.molbev.a003951. PMID 11504843.
26. Mattick, JS (2001). "Noncoding RNAs: the architects of eukaryotic complexity". EMBO Reports. 2 (11): 986–91. doi:10.1093/embo-reports/kve230. PMC 1084129 . PMID 11713189.
27. Mattick JS (October 2003). "Challenging the dogma: the hidden layer of non-protein-coding RNAs in complex organisms" (PDF). BioEssays. 25 (10): 930–9. doi:10.1002/bies.10332. PMID 14505360.
28. Mattick JS (October 2004). "The hidden genetic program of complex organisms". Scientific American. 291 (4): 60–7. doi:10.1038/scientificamerican1004-60. PMID 15487671.[dead link]
29. Wirta W (2006). Mining the transcriptome – methods and applications. Stockholm: School of Biotechnology, Royal Institute of Technology. ISBN 91-7178-436-5. OCLC 185406288.
30. Rossi JJ (2004). "Ribozyme diagnostics comes of age". Chemistry & Biology. 11 (7): 894–95. doi:10.1016/j.chembiol.2004.07.002. PMID 15271347.
31. Storz, G (17 May 2002). "An expanding universe of noncoding RNAs.". Science. 296 (5571): 1260–3. doi:10.1126/science.1072249. PMID 12016301.
32. Fatica, A; Bozzoni, I (January 2014). "Long non-coding RNAs: new players in cell differentiation and development.". Nature Reviews Genetics. 15 (1): 7–21. doi:10.1038/nrg3606. PMID 24296535.
33. Chen, Q; Yan, M; Cao, Z; Li, X; Zhang, Y; Shi, J; Feng, GH; Peng, H; Zhang, X; Zhang, Y; Qian, J; Duan, E; Zhai, Q; Zhou, Q (22 January 2016). "Sperm tsRNAs contribute to intergenerational inheritance of an acquired metabolic disorder.". Science. 351 (6271): 397–400. doi:10.1126/science.aad7977. PMID 26721680.
34. Wei, H; Zhou, B; Zhang, F; Tu, Y; Hu, Y; Zhang, B; Zhai, Q (2013). "Profiling and identification of small rDNA-derived RNAs and their potential biological

functions.". PLOS ONE. 8 (2): e56842. doi:10.1371/journal.pone.0056842. PMC 3572043 . PMID 23418607.

35. Gueneau de Novoa P, Williams KP; Williams (2004). "The tmRNA website: reductive evolution of tmRNA in plastids and other endosymbionts". Nucleic Acids Res. 32 (Database issue): D104–8. doi:10.1093/nar/gkh102. PMC 308836 . PMID 14681369.
36. Carthew, RW; Sontheimer, EJ. (February 2009). "Origins and Mechanisms of miRNAs and siRNAs.". Cell. 136: 642–55. doi:10.1016/j.cell.2009.01.035. PMC 2675692 . PMID 19239886.
37. Liang, Kung-Hao; Yeh, Chau-Ting (2013). "A gene expression restriction network mediated by sense and antisense Alu sequences located on protein-coding messenger RNAs.". BMC Genomics. 14: 325. doi:10.1186/1471-2164-14-325. PMC 3655826 . PMID 23663499. Retrieved 2013-05-11.
38. Wu L; Belasco JG (January 2008). "Let me count the ways: mechanisms of gene regulation by miRNAs and siRNAs". Mol. Cell. 29 (1): 1–7. doi:10.1016/j.molcel.2007.12.010. PMID 18206964.
39. Matzke MA; Matzke AJM (2004). "Planting the seeds of a new paradigm". PLoS Biology. 2 (5): e133. doi:10.1371/journal.pbio.0020133. PMC 406394 . PMID 15138502.
40. Vazquez F; Vaucheret H; Rajagopalan R; Lepers C; Gasciolli V; Mallory AC; Hilbert J; Bartel DP; Crété P (2004). "Endogenous trans-acting siRNAs regulate the accumulation of Arabidopsis mRNAs". Molecular Cell. 16 (1): 69–79. doi:10.1016/j.molcel.2004.09.028. PMID 15469823.
41. Watanabe T, Totoki Y, Toyoda A, et al. (May 2008). "Endogenous siRNAs from naturally formed dsRNAs regulate transcripts in mouse oocytes". Nature. 453 (7194): 539–43. Bibcode:2008Natur.453..539W. doi:10.1038/nature06908. PMID 18404146.
42. Sontheimer EJ; Carthew RW (July 2005). "Silence from within: endogenous siRNAs and miRNAs". Cell. 122 (1): 9–12. doi:10.1016/j.cell.2005.06.030. PMID 16009127.
43. Doran G (2007). "RNAi – Is one suffix sufficient?". Journal of RNAi and Gene Silencing. 3 (1): 217–19.
44. Pushparaj PN; Aarthi JJ; Kumar SD; Manikandan J (2008). "RNAi and RNAa — The Yin and Yang of RNAome". Bioinformation. 2 (6): 235–7. doi:10.6026/97320630002235. PMC 2258431 . PMID 18317570.
45. Horwich MD; Li C; Matranga C; Vagin V; Farley G; Wang P; Zamore PD (2007). "The Drosophila RNA methyltransferase, DmHen1, modifies germline piRNAs and single-stranded siRNAs in RISC". Current Biology. 17 (14): 1265–72. doi:10.1016/j.cub.2007.06.030. PMID 17604629.
46. Girard A; Sachidanandam R; Hannon GJ; Carmell MA (2006). "A germline-specific class of small RNAs binds mammalian Piwi proteins". Nature. 442 (7099): 199–202. Bibcode:2006Natur.442..199G. doi:10.1038/nature04917. PMID 16751776.

47. Horvath P; Barrangou R (2010). "CRISPR/Cas, the Immune System of Bacteria and Archaea". Science. 327 (5962): 167–70. Bibcode:2010Sci...327..167H. doi:10.1126/science.1179555. PMID 20056882.

48. Wagner EG; Altuvia S; Romby P (2002). "Antisense RNAs in bacteria and their genetic elements". Adv Genet. Advances in Genetics. 46: 361–98. doi:10.1016/S0065-2660(02)46013-0. ISBN 9780120176465. PMID 11931231.

49. Gilbert SF (2003). Developmental Biology (7th ed.). Sinauer. pp. 101–3. ISBN 0-87893-258-5. OCLC 154656422.

50. Amaral PP; Mattick JS (October 2008). "Noncoding RNA in development". Mammalian genome : official journal of the International Mammalian Genome Society. 19 (7–8): 454–92. doi:10.1007/s00335-008-9136-7. PMID 18839252.

51. Heard E; Mongelard F; Arnaud D; Chureau C; Vourc'h C; Avner P (1999). "Human XIST yeast artificial chromosome transgenes show partial X inactivation center function in mouse embryonic stem cells". Proc. Natl. Acad. Sci. USA. 96 (12): 6841–46. Bibcode:1999PNAS...96.6841H. doi:10.1073/pnas.96.12.6841. PMC 22003 . PMID 10359800.

52. Batey RT (2006). "Structures of regulatory elements in mRNAs". Curr. Opin. Struct. Biol. 16 (3): 299–306. doi:10.1016/j.sbi.2006.05.001. PMID 16707260.

53. Scotto L; Assoian RK (June 1993). "A GC-rich domain with bifunctional effects on mRNA and protein levels: implications for control of transforming growth factor beta 1 expression". Mol. Cell. Biol. 13 (6): 3588–97. PMC 359828 . PMID 8497272.

54. Steitz TA; Steitz JA (1993). "A general two-metal-ion mechanism for catalytic RNA". Proc. Natl. Acad. Sci. U.S.A. 90 (14): 6498–502. Bibcode:1993PNAS...90.6498S. doi:10.1073/pnas.90.14.6498. PMC 46959 . PMID 8341661.

55. Xie J; Zhang M; Zhou T; Hua X; Tang L; Wu W (2007). "Sno/scaRNAbase: a curated database for small nucleolar RNAs and cajal body-specific RNAs". Nucleic Acids Res. 35 (Database issue): D183–7. doi:10.1093/nar/gkl873. PMC 1669756 . PMID 17099227.

56. Omer AD; Ziesche S; Decatur WA; Fournier MJ; Dennis PP (2003). "RNA-modifying machines in archaea". Molecular Microbiology. 48 (3): 617–29. doi:10.1046/j.1365-2958.2003.03483.x. PMID 12694609.

57. Cavaillé J; Nicoloso M; Bachellerie JP (1996). "Targeted ribose methylation of RNA in vivo directed by tailored antisense RNA guides". Nature. 383 (6602): 732–5. Bibcode:1996Natur.383..732C. doi:10.1038/383732a0. PMID 8878486.

58. Kiss-László Z; Henry Y; Bachellerie JP; Caizergues-Ferrer M; Kiss T (1996). "Site-specific ribose methylation of preribosomal RNA: a novel function for small nucleolar RNAs". Cell. 85 (7): 1077–88. doi:10.1016/S0092-8674(00)81308-2. PMID 8674114.

59. Daròs JA; Elena SF; Flores R (2006). "Viroids: an Ariadne's thread into the RNA labyrinth". EMBO Rep. 7 (6): 593–8. doi:10.1038/sj.embor.7400706. PMC 1479586 . PMID 16741503.

60. Kalendar R; Vicient CM; Peleg O; Anamthawat-Jonsson K; Bolshoy A; Schulman AH (2004). "Large retrotransposon derivatives: abundant, conserved but nonautonomous retroelements of barley and related genomes". Genetics. 166 (3): 1437–50. doi:10.1534/genetics.166.3.1437. PMC 1470764 . PMID 15082561.
61. Podlevsky JD; Bley CJ; Omana RV; Qi X; Chen JJ (2008). "The telomerase database". Nucleic Acids Res. 36 (Database issue): D339–43. doi:10.1093/nar/gkm700. PMC 2238860 . PMID 18073191.
62. Blevins, T.; Rajeswaran, R.; Shivaprasad, PV.; Beknazariants, D.; Si-Ammour, A.; Park, HS.; Vazquez, F.; Robertson, D.; Meins, F. (2006). "Four plant Dicers mediate viral small RNA biogenesis and DNA virus induced silencing". Nucleic Acids Res. 34 (21): 6233–46. doi:10.1093/nar/gkl886. PMC 1669714 . PMID 17090584.
63. Jana S; Chakraborty C; Nandi S; Deb JK (2004). "RNA interference: potential therapeutic targets". Appl. Microbiol. Biotechnol. 65 (6): 649–57. doi:10.1007/s00253-004-1732-1. PMID 15372214.
64. Schultz U; Kaspers B; Staeheli P (2004). "The interferon system of non-mammalian vertebrates". Dev. Comp. Immunol. 28 (5): 499–508. doi:10.1016/j.dci.2003.09.009. PMID 15062646.
65. Whitehead, K. A.; Dahlman, J. E.; Langer, R. S.; Anderson, D. G. (2011). "Silencing or Stimulation? SiRNA Delivery and the Immune System". Annual Review of Chemical and Biomolecular Engineering. 2: 77–96. doi:10.1146/annurev-chembioeng-061010-114133. PMID 22432611.
66. Dahm R (2005). "Friedrich Miescher and the discovery of DNA". Developmental Biology. 278 (2): 274–88. doi:10.1016/j.ydbio.2004.11.028. PMID 15680349.
67. Caspersson T; Schultz J (1939). "Pentose nucleotides in the cytoplasm of growing tissues". Nature. 143 (3623): 602–3. Bibcode:1939Natur.143..602C. doi:10.1038/143602c0.
68. Ochoa S (1959). "Enzymatic synthesis of ribonucleic acid" (PDF). Nobel Lecture.
69. Rich A; Davies, D (1956). "A New Two-Stranded Helical Structure: Polyadenylic Acid and Polyuridylic Acid". Journal of the American Chemical Society. 78 (14): 3548–3549. doi:10.1021/ja01595a086.
70. Holley RW, et al. (1965). "Structure of a ribonucleic acid". Science. 147 (3664): 1462–65. Bibcode:1965Sci...147.1462H. doi:10.1126/science.147.3664.1462. PMID 14263761.
71. Siebert S (2006). "Common sequence structure properties and stable regions in RNA secondary structures" (PDF). Dissertation, Albert-Ludwigs-Universität, Freiburg im Breisgau. p. 1. Archived from the original (PDF) on March 9, 2012.
72. Szathmáry E (1999). "The origin of the genetic code: amino acids as cofactors in an RNA world". Trends Genet. 15 (6): 223–9. doi:10.1016/S0168-9525(99)01730-8. PMID 10354582.
73. Fiers W, et al. (1976). "Complete nucleotide-sequence of bacteriophage MS2-RNA: primary and secondary structure of replicase gene". Nature. 260 (5551): 500–7. Bibcode:1976Natur.260..500F. doi:10.1038/260500a0. PMID 1264203.

74. Napoli C; Lemieux C; Jorgensen R (1990). "Introduction of a chimeric chalcone synthase gene into petunia results in reversible co-suppression of homologous genes in trans". Plant Cell. 2 (4): 279–89. doi:10.1105/tpc.2.4.279. PMC 159885 . PMID 12354959.
75. Dafny-Yelin M; Chung SM; Frankman EL; Tzfira T (December 2007). "pSAT RNA interference vectors: a modular series for multiple gene down-regulation in plants". Plant Physiol. 145 (4): 1272–81. doi:10.1104/pp.107.106062. PMC 2151715 . PMID 17766396.
76. Ruvkun G (2001). "Glimpses of a tiny RNA world". Science. 294 (5543): 797–99. doi:10.1126/science.1066315. PMID 11679654.
77. Fichou Y; Férec C (2006). "The potential of oligonucleotides for therapeutic applications". Trends in Biotechnology. 24 (12): 563–70. doi:10.1016/j.tibtech.2006.10.003. PMID 17045686.
78. Marlaire, Ruth (3 March 2015). "NASA Ames Reproduces the Building Blocks of Life in Laboratory". NASA. Retrieved 5 March 2015.
79. Plescia, O J; Palczuk, N C; Cora-Figueroa, E; Mukherjee, A; Braun, W (October 1965). "Production of antibodies to soluble RNA (sRNA)". Proc. Natl. Acad. Sci. USA. 54 (4): 1281–1285. doi:10.1073/pnas.54.4.1281. PMC 219862 . PMID 5219832.
80. Sharp, Stephen J; Schaack, Jerome; Cooley, Lynn; Burke, Deborah J; Soll, Dieter (1985). "Structure and Transcription of Eukaryotic tRNA Genes". CRC Critical Reviews in Biochemistry. 19 (2): 107–144. doi:10.3109/10409238509082541. PMID 3905254.
81. Crick F (1968). "The origin of the genetic code". J Mol Biol. 38 (3): 367–379. doi:10.1016/0022-2836(68)90392-6. PMID 4887876.
82. Stryer L, Berg JM, Tymoczko JL (2002). Biochemistry (5th ed.). San Francisco: W.H. Freeman. ISBN 0-7167-4955-6.
83. Itoh, Yuzuru; Sekine, Shun-ichi Sekine; Suetsugu, Shiro; Yokoyama, Shigeyuki (6 May 2013). "Tertiary structure of bacterial selenocysteine tRNA". Nucleic Acids Research. 41 (13): 6729–6738. doi:10.1093/nar/gkt321. PMC 3711452 . PMID 23649835. Retrieved 23 November 2014.
84. Goodenbour, J. M.; Pan, T. (29 October 2006). "Diversity of tRNA genes in eukaryotes" (PDF). Nucleic Acids Research. 34 (21): 6137–6146. doi:10.1093/nar/gkl725. PMC 1693877 . PMID 17088292. Retrieved 23 November 2014.
85. Jahn, Martina; Rogers, M. John; Söll, Dieter (18 July 1991). "Anticodon and acceptor stem nucleotides in tRNAGln are major recognition elements for E. coli glutaminyl-tRNA synthetase". Nature. 352 (6332): 258–260. doi:10.1038/352258a0. PMID 1857423. Retrieved 23 November 2014.
86. Ibba, Michael; Söll, Dieter (June 2000). "Aminoacyl-tRNA Synthesis". Annual Review of Biochemistry. 69 (1): 617–650. doi:10.1146/annurev.biochem.69.1.617. PMID 10966471. Retrieved 23 November 2014.
87. Sprinzl, M., and Cramer, F. (1979) Prog. Nucleic Acids Res. Mol. Biol. 22, 1–16

88. Green, R., and Noller, H. F. (1997) Annu. Rev. Biochem. 66, 679–716

89. Aebi M, Kirchner G, Chen JY, et al. (September 1990). "Isolation of a temperature-sensitive mutant with an altered tRNA nucleotidyltransferase and cloning of the gene encoding tRNA nucleotidyltransferase in the yeast Saccharomyces cerevisiae". J. Biol. Chem. 265 (27): 16216–16220. PMID 2204621.

90. McCloskey, James A.; Nishimura, Susumu (November 1977). "Modified nucleosides in transfer RNA". Accounts of Chemical Research. 10 (11): 403–410. doi:10.1021/ar50119a004. Retrieved 23 November 2014.

91. Felsenfeld G, Cantoni G; Cantoni (1964). "Use of thermal denaturation studies to investigate the base sequence of yeast serine sRNA". Proc Natl Acad Sci USA. 51 (5): 818–26. Bibcode:1964PNAS...51..818F. doi:10.1073/pnas.51.5.818. PMC 300168 . PMID 14172997.

92. Suzuki, T; Suzuki, T (June 2014). "A complete landscape of post-transcriptional modifications in mammalian mitochondrial tRNAs.". Nucleic Acids Research. 42 (11): 7346–57. doi:10.1093/nar/gku390. PMID 24831542.

93. Lodish H, Berk A, Matsudaira P, Kaiser CA, Krieger M, Scott MP, Zipursky SL, Darnell J. (2004). Molecular Biology of the Cell. WH Freeman: New York, NY. 5th ed.

94. Schimmel P, Giege R, Moras D, Yokoyama S; Giege; Moras; Yokoyama (1993). "An operational RNA code for amino acids and possible relationship to genetic code". Proc. Natl. Acad. Sci. USA. 90 (19): 8763–876. Bibcode:1993PNAS...90.8763S. doi:10.1073/pnas.90.19.8763.

95. Dunkle JA, Wang L, Feldman MB, Pulk A, Chen VB, Kapral GJ, Noeske J, Richardson JS, Blanchard SC, Cate JH; Wang; Feldman; Pulk; Chen; Kapral; Noeske; Richardson; Blanchard; Cate (2011). "Structures of the bacterial ribosome in classical and hybrid states of tRNA binding". Science. 332 (6032): 981–984. Bibcode:2011Sci...332..981D. doi:10.1126/science.1202692. PMC 3176341 . PMID 21596992.

96. Konevega, AL; Soboleva, NG; Makhno, VI; Semenkov, YP; Wintermeyer, W; Rodnina, MV; Katunin, VI (Jan 2004). "Purine bases at position 37 of tRNA stabilize codon-anticodon interaction in the ribosomal A site by stacking and Mg2+-dependent interactions". RNA. 10 (1): 90–101. doi:10.1261/rna.5142404. PMC 1370521 . PMID 14681588.

97. Agirrezabala X, Frank J; Frank (2009). "Elongation in translation as a dynamic interaction among the ribosome, tRNA, and elongation factors EF-G and EF-Tu". Q Rev Biophys. 42 (3): 159–200. doi:10.1017/S0033583509990060. PMC 2832932 . PMID 20025795.

98. Allen GS, Zavialov A, Gursky R, Ehrenberg M, Frank J; Zavialov; Gursky; Ehrenberg; Frank (2005). "The cryo-EM structure of a translation initiation complex from Escherichia coli". Cell. 121 (5): 703–712. doi:10.1016/j.cell.2005.03.023. PMID 15935757.

99. WormBase web site, http://www.wormbase.org, release WS187, date 25-Jan-2008.

100. Spieth, J; Lawson, D (Jan 2006). "Overview of gene structure". WormBook: 1–10. doi:10.1895/wormbook.1.65.1. PMID 18023127.

101. Hartwell LH, Hood L, Goldberg ML, Reynolds AE, Silver LM, Veres RC. (2004). Genetics: From Genes to Genomes 2nd ed. McGraw-Hill: New York, NY. p 264.

102. Ensembl release 70 - Jan 2013 http://www.ensembl.org/Homo_sapiens/Info/StatsTable?db=core

103. Lander E.; et al. (2001). "Initial sequencing and analysis of the human genome". Nature. 409 (6822): 860–921. doi:10.1038/35057062. PMID 11237011.

104. Rogers Theresa E.; et al. (2012). "A Pseudo-tRNA Modulates Antibiotic Resistance in Bacillus cereus". PLoS ONE. 7 (7): e41248. doi:10.1371/journal.pone.0041248. PMC 3399842 . PMID 22815980.

105. Telonis Aristeidis G.; et al. (2014). "Nuclear and Mitochondrial tRNA-lookalikes in the Human Genome". Frontiers in Genetics. 5: 00344. doi:10.3389/fgene.2014.00344. PMC 4189335 . PMID 25339973.

106. Ramos A.; et al. (2011). "Nuclear Insertions of Mitochondrial Origin: Database Updating and Usefulness in Cancer Studies". Mitochondrion. 11 (6): 946–53. doi:10.1016/j.mito.2011.08.009. PMID 21907832.

107. Ibid. p 529.

108. Novoa, Eva Maria; Pavon-Eternod, Mariana; Pan, Tao; Ribas de Pouplana, Lluís (March 2012). "A Role for tRNA Modifications in Genome Structure and Codon Usage". Cell. 149 (1): 202–213. doi:10.1016/j.cell.2012.01.050. PMID 22464330. Retrieved 23 November 2014.

109. Gebetsberger Jennifer; et al. (2013). "Slicing tRNAs to boost functional ncRNA diversity". RNA biology. 10: 1798–1806. doi:10.4161/rna.27177. PMC 3917982 . PMID 24351723.

110. Shigematsu Megumi; et al. (2014). "Tranfer RNA as a source of small functional RNA". Journal of Molecular Biology and Molecular Imaging. 1: 8.

111. ^ Jump up to:a b Sobala Andrew; et al. (2011). "Transfer RNA-derived fragments: origins, processing, and functions". Wiley interdisciplinary reviews. 2: 853–862. doi:10.1002/wrna.96. PMID 21976287.

112. Keam Simon P; et al. (2015). "tRNA-Derived Fragments (tRFs): Emerging New Roles for an Ancient RNA in the Regulation of Gene Expression". Life (Basel). 5: 1638–1651. doi:10.3390/life5041638. PMC 4695841 . PMID 26703738.

113. Telonis Aristeidis G; et al. (2015). "Dissecting tRNA-derived fragment complexities using personalized transcriptomes reveals novel fragment classes and unexpected dependencies". Oncotarget. 6: 24797–822. doi:10.18632/oncotarget.4695. PMC 4694795 . PMID 26325506.

114. Kumar Pankaj; et al. (2014). "Meta-analysis of tRNA derived RNA fragments reveals that they are evolutionarily conserved and associate with AGO proteins to recognize specific RNA targets". BMC Biol. 12: 78. doi:10.1186/s12915-014-0078-0. PMC 4203973 . PMID 25270025.

115. Honda Shozo; et al. (2015). "Sex hormone-dependent tRNA halves enhance cell proliferation in breast and prostate cancers". Proc Natl Acad Sci USA. 112: E3816–25. doi:10.1073/pnas.1510077112. PMC 4517238 . PMID 26124144.

116. Shigematsu Megumi; et al. (2015). "tRNA-Derived Short Non-coding RNA as Interacting Partners of Argonaute Proteins". Gene Regul Syst Bio. 9: 27–33. doi:10.4137/GRSB.S29411. PMC 4567038 . PMID 26401098.

117. Emara Mohamed M; et al. (2010). "Angiogenin-induced tRNA-derived stress-induced RNAs promote stress-induced stress granule assembly". J Biol Chem. 285: 10959–68. doi:10.1074/jbc.M109.077560. PMC 2856301 . PMID 20129916.

118. Goodarzi Hani; et al. (2015). "Endogenous tRNA-derived fragments suppress breast cancer progression via YBX1 displacement". Cell. 161: 790–802. doi:10.1016/j.cell.2015.02.053. PMC 4457382 . PMID 25957686.

119. Ivanov Pavel; et al. (2011). "Angiogenin-induced tRNA fragments inhibit translation initiation". Mol Cell. 43: 613–23. doi:10.1016/j.molcel.2011.06.022. PMC 3160621 . PMID 21855800.

120. Selitsky Sara R; et al. (2015). "Small tRNA-derived RNAs are increased and more abundant than microRNAs in chronic hepatitis B and C". Sci Rep. 5: 7675. doi:10.1038/srep07675. PMC 4286764 . PMID 25567797.

121. Sharma Upasna; et al. (2016). "Biogenesis and function of tRNA fragments during sperm maturation and fertilization in mammals". Science. 351: 391–6. doi:10.1126/science.aad6780. PMC 4888079 . PMID 26721685.

122. Casas Eduardo; et al. (2015). "Characterization of circulating transfer RNA-derived RNA fragments in cattle". Front Genet. 6: 271. doi:10.3389/fgene.2015.00271. PMC 4547532 . PMID 26379699.

123. Hirose Yuka; et al. (2015). "Precise mapping and dynamics of tRNA-derived fragments (tRFs) in the development of Triops cancriformis (tadpole shrimp)". BMC Genet. 16: 83. doi:10.1186/s12863-015-0245-5. PMC 4501094 . PMID 26168920.

124. Karaiskos Spyros; et al. (2015). "Age-driven modulation of tRNA-derived fragments in Drosophila and their potential targets". Biol Direct. 10: 51. doi:10.1186/s13062-015-0081-6. PMC 4572633 . PMID 26374501.

125. Pliatsika Venetia; et al. (2016). "MINTbase: a framework for the interactive exploration of mitochondrial and nuclear tRNA fragments". Bioinformatics.

126. Kumar Panjav; et al. (2014). "tRFdb: a database for transfer RNA fragments". Nucleic Acids Res. 43: D141–D145. doi:10.1093/nar/gku1138. PMC 4383946 . PMID 25392422.

127. White RJ (1997). "Regulation of RNA polymerases I and III by the retinoblastoma protein: a mechanism for growth control?". Trends in Biochemical Sciences. 22 (3): 77–80. doi:10.1016/S0968-0004(96)10067-0. PMID 9066256.

128. Sharp, Stephen; Dingermann, Theodor; Söll, Dieter (1982). "The minimum intragenic sequences required for promotion of eukaryotic tRNA gene

transcription" (PDF). Nucleic Acids Research. 10 (18): 5393–5406. doi:10.1093/nar/10.18.5393. PMC 320884 . PMID 6924209. Retrieved 23 November 2014.

129. Dieci G, Fiorino G, Castelnuovo M, Teichmann M, Pagano A; Fiorino; Castelnuovo; Teichmann; Pagano (December 2007). "The expanding RNA polymerase III transcriptome". Trends Genet. 23 (12): 614–22. doi:10.1016/j.tig.2007.09.001. PMID 17977614.
130. Tocchini-Valentini, Giuseppe D.; Fruscoloni, Paolo; Tocchini-Valentini, Glauco P. (12 November 2009). "Processing of multiple-intron-containing pretRNA". Proceedings of the National Academy of Sciences. 106 (48): 20246–20251. doi:10.1073/pnas.0911658106. PMC 2787110 . PMID 19910528.
131. Abelson J, Trotta CR, Li H; Trotta; Li (1998). "tRNA Splicing". J Biol Chem. 273 (21): 12685–12688. doi:10.1074/jbc.273.21.12685. PMID 9582290.
132. Soma, Akiko (2014). "Circularly permuted tRNA genes: their expression and implications for their physiological relevance and development". Frontiers in Genetics. 5. doi:10.3389/fgene.2014.00063. ISSN 1664-8021.
133. Frank DN, Pace NR; Pace (1998). "Ribonuclease P: unity and diversity in a tRNA processing ribozyme". Annu. Rev. Biochem. 67 (1): 153–80. doi:10.1146/annurev.biochem.67.1.153. PMID 9759486.
134. Ceballos M, Vioque A; Vioque (2007). "tRNase Z". Protein Pept. Lett. 14 (2): 137–45. doi:10.2174/092986607779816050. PMID 17305600.
135. Randau L, Schröder I, Söll D; Schröder; Söll (May 2008). "Life without RNase P". Nature. 453 (7191): 120–3. Bibcode:2008Natur.453..120R. doi:10.1038/nature06833. PMID 18451863.
136. Weiner AM (October 2004). "tRNA maturation: RNA polymerization without a nucleic acid template". Curr. Biol. 14 (20): R883–5. doi:10.1016/j.cub.2004.09.069. PMID 15498478.
137. Kutay, U. .; Lipowsky, G. .; Izaurralde, E. .; Bischoff, F. .; Schwarzmaier, P. .; Hartmann, E. .; Görlich, D. . (1998). "Identification of a tRNA-Specific Nuclear Export Receptor". Molecular Cell. 1 (3): 359–369. doi:10.1016/S1097-2765(00)80036-2. PMID 9660920.
138. Arts, G. J.; Fornerod, M. .; Mattaj, L. W. (1998). "Identification of a nuclear export receptor for tRNA". Current Biology. 8 (6): 305–314. doi:10.1016/S0960-9822(98)70130-7. PMID 9512417.
139. Arts, G. -J.; Kuersten, S.; Romby, P.; Ehresmann, B.; Mattaj, I. W. (1998). "The role of exportin-t in selective nuclear export of mature tRNAs". The EMBO Journal. 17 (24): 7430–7441. doi:10.1093/emboj/17.24.7430. PMC 1171087 . PMID 9857198.
140. Yoshihisa, T.; Yunoki-Esaki, K.; Ohshima, C.; Tanaka, N.; Endo, T. (2003). "Possibility of cytoplasmic pre-tRNA splicing: the yeast tRNA splicing endonuclease mainly localizes on the mitochondria". Molecular Biology of the Cell. 14 (8): 3266–3279. doi:10.1091/mbc.E02-11-0757. PMC 181566 . PMID 12925762.

141. Brian F.C. Clark (October 2006). "The crystal structure of tRNA"(PDF). J. Biosci. 31 (4): 453–7. doi:10.1007/BF02705184. PMID 17206065.

142. HOLLEY RW; APGAR J; EVERETT GA; et al. (March 1965). "STRUCTURE OF A RIBONUCLEIC ACID". Science. 147 (3664): 1462–5. Bibcode:1965Sci...147.1462H. doi:10.1126/science.147.3664.1462. PMID 14263761. Retrieved 2010-09-03.

143. http://www.nytimes.com/1991/07/04/obituaries/robert-m-bock-67-biologist-and-a-dean.html

144. "The Nobel Prize in Physiology or Medicine 1968". Nobel Foundation. Retrieved 2007-07-28.

145. Ladner JE; Jack A; Robertus JD; et al. (November 1975). "Structure of yeast phenylalanine transfer RNA at 2.5 A resolution". Proc. Natl. Acad. Sci. U.S.A. 72 (11): 4414–8. Bibcode:1975PNAS...72.4414L. doi:10.1073/pnas.72.11.4414. PMC 388732 . PMID 1105583.

146. Kim SH; Quigley GJ; Suddath FL; et al. (1973). "Three-dimensional structure of yeast phenylalanine transfer RNA: folding of the polynucleotide chain". Science. 179 (4070): 285–8. Bibcode:1973Sci...179..285K. doi:10.1126/science.179.4070.285. PMID 4566654.

147. Krieger M, Scott MP, Matsudaira PT, Lodish HF, Darnell JE, Lawrence Z, Kaiser C, Berk A (2004). "Section 4.1: Structure of Nucleic Acids". Molecular cell biology. New York: W.H. Freeman and CO. ISBN 0-7167-4366-3.

148. "Structure of Nucleic Acids". SparkNotes.

149. Anthony-Cahill SJ, Mathews CK, van Holde KE, Appling DR (2012). Biochemistry (4th Edition). Englewood Cliffs, N.J: Prentice Hall. ISBN 0-13-800464-1.

150. Alberts B, Johnson A, Lewis J, Raff M, Roberts K, Wlater P (2002). Molecular Biology of the Cell (4th ed.). New York NY: Garland Science. ISBN 0-8153-3218-1.

151. Mao C (Dec 2004). "The emergence of complexity: lessons from DNA". PLoS Biology. 2 (12): e431. doi:10.1371/journal.pbio.0020431. PMC 535573 . PMID 15597116.

152. Katsuyuki, Aoki; Kazutaka, Murayama; Hu, Ning-Hai (2016). "Chapter 3, section3. Nucleic Acid Constituent complexes". In Astrid, Sigel; Helmut, Sigel; Roland K.O., Sigel. The Alkali Metal Ions: Their Role in Life. Metal Ions in Life Sciences. 16. Springer. pp. 43–66. doi:10.1007/978-3-319-21756-7_3.

153. Tinoco I, Bustamante C (Oct 1999). "How RNA folds". Journal of Molecular Biology. 293 (2): 271–81. doi:10.1006/jmbi.1999.3001. PMID 10550208.

154. "RNA structure (Molecular Biology)".

155. Hollyfield JG, Besharse JC, Rayborn ME (Dec 1976). "The effect of light on the quantity of phagosomes in the pigment epithelium". Experimental Eye Research. 23 (6): 623–35. doi:10.1016/0014-4835(76)90221-9. PMID 1087245.

156. Rietveld K, Van Poelgeest R, Pleij CW, Van Boom JH, Bosch L (Mar 1982). "The tRNA-like structure at the 3' terminus of turnip yellow mosaic virus RNA.

Differences and similarities with canonical tRNA". Nucleic Acids Research. 10 (6): 1929–46. doi:10.1093/nar/10.6.1929. PMC 320581 . PMID 7079175.

157. Staple DW, Butcher SE (Jun 2005). "Pseudoknots: RNA structures with diverse functions". PLoS Biology. 3 (6): e213. doi:10.1371/journal.pbio.0030213. PMC 1149493 . PMID 15941360.

158. Sperschneider J, Datta A, Wise MJ (Dec 2012). "Predicting pseudoknotted structures across two RNA sequences". Bioinformatics. 28 (23): 3058–65. doi:10.1093/bioinformatics/bts575. PMC 3516145 . PMID 23044552.

159. Dickerson RE, Drew HR, Conner BN, Wing RM, Fratini AV, Kopka ML (Apr 1982). "The anatomy of A-, B-, and Z-DNA". Science. 216 (4545): 475–85. doi:10.1126/science.7071593. PMID 7071593.

160. Chen X; Ramakrishnan B; Sundaralingam M (1995). "Crystal structures of B-form DNA-RNA chimers complexed with distamycin". Nature Structural Biology. 2 (9): 733–735. doi:10.1038/nsb0995-733.

161. Sedova A, Banavali NK (Oct 2015). "RNA approaches the B-form in stacked single strand dinucleotide contexts". Biopolymers. 105 (2). doi:10.1002/bip.22750. PMID 26443416.

162. Mirkin SM (2001). "DNA Topology: Fundamentals". Encyclopedia of Life Sciences. doi:10.1038/npg.els.0001038. ISBN 0470016175.

163. "Strucual Biochemistry/Nucleic Acid/DNA/DNA Structure". Retrieved 11 December 2012.

CHAPTER 5

Protein Biosynthesis

Chapter Contents

- **Introduction**
- **Proteolysis**
- **Post-Translational Modification**
- **Protein Folding**

Protein synthesis is the process by which individual cells construct proteins. Both deoxyribonucleic acid (DNA) and all types of ribonucleic acid (RNA) are involved in this process. Enzymes in the cell's nucleus begin the process of synthesizing protein by unwinding the needed section of DNA, so that RNA can be made. The RNA forms as a copy of one side of the DNA strand, and is sent to other areas of the cell to aid in the bringing together of different amino acids that form proteins. Protein synthesis is so called because proteins are "synthesized" through mechanical and chemical processes in the cell.

INTRODUCTION

Protein synthesis is the process whereby biological cells generate new proteins; it is balanced by the loss of cellular proteins via degradation or export. Translation, the assembly of amino acids by ribosomes, is an essential part of the biosynthetic pathway, along with generation of messenger RNA (mRNA), aminoacylation of transfer RNA (tRNA), co-translational transport, and post-translational modification. Protein biosynthesis is strictly regulated at multiple steps. They are principally during transcription (phenomena of RNA synthesis from DNA template) and translation (phenomena of amino acid assembly from RNA).

The cistron DNA is transcribed into the first of a series of RNA intermediates. The last version is used as a template in synthesis of a polypeptide chain. Protein will often be synthesized directly from genes by translating mRNA. However, when a protein must be available on short notice or in large quantities, a protein precursor is produced. A proprotein is an inactive protein containing one or more inhibitory peptides that can be activated when the inhibitory sequence is removed by proteolysis during posttranslational modification. A preprotein is a form that contains a signal sequence (an N-terminal signal peptide) that specifies its insertion into or through membranes, i.e., targets them for secretion. The signal peptide is cleaved off in the endoplasmic reticulum. Preproproteins have both sequences (inhibitory and signal) still present.

In protein synthesis, a succession of tRNA molecules charged with appropriate amino acids are brought together with an mRNA molecule and matched up by base-pairing through the anti-codons of the tRNA with successive codons of the mRNA. The amino acids are then linked together to extend the growing protein chain, and the tRNAs, no longer carrying amino acids, are released. This whole complex of processes is carried out by the ribosome, formed of two main chains of RNA, called ribosomal RNA (rRNA), and more than 50 different proteins. The ribosome latches onto the end of an mRNA molecule and moves along it, capturing loaded tRNA molecules and joining together their amino acids to form a new protein chain.

Protein biosynthesis, although very similar, is different for prokaryotes and eukaryotes.

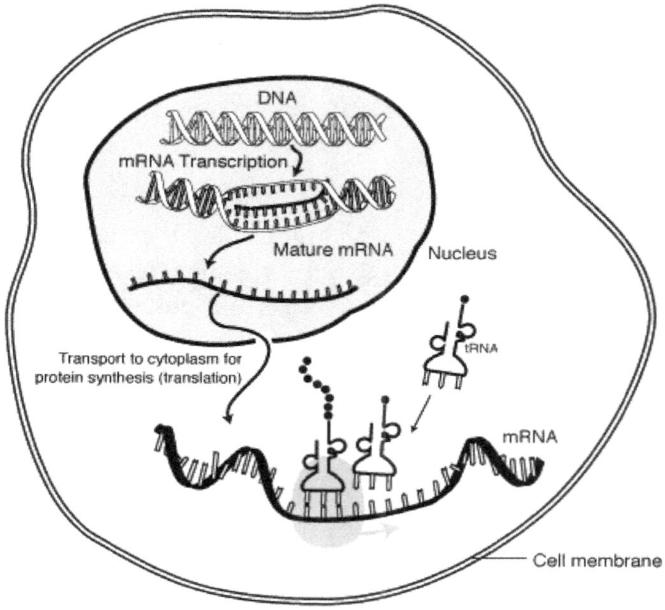

RNA is transcribed in the nucleus; once completely processed, it is transported to the cytoplasm and translated by the ribosome (shown in very pale grey behind the tRNA).

Transcription

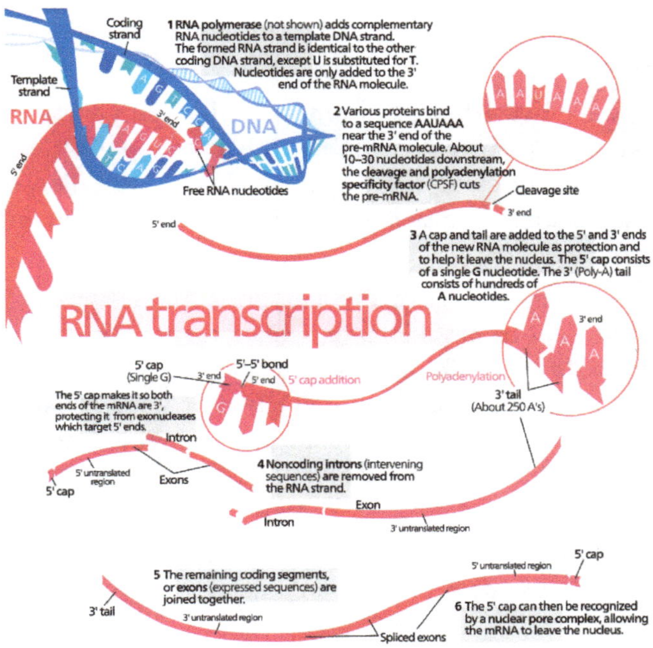

Diagram showing the process of transcription

In transcription an mRNA chain is generated, with one strand of the DNA double helix in the genome as a template. This strand is called the template strand. Transcription can be divided into 3 stages: initiation, elongation, and termination, each regulated by a large number of proteins such as transcription factors and coactivators that ensure that the correct gene is transcribed.

Transcription occurs in the cell nucleus, where the DNA is held. The DNA structure of the cell is made up of two helixes made up of sugar and phosphate held together by hydrogen bonds between the bases of opposite strands. The sugar and the phosphate in each strand are joined together by stronger phosphodiester covalent bonds. The DNA is "unzipped" (disruption of hydrogen bonds between different single strands) by the enzyme helicase, leaving the single nucleotide chain open to be copied. RNA polymerase reads the DNA strand from the 3-prime (3') end to the 5-prime (5') end, while it synthesizes a single strand of messenger RNA in the 5'-to-3' direction. The general RNA structure is very similar to the DNA structure, but in RNA the nucleotide uracil takes the place that thymine occupies in DNA. The single strand of mRNA leaves the nucleus through nuclear pores, and migrates into the cytoplasm.

The first product of transcription differs in prokaryotic cells from that of eukaryotic cells, as in prokaryotic cells the product is mRNA, which needs no post-transcriptional modification, whereas, in eukaryotic cells, the first product is called primary transcript, that needs post-transcriptional modification (capping with 7-methyl-guanosine, tailing with a poly A tail) to give hnRNA (heterogeneous nuclear RNA). hnRNA then undergoes splicing of introns (noncoding parts of the gene) via spliceosomes to produce the final mRNA.

PROTEOLYSIS

Proteolysis is the breakdown of proteins into smaller polypeptides or amino acids. Uncatalysed, the hydrolysis of peptide bonds is extremely slow, taking hundreds of years. Proteolysis is typically catalysed by cellular enzymes called proteases, but may also occur by intra-molecular digestion. Low pH or high temperatures can also cause proteolysis non-enzymatically.

Proteolysis in organisms serves many purposes; for example, digestive enzymes break down proteins in food to provide amino acids for the organism, while proteolytic processing of a polypeptide chain after its synthesis may be necessary for the production of an active protein. It is also important in the regulation of some physiological and cellular processes, as well as preventing the accumulation of unwanted or abnormal proteins in cells. Consequently, dis-regulation of proteolysis can cause diseases and is used in some venoms to damage their prey.

Proteolysis is important as an analytical tool for studying proteins in the laboratory, as well as industrially, for example in food processing and stain removal.

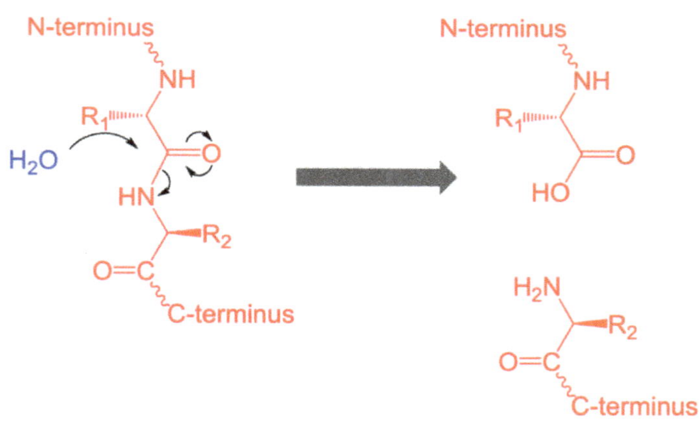

The hydrolysis of a protein (red) by the nucleophilic attack of water (blue). The uncatalysed half-life is several hundred years.

Biological Functions

Post-Translational Proteolytic Processing

Limited proteolysis of a polypeptide during or after translation in protein synthesis often occurs for many proteins. This may involve removal of the N-terminal methionine, signal peptide, and/or the conversion of an inactive or non-functional protein to an active one.

The precursor to the final functional form of protein is termed proprotein, and these proproteins may be first synthesized as preproprotein. For example, albumin is first synthesized as preproalbumin and contains an uncleaved signal peptide. This forms the proalbumin after the signal peptide is cleaved, and a further processing to remove the N-terminal 6-residue propeptide yields the mature form of the protein.

Removal of N-Terminal Methionine

The initiating methionine (and, in prokaryotes, fMet) may be removed during translation of the nascent protein. For *E. coli*, fMet is efficiently removed if the second residue is small and uncharged, but not if the second residue is bulky and charged. In both prokaryotes and eukaryotes, the exposed N-terminal residue may determine the half-life of the protein according to the N-end rule.

Removal of the Signal Sequence

Proteins that are to be targeted to a particular organelle or for secretion have an N-terminal signal peptide that directs the protein to its final destination. This signal peptide is removed by proteolysis after their transport through a membrane.

Cleavage of Polyproteins

Some proteins and most eukaryotic polypeptide hormones are synthesized as a large precursor polypeptide known as a polyprotein that requires proteolytic cleavage into individual smaller polypeptide chains. The polyprotein pro-opiomelanocortin (POMC) contains many polypeptide hormones. The cleavage pattern of POMC, however, may vary between different tissues, yielding different sets of polypeptide hormones from the same polyprotein.

Many viruses also produce their proteins initially as a single polypeptide chain that were translated from a polycistronic mRNA. This polypeptide is subsequently cleaved into individual polypeptide chains.

Cleavage of Precursor Proteins

Many proteins and hormones are synthesized in the form of their precursors - zymogens, proenzymes, and prehormones. These proteins are cleaved to form their final active structures. Insulin, for example, is synthesized as preproinsulin, which yields proinsulin after the signal peptide has been cleaved. The proinsulin is then cleaved at two positions to yield two polypeptide chains linked by two disulfide bonds. Removal of two C-terminal residues from the B-chain then yields the mature insulin. Protein folding occurs in the single-chain Proinsulin form which facilitates formation of the ultimately inter-peptide disulfide bonds, and the ultimately intra-peptide disulfide bond, found in the native structure of insulin.

Proteases in particular are synthesized in the inactive form so that they may be safely stored in cells, and ready for release in sufficient quantity when required. This is to ensure that the protease is activated only in the correct location or context, as inappropriate activation of these proteases can be very destructive for an organism. Proteolysis of the zymogen yields an active protein; for example, when trypsinogen is cleaved to form trypsin, a slight rearrangement of the protein structure that completes the active site of the protease occurs, thereby activating the protein.

Proteolysis can, therefore, be a method of regulating biological processes by turning inactive proteins into active ones. A good example is the blood clotting cascade whereby an initial event triggers a cascade of sequential proteolytic activation of many specific proteases, resulting in blood coagulation. The complement system of the immune response also involves a complex sequential proteolytic activation and interaction that result in an attack on invading pathogens.

Protein Degradation

Protein degradation may take place intracellularly or extracellularly. In digestion of food, digestive enzymes may be released into the environment for extracellular digestion whereby proteolytic cleavage breaks down proteins into smaller peptides and amino acids so that they may be absorbed and used by an organism. In animals the food may be processed extracellularly in specialized digestive organs or guts, but in many bacteria the food may be internalized into the cell via phagocytosis. Microbial degradation of protein in the environment can be regulated by nutrient availability. For example, limitation for major elements in proteins (carbon, nitrogen, and sulfur) has been shown to induce proteolytic activity in the fungus *Neurospora crassa* as well as in whole communities of soil organisms.

Proteins in cells are also constantly being broken down into amino acids. This intracellular degradation of protein serves a number of functions: It removes damaged and abnormal protein and prevent their accumulation, and it also serves to regulate cellular processes by removing enzymes and regulatory proteins that are no longer needed. The amino acids may then be reused for protein synthesis.

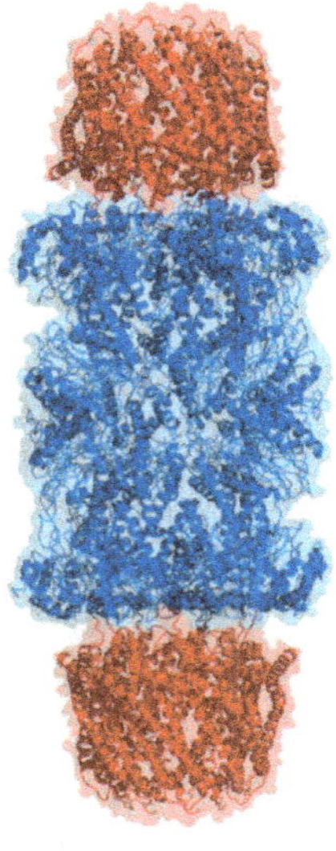

Structure of a proteasome. Its active sites are inside the tube (blue) where proteins are degraded.

Lysosome and Proteasome

The intracellular degradation of protein may be achieved in two ways - proteolysis in lysosome, or a ubiquitin-dependent process that targets unwanted proteins to proteasome. The autophagy-lysosomal pathway is normally a non-selective process, but it may become selective upon starvation whereby proteins with peptide sequence KFERQ or similar are selectively broken down. The lysosome contains a large number of proteases such as cathepsins.

The ubiquitin-mediated process is selective. Proteins marked for degradation are covalently linked to ubiquitin. Many molecules of ubiquitin may be linked in tandem to a protein destined for degradation. The polyubiquinated protein is targeted to an ATP-dependent protease complex, the proteasome. The ubiquitin is released and reused, while the targeted protein is degraded.

Rate of Intracellular Protein Degradation

Different proteins are degraded at different rates. Abnormal proteins are quickly degraded, whereas the rate of degradation of normal proteins may vary widely depending on their functions. Enzymes at important metabolic control points may be degraded much faster than those enzymes whose activity is largely constant under all physiological conditions. One of the most rapidly degraded proteins is ornithine decarboxylase, which has a half-life of 11 minutes. In contrast, other proteins like actin and myosin have a half-life of a month or more, while, in essence, haemoglobin lasts for the entire life-time of an erythrocyte.

The N-end rule may partially determine the half-life of a protein, and proteins with segments rich in proline, glutamic acid, serine, and threonine (the so-called PEST proteins) have short half-life. Other factors suspected to affect degradation rate include the rate deamination of glutamine and asparagine and oxidation of cystein, histidine, and methionine, the absence of stabilizing ligands, the presence of attached carbohydrate or phosphate groups, the presence of free α-amino group, the negative charge of protein, and the flexibility and stability of the protein. Proteins with larger degrees of intrinsic disorder also tend to have short cellular half-life, with disordered segments having been proposed to facilitate efficient initiation of degradation by the proteasome.

The rate of proteolysis may also depend on the physiological state of the organism, such as its hormonal state as well as nutritional status. In time of starvation, the rate of protein degradation increases.

Digestion

In human digestion, proteins in food are broken down into smaller peptide chains by digestive enzymes such as pepsin, trypsin, chymotrypsin, and elastase, and into amino acids by various enzymes such as carboxypeptidase, aminopeptidase, and dipeptidase. It is necessary to break down proteins into small peptides (tripeptides and dipeptides) and amino acids so they can be absorbed by the intestines, and the absorbed tripeptides and dipeptides are also further broken into amino acids intracellularly before they enter the bloodstream. Different enzymes have different specificity for their substrate; trypsin, for example, cleaves the peptide bond after a positively charged residue (arginine and lysine);

chymotrypsin cleaves the bond after an aromatic residue (phenylalanine, tyrosine, and tryptophan); elastase cleaves the bond after a small non-polar residue such as alanine or glycine.

In order to prevent inappropriate or premature activation of the digestive enzymes (they may, for example, trigger pancreatic self-digestion causing pancreatitis), these enzymes are secreted as inactive zymogen. The precursor of pepsin, pepsinogen, is secreted by the stomach, and is activated only in the acidic environment found in stomach. The pancreas secretes the precursors of a number of proteases such as trypsin and chymotrypsin. The zymogen of trypsin is trypsinogen, which is activated by a very specific protease, enterokinase, secreted by the mucosa of the duodenum. The trypsin, once activated, can also cleave other trypsinogens as well as the precursors of other proteases such as chymotrypsin and carboxypeptidase to activate them.

In bacteria, a similar strategy of employing an inactive zymogen or prezymogen is used. Subtilisin, which is produced by *Bacillus subtilis*, is produced as preprosubtilisin, and is released only if the signal peptide is cleaved and autocatalytic proteolytic activation has occurred.

Cellular Regulation

Proteolysis is also involved in the regulation of many cellular processes by activating or deactivating enzymes, transcription factors, and receptors, for example in the biosynthesis of cholesterol, or the mediation of thrombin signalling through protease-activated receptors.

Some enzymes at important metabolic control points such as ornithine decarboxylase is regulated entirely by its rate of synthesis and its rate of degradation. Other rapidly degraded proteins include the protein products of proto-oncogenes, which play central roles in the regulation of cell growth.

Cell Cycle Regulation

Cyclins are a group of proteins that activate kinases involved in cell division. The degradation of cyclins is the key step that governs the exit from mitosis and progress into the next cell cycle. Cyclins accumulate in the course the cell cycle, then abruptly disappear just before the anaphase of mitosis. The cyclins are removed via a ubiquitin-mediated proteolytic pathway.

Apoptosis

Caspases are an important group of proteases involved in apoptosis. The precursors of caspase, procaspase, may be activated by proteolysis through its association with a protein complex that forms apoptosome, or by granzyme B, or via the death receptor pathways.

Proteolysis and Diseases

Abnormal proteolytic activity is associated with many diseases. In pancreatitis, leakage of proteases and their premature activation in the pancreas results in the self-digestion of the pancreas. People with diabetes mellitus may have increased lysosomal activity and the degradation of some proteins can increase significantly. Chronic inflammatory diseases such as rheumatoid arthritis may involve the release of lysosomal enzymes into extracellular space that break down surrounding tissues. Abnormal proteolysis and generation of peptides that aggregate in cells and their ineffective removal may result in many age-related neurological diseases such as Alzheimer's.

Proteases may be regulated by antiproteases or protease inhibitors, and imbalance between proteases and antiproteases can result in diseases, for example, in the destruction of lung tissues in emphysema brought on by smoking tobacco. Smoking is thought to increase the neutrophils and macrophages in the lung which release excessive amount of proteolytic enzymes such as elastase, such that they can no longer be inhibited by serpins such as α_1-antitrypsin, thereby resulting in the breaking down of connective tissues in the lung. Other proteases and their inhibitors may also be involved in this disease, for example matrix metalloproteinases (MMPs) and tissue inhibitors of metalloproteinases (TIMPs).

Other diseases linked to aberrant proteolysis include muscular dystrophy, degenerative skin disorders, respiratory and gastrointestinal diseases, and malignancy.

Non-Enzymatic Proteolysis

Protein backbones are very stable in water at neutral pH and room temperature, although the rate of hydrolysis of different peptide bonds can vary. The half life of a peptide bond under normal conditions can range from 7 years to 350 years, even higher for peptides protected by modified terminus or within the protein interior. The rate of proteolysis however can be significantly increased by extremes of pH and heat.

Strong mineral acids can readily hydrolyse the peptide bonds in a protein (acid hydrolysis). The standard way to hydrolyze a protein or peptide into its constituent amino acids for analysis is to heat it to 105 °C for around 24 hours in 6M hydrochloric acid. However, some proteins are resistant to acid hydrolysis. One well-known example is ribonuclease A, which can be purified by treating crude extracts with hot sulphuric acid so that other proteins become degraded while ribonuclease A is left intact.

Certain chemicals cause proteolysis only after specific residues, and these can be used to selectively break down a protein into smaller polypeptides for laboratory analysis. For example, cyanogen bromide cleaves the peptide bond after a methionine. Similar methods may be used to specifically cleave tryptophanyl, aspartyl, cysteinyl, and asparaginyl peptide bonds. Acids such as trifluoroacetic acid and formic acid may also be used.

Like other biomolecules, proteins can also be broken down by high heat alone. At 250 °C, the peptide bond may be easily hydrolyzed, with its half-life dropping to about a minute. Protein may also be broken down without hydrolysis through pyrolysis; small heterocyclic compounds may start to form upon degradation, above 500 °C, polycyclic aromatic hydrocarbon may also form, which is of interest in the study of generation of carcinogens in tobacco smoke and cooking at high heat.

Laboratory Applications

Proteolysis is also used in research and diagnostic applications:

- Cleavage of fusion protein so that the fusion partner and protein tag used in protein expression and purification may be removed. The proteases used have high degree of specificity, such as thrombin, enterokinase, and TEV protease, so that only the targeted sequence may be cleaved.
- Complete inactivation of undesirable enzymatic activity or removal of unwanted proteins. For example, proteinase K, a broad-spectrum proteinase stable in urea and SDS, is often used in the preparation of nucleic acids to remove unwanted nuclease contaminants that may otherwise degrade the DNA or RNA.
- Partial inactivation, or changing the functionality, of specific protein. For example, treatment of DNA polymerase I with subtilisin yields the Klenow fragment, which retains its polymerase function but lacks 5›-exonuclease activity.
- Digestion of proteins in solution for proteome analysis by liquid chromatography-mass spectrometry (LC-MS). This may also be done by in-gel digestion of proteins after separation by gel electrophoresis for the identification by mass spectrometry.
- Analysis of the stability of folded domain under a wide range of conditions.
- Increasing success rate of crystallisation projects
- Production of digested protein used in growth media to culture bacteria and other organisms, e.g. tryptone in Lysogeny Broth.

Protease Enzymes

Proteases may be classified according to the catalytic group involved in its active site.

- Cysteine protease
- Serine protease
- Threonine protease
- Aspartic protease
- Glutamic protease
- Metalloprotease
- Asparagine peptide lyase

Venoms

Certain types of venom, such as those produced by venomous snakes, can also cause proteolysis. These venoms are, in fact, complex digestive fluids that begin their work outside of the body. Proteolytic venoms cause a wide range of toxic effects, including effects that are:

- cytotoxic (cell-destroying)

- hemotoxic (blood-destroying)
- myotoxic (muscle-destroying)
- hemorrhagic (bleeding)

POST-TRANSLATIONAL MODIFICATION

Post-translational modification (PTM) refers to the covalent and generally enzymatic modification of proteins during or after protein biosynthesis. Proteins are synthesized by ribosomes translating mRNA into polypeptide chains, which may then undergo PTM to form the mature protein product. PTMs are important components in cell signaling.

Post-translational modifications can occur on the amino acid side chains or at the protein›s C- or N- termini. They can extend the chemical repertoire of the 20 standard amino acids by introducing new functional groups such as phosphate, acetate, amide groups, or methyl groups. Phosphorylation is a very common mechanism for regulating the activity of enzymes and is the most common post-translational modification. Many eukaryotic proteins also have carbohydrate molecules attached to them in a process called glycosylation, which can promote protein folding and improve stability as well as serving regulatory functions. Attachment of lipid molecules, known as lipidation, often targets a protein or part of a protein attached to the cell membrane.

Other forms of post-translational modification consist of cleaving peptide bonds, as in processing a propeptide to a mature form or removing the initiator methionine residue. The formation of disulfide bonds from cysteine residues may also be referred to as a post-translational modification. For instance, the peptide hormone insulin is cut twice after disulfide bonds are formed, and a propeptide is removed from the middle of the chain; the resulting protein consists of two polypeptide chains connected by disulfide bonds.

Some types of post-translational modification are consequences of oxidative stress. Carbonylation is one example that targets the modified protein for degradation and can result in the formation of protein aggregates. Specific amino acid modifications can be used as biomarkers indicating oxidative damage.

Sites that often undergo post-translational modification are those that have a functional group that can serve as a nucleophile in the reaction: the hydroxyl groups of serine, threonine, and tyrosine; the amine forms of lysine, arginine, and histidine; the thiolate anion of cysteine; the carboxylates of aspartate and glutamate; and the N- and C-termini. In addition, although the amides of asparagine and glutamine are weak nucleophiles, both can serve as attachment points for glycans. Rarer modifications can occur at oxidized methionines and at some methylenes in side chains.

Post-translational modification of proteins can be experimentally detected by a variety of techniques, including mass spectrometry, Eastern blotting, and Western blotting.

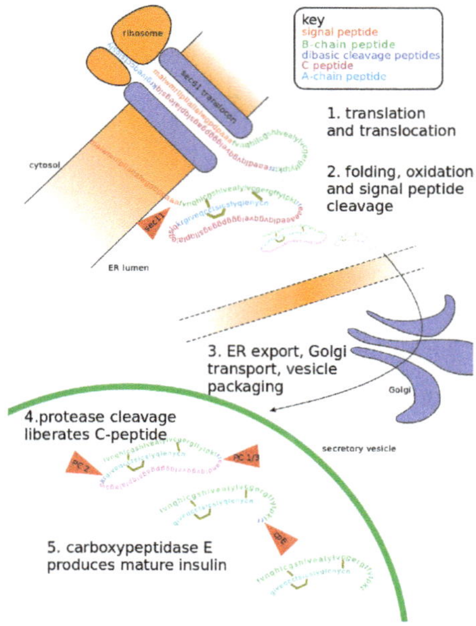

Post-translational modification of insulin. At the top, the ribosome translates a mRNA sequence into a protein, insulin, and passes the protein through the endoplasmic reticulum, where it is cut, folded and held in shape by disulfide (-S-S-) bonds. Then the protein passes through the golgi apparatus, where it is packaged into a vesicle. In the vesicle, more parts are cut off, and it turns into mature insulin.

PTMs Involving Addition of Functional Groups

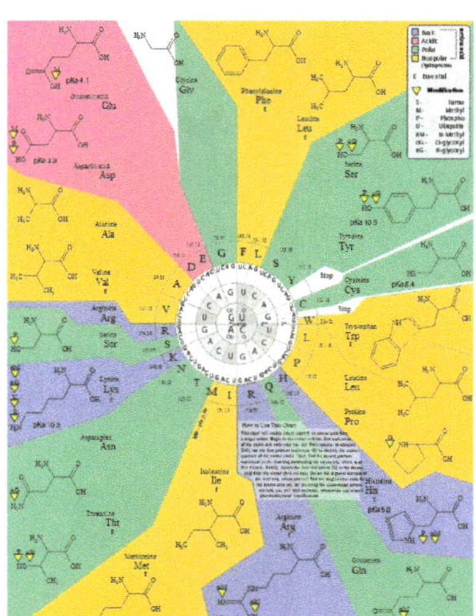

The genetic code diagram showing the amino acid residues as target of modification.

Addition by an Enzyme In Vivo

Hydrophobic Groups for Membrane Localization

- myristoylation, attachment of myristate, a C_{14} saturated acid
- palmitoylation, attachment of palmitate, a C_{16} saturated acid
- isoprenylation or prenylation, the addition of an isoprenoid group (e.g. farnesol and geranylgeraniol)
    - farnesylation
    - geranylgeranylation
- glypiation, glycosylphosphatidylinositol (GPI) anchor formation via an amide bond to C-terminal tail

Cofactors for Enhanced Enzymatic Activity

- lipoylation, attachment of a lipoate (C_8) functional group
- flavin moiety (FMN or FAD) may be covalently attached
- heme C attachment via thioether bonds with cysteins
- phosphopantetheinylation, the addition of a 4›-phosphopantetheinyl moiety from coenzyme A, as in fatty acid, polyketide, non-ribosomal peptide and leucine biosynthesis
- retinylidene Schiff base formation

Modifications of Translation Factors

- diphthamide formation (on a histidine found in eEF2)
- ethanolamine phosphoglycerol attachment (on glutamate found in eEF1α)
- hypusine formation (on conserved lysine of eIF5A (eukaryotic) and aIF5A (archaeal))

Smaller Chemical Groups

- acylation, e.g. *O*-acylation (esters), *N*-acylation (amides), *S*-acylation (thioesters)
    - acetylation, the addition of an acetyl group, either at the N-terminus of the protein or at lysine residues. *histone acetylation*. The reverse is called deacetylation.
    - formylation
- alkylation, the addition of an alkyl group, e.g. methyl, ethyl
    - methylation the addition of a methyl group, usually at lysine or arginine residues. The reverse is called demethylation.

- amide bond formation
    - amidation at C-terminus
    - amino acid addition
        - arginylation, a tRNA-mediation addition
        - polyglutamylation, covalent linkage of glutamic acid residues to the N-terminus of tubulin and some other proteins. (See tubulin polyglutamylase)
        - polyglycylation, covalent linkage of one to more than 40 glycine residues to the tubulin C-terminal tail
- butyrylation
- gamma-carboxylation dependent on Vitamin K
- glycosylation, the addition of a glycosyl group to either arginine, asparagine, cysteine, hydroxylysine, serine, threonine, tyrosine, or tryptophan resulting in a glycoprotein. Distinct from glycation, which is regarded as a nonenzymatic attachment of sugars.
    - polysialylation, addition of polysialic acid, PSA, to NCAM
- malonylation
- hydroxylation: addition of an oxygen atom to the side-chain of Pro or Lys residues.
- iodination (e.g. of thyroglobulin)
- nucleotide addition such as ADP-ribosylation
- phosphate ester (*O*-linked) or phosphoramidate (*N*-linked) formation
    - phosphorylation, the addition of a phosphate group, usually to serine, threonine, and tyrosine (*O*-linked), or histidine (*N*-linked)
    - adenylylation, the addition of an adenylyl moiety, usually to tyrosine (*O*-linked), or histidine and lysine (*N*-linked)
- propionylation
- pyroglutamate formation
- *S*-glutathionylation
- *S*-nitrosylation
- *S*-sulfenylation (*aka* *S*-sulphenylation), reversible covalent attachment of hydroxide to the thiol group of cysteine residues
- succinylation addition of a succinyl group to lysine
- sulfation, the addition of a sulfate group to a tyrosine.

Non-Enzymatic Additions in vivo

- glycation, the addition of a sugar molecule to a protein without the controlling action of an enzyme.

- carbamylation the addition of Isocyanic acid to an N-terminus of either lysine, histidine, taurine, arginine, or cysteine.
- carbonylation the addition of carbon monoxide to other organic/inorganic compounds.

Non-Enzymatic Additions in Vitro

- biotinylation: covalent attachment of a biotin moiety using a biotinylation reagent, typically for the purpose of labeling a protein.
- oxidation: addition of one or more Oxygen atoms to a susceptible side-chain, principally of Met, Trp, His or Cys residues. Formation of disulfide bonds between Cys residues.
- pegylation: covalent attachment of polyethylene glycol (PEG) using a pegylation reagent, typically to the N-terminus or the side-chains of Lys residues. Pegylation is used to improve the efficacy of protein pharmaceuticals.

Other Proteins or Peptides

- ISGylation, the covalent linkage to the ISG15 protein (Interferon-Stimulated Gene 15)
- SUMOylation, the covalent linkage to the SUMO protein (Small Ubiquitin-related MOdifier)
- ubiquitination, the covalent linkage to the protein ubiquitin.
- Neddylation, the covalent linkage to Nedd
- Pupylation, the covalent linkage to the Prokaryotic ubiquitin-like protein

Chemical Modification of Amino Acids

- citrullination, or **deimination**, the conversion of arginine to citrulline
- deamidation, the conversion of glutamine to glutamic acid or asparagine to aspartic acid
- eliminylation, the conversion to an alkene by beta-elimination of phosphothreonine and phosphoserine, or dehydration of threonine and serine, as well as by decarboxylation of cysteine
- carbamylation of the N-terminus (an α-amine) and the side-chain amino groups of lysine and, usually to a lesser extent, arginine

Structural Changes

- disulfide bridges, the covalent linkage of two cysteine amino acids
- proteolytic cleavage, cleavage of a protein at a peptide bond
- racemization

- of proline by prolyl isomerase
- of serine by protein-serine epimerase
- of alanine in dermorphin, a frog opioid peptide
- of methionine in deltorphin, also a frog opioid peptide

- protein splicing, self-catalytic removal of inteins analogous to mRNA processing

Statistics

In 2011, statistics of each post-translational modification experimentally and putatively detected have been compiled using proteome-wide information from the Swiss-Prot database. The 10 most common experimentally found modifications were as follows:

Frequency	Modification
58383	Phosphorylation
6751	Acetylation
5526	N-linked glycosylation
2844	Amidation
1619	Hydroxylation
1523	Methylation
1133	O-linked glycosylation
878	Ubiquitylation
826	Pyrrolidone Carboxylic Acid
504	Sulfation

PROTEIN FOLDING

Protein folding is the physical process by which a protein chain acquires its native 3-dimensional structure, a conformation that is usually biologically functional, in an expeditious and reproducible manner. It is the physical process by which a polypeptide folds into its characteristic and functional three-dimensional structure from random coil. Each protein exists as an unfolded polypeptide or random coil when translated from a sequence of mRNA to a linear chain of amino acids. This polypeptide lacks any stable (long-lasting) three-dimensional structure (the left hand side of the first figure). As the polypeptide chain is being synthesized by the ribosome, the linear chain begins to fold into its three dimensional structure. Folding begins to occur even during translation of the polypeptide chain. Amino acids interact with each other to produce a well-defined three-dimensional structure, the folded protein (the right hand side of the figure), known as the native state. The resulting three-dimensional structure is determined by the amino acid sequence or primary structure (Anfinsen's dogma). The energy landscape describes the folding pathways in which the unfolded protein is able to assume its native state. Experiments beginning in the 1980s indicate the codon for an amino acid can also influence protein structure.

The correct three-dimensional structure is essential to function, although some parts of functional proteins may remain unfolded, so that protein dynamics is important. Failure to fold into native structure generally produces inactive proteins, but in some instances misfolded proteins have modified or toxic functionality. Several neurodegenerative and other diseases are believed to result from the accumulation of amyloid fibrils formed by misfolded proteins. Many allergies are caused by incorrect folding of some proteins, because the immune system does not produce antibodies for certain protein structures.

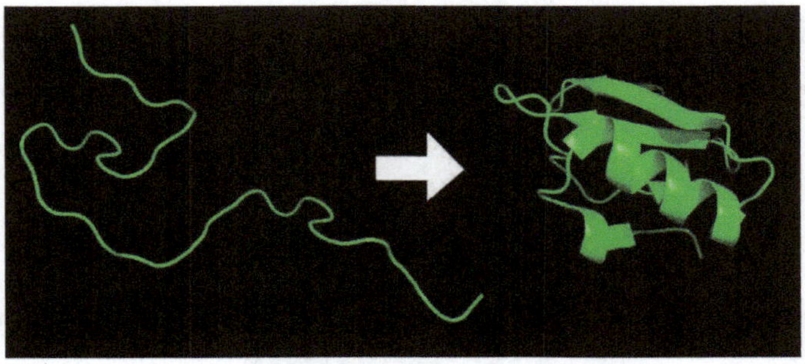

Protein before and after folding.

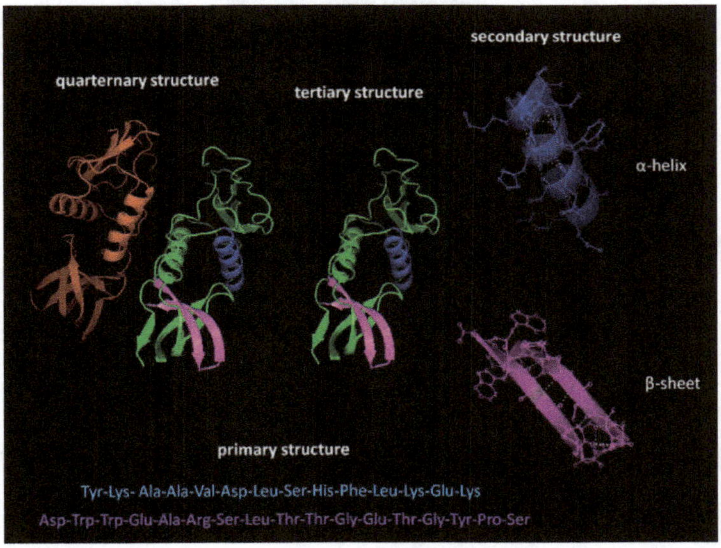

Results of protein folding.

Process of Protein Folding

Primary Structure

The primary structure of a protein, its linear amino-acid sequence, determines its native conformation. The specific amino acid residues and their position in the polypeptide chain are the determining factors for which portions of the protein fold closely together and form its three dimensional conformation. The amino acid composition is not as important as the sequence. The essential fact of folding, however, remains that the amino acid sequence of each protein contains the information that specifies both the native structure and the pathway to attain that state. This is not to say that nearly identical amino acid sequences always fold similarly. Conformations differ based on environmental factors as well; similar proteins fold differently based on where they are found.

Secondary Structure

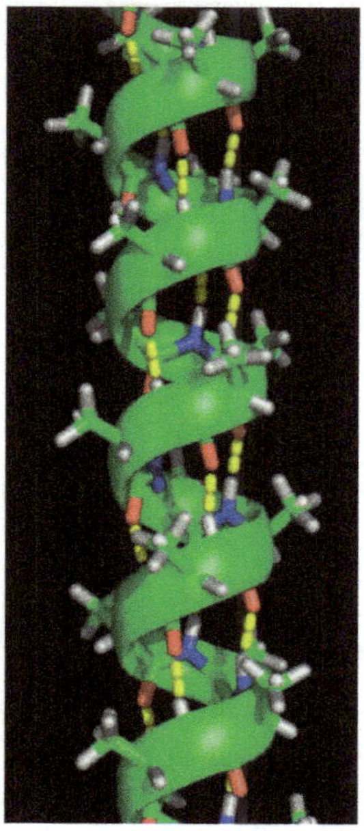

The alpha helix spiral formation.

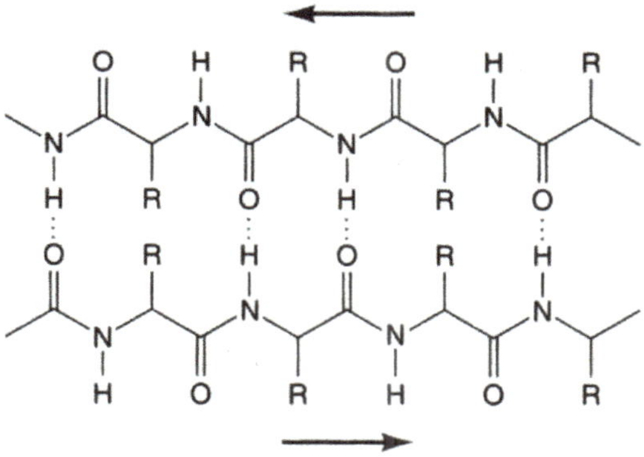

An anti-parallel beta pleated sheet displaying hydrogen bonding within the backbone.

Formation of a secondary structure is the first step in the folding process that a protein takes to assume its native structure. Characteristic of secondary structure is the structures known as alpha helices and beta sheets that fold rapidly because they are stabilized by intramolecular hydrogen bonds, as was first characterized by Linus Pauling. Formation of intramolecular hydrogen bonds provides another important contribution to protein stability. Alpha helices are formed by hydrogen bonding of the backbone to form a spiral shape (refer to figure on the right). The beta pleated sheet is a structure that forms with the backbone bending over itself to form the hydrogen bonds (as displayed in the figure to the left). The hydrogen bonds are between the amide hydrogen and carbonyl carbon of the peptide bond. There exists anti-parallel beta pleated sheets and parallel beta pleated sheets where the stability of the hydrogen bonds is stronger in the anti-parallel beta sheet as it hydrogen bonds with the ideal 180 degree angle compared to the slanted hydrogen bonds formed by parallel sheets. The alpha helices and beta pleated sheets can be amphipathic in nature, or contain a hydrophilic portion and a hydrophobic portion. This property of these secondary structures aids in the folding of the protein as it aligns the helices and sheets in such a way where the hydrophilic sides are facing the aqueous environment surrounding the protein and the hydrophobic sides are facing the hydrophobic core of the protein. Secondary structure hierarchically gives way to tertiary structure formation.

Tertiary Structure

Once the protein's tertiary structure is formed and stabilized by the hydrophobic interactions, there may also be covalent bonding in the form of disulfide bridges formed between two cysteine residues. Tertiary structure of a protein involves a single polypeptide chain; however, additional interactions of folded polypeptide chains give rise to quaternary structure formation.

Quaternary Structure

Tertiary structure may give way to the formation of quaternary structure in some proteins, which usually involves the "assembly" or "coassembly" of subunits that have already folded; in other words, multiple polypeptide chains could interact to form a fully functional quaternary protein.

Driving force of Protein Folding

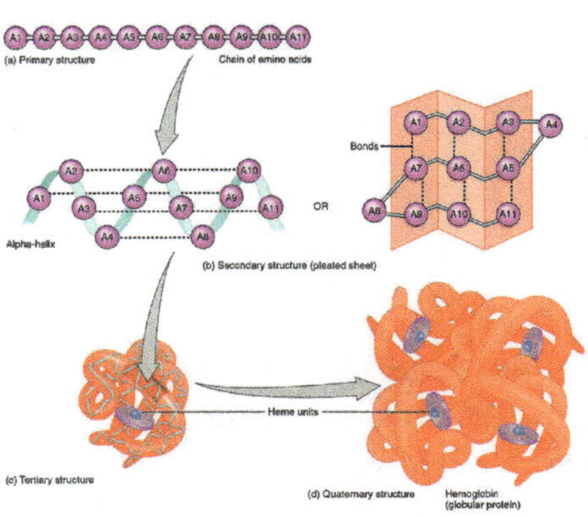

All forms of protein structure summarized.

Folding is a spontaneous process independent of energy inputs from nucleoside triphosphates. The passage of the folded state is mainly guided by hydrophobic interactions, formation of intramolecular hydrogen bonds, van der Waals forces, and it is opposed by conformational entropy. The process of folding often begins co-translationally, so that the N-terminus of the protein begins to fold while the C-terminal portion of the protein is still being synthesized by the ribosome; however, a protein molecule may fold spontaneously during or after biosynthesis. While these macromolecules may be regarded as "folding themselves", the process also depends on the solvent (water or lipid bilayer), the concentration of salts, the pH, the temperature, the possible presence of cofactors and of molecular chaperones. Proteins will have limitations on their folding abilities by the restricted bending angles or conformations that are possible. These allowable angles of protein folding are described with a two-dimensional plot known as the Ramachandran plot, depicted with psi and phi angles of allowable rotation.

Hydrophobic Effect

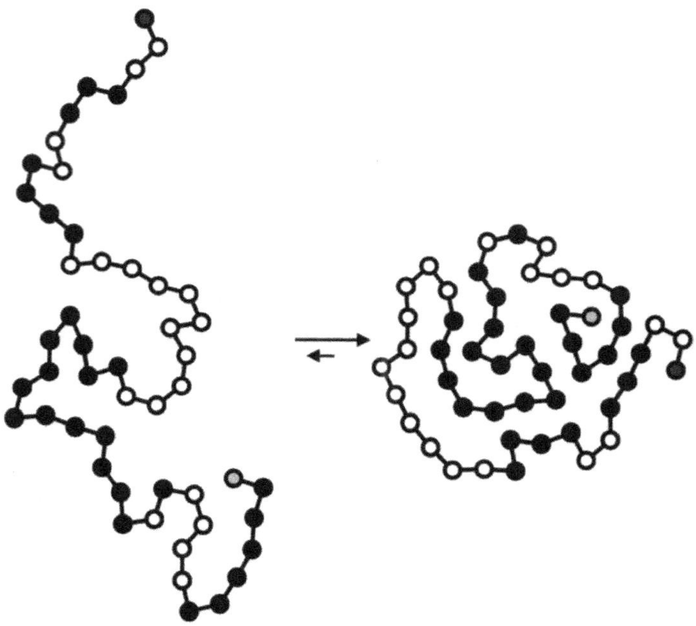

Hydrophobic collapse. In the compact fold (to the right), the hydrophobic amino acids (shown as black spheres) collapse toward the center to become shielded from aqueous environment.

Protein folding must be thermodynamically favorable within a cell in order for it to be a spontaneous reaction. Since it is known that protein folding is a spontaneous reaction, then it must assume a negative Gibbs free energy value. Gibbs free energy in protein folding is directly related to enthalpy and entropy. For a negative delta G to arise and for protein folding to become thermodynamically favorable, then either enthalpy, entropy, or both terms must be favorable.

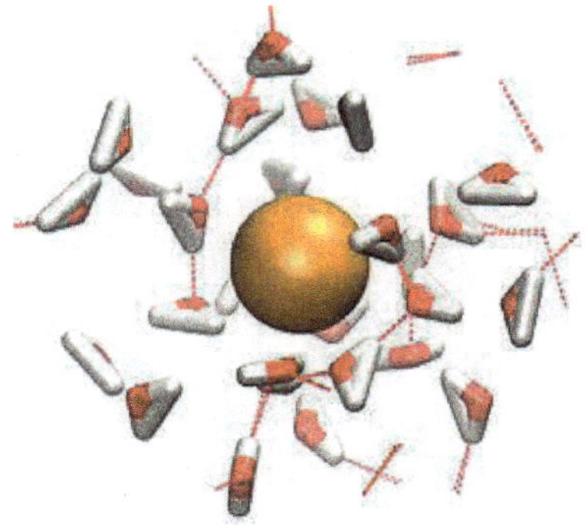

Entropy is decreased as the water molecules become more orderly near the hydrophobic solute.

Minimizing the number of hydrophobic side-chains exposed to water is an important driving force behind the folding process.The hydrophobic effect is the phenomenon in which the hydrophobic chains of a protein collapse into the core of the protein (away from the hydrophilic environment). In an aqueous environment, the water molecules tend to aggregate around the hydrophobic regions or side chains of the protein, creating water shells of ordered water molecules. An ordering of water molecules around a hydrophobic region increases order in a system and therefore contributes a negative change in entropy (less entropy in the system). The water molecules are fixed in these water cages which drives the hydrophobic collapse, or the inward folding of the hydrophobic groups. The hydrophobic collapse introduces entropy back to the system via the breaking of the water cages which frees the ordered water molecules. The multitude of hydrophobic groups interacting within the core of the globular folded protein contributes a significant amount to protein stability after folding, because of the vastly accumulated van der Waals forces (specifically London Dispersion forces). The hydrophobic effect exists as a driving force in thermodynamics only if there is the presence of an aqueous medium with an amphiphilic molecule containing a large hydrophobic region. The strength of hydrogen bonds depends on their environment; thus, H-bonds enveloped in a hydrophobic core contribute more than H-bonds exposed to the aqueous environment to the stability of the native state.

Chaperones

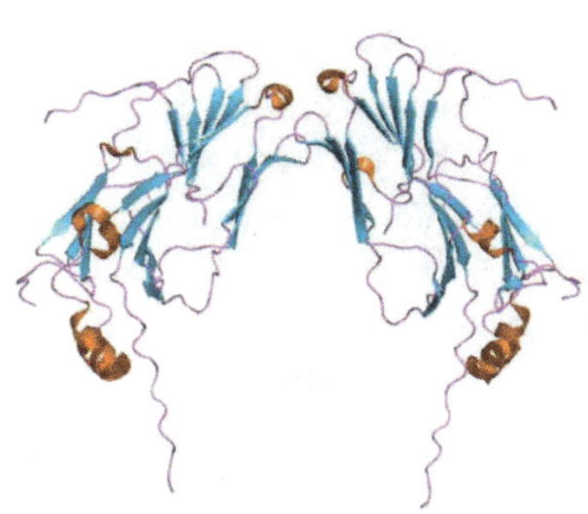

Example of a small eukaryotic heat shock protein.

Chaperones are a class of proteins that aid in the correct folding of other proteins *in vivo*. Chaperones exist in all cellular compartments and interact with the polypeptide chain in order to allow the native three-dimensional conformation of the protein to form; however, chaperones themselves are not included in the final structure of the protein they are assisting in. Chaperones may assist in folding even when the nascent polypeptide is being synthesized by the ribosome. Molecular chaperones operate by binding to stabilize an otherwise unstable structure of a protein in its folding pathway, but chaperones do not contain the necessary information to know the correct native structure of the protein they are aiding; rather, chaperones work by preventing incorrect folding conformations. In this way, chaperones do not actually increase the rate of individual steps involved in the folding pathway toward the native structure; instead, they work by reducing possible unwanted aggregations of the polypeptide chain that might otherwise slow down the search for the proper intermediate and they provide a more efficient pathway for the polypeptide chain to assume the correct conformations. Chaperones are not to be confused with folding catalysts, which actually do catalyze the otherwise slow steps in the folding pathway. Examples of folding catalysts are protein disulfide isomerases and peptidyl-prolyl isomerases that may be involved in formation of disulfide bonds or interconversion between cis and trans stereoisomers, respectively. Chaperones are shown to be critical in the process of protein folding *in vivo* because they provide the protein with the aid needed to assume its proper alignments and conformations efficiently enough to become "biologically relevant". This means that the polypeptide chain could theoretically fold into its native structure without the aid of chaperones, as demonstrated by protein folding experiments conducted *in vitro*; however, this process proves to be too inefficient or too slow to exist in biological systems; therefore, chaperones are necessary for protein folding *in vivo*. Along with its role in aiding native structure formation, chaperones are shown to be involved in various roles such as protein transport, degradation, and even allow denatured proteins exposed to certain external denaturant factors an opportunity to refold into their correct native structures.

A fully denatured protein lacks both tertiary and secondary structure, and exists as a so-called random coil. Under certain conditions some proteins can refold; however, in many cases, denaturation is irreversible. Cells sometimes protect their proteins against the denaturing influence of heat with enzymes known as heat shock proteins (a type of chaperone), which assist other proteins both in folding and in remaining folded. Some proteins never fold in cells at all except with the assistance of chaperones which either isolate individual proteins so that their folding is not interrupted by interactions with other proteins or help to unfold misfolded proteins, allowing them to refold into the correct native structure. This function is crucial to prevent the risk of precipitation into insoluble amorphous aggregates. The external factors involved in protein denaturation or disruption of the native state include temperature, external fields (electric, magnetic), molecular crowding, and even the limitation of space, which can have a big influence on the folding of proteins. High concentrations of solutes, extremes of pH, mechanical forces, and the presence of chemical denaturants can contribute to protein denaturation, as well. These individual factors are categorized together as stresses. Chaperones are shown to exist in increasing concentrations during times of cellular stress and help the proper folding of emerging proteins as well as denatured or misfolded ones.

Under some conditions proteins will not fold into their biochemically functional forms. Temperatures above or below the range that cells tend to live in will cause thermally unstable proteins to unfold or denature (this is why boiling makes an egg white turn opaque). Protein thermal stability is far from constant, however; for example, hyperthermophilic bacteria have been found that grow at temperatures as high as 122 °C, which of course requires that their full complement of vital proteins and protein assemblies be stable at that temperature or above.

Computational Methods for Studying Protein Folding

The study of protein folding includes three main aspects related to the prediction of protein stability, kinetics, and structure. A recent review summarizes the available computational methods for protein folding.

Energy Landscape of Protein Folding

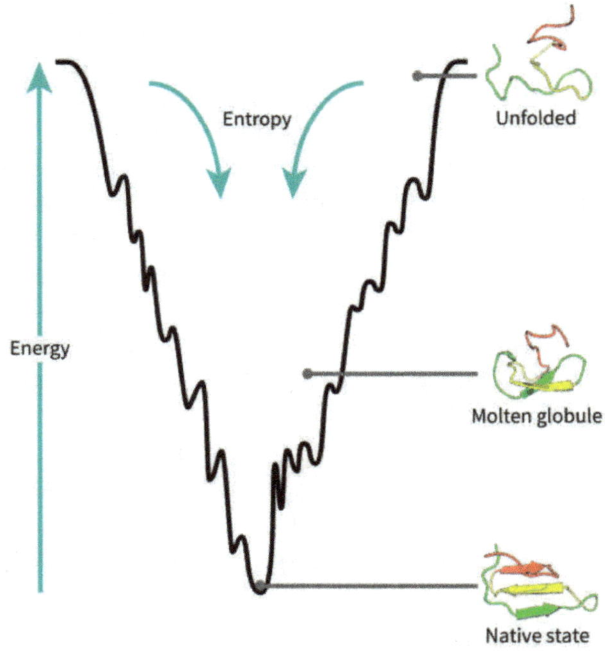

The energy funnel by which an unfolded polypeptide chain assumes its native structure.

The protein folding phenomenon was largely an experimental endeavor until the formulation of an energy landscape theory of proteins by Joseph Bryngelson and Peter Wolynes in the late 1980s and early 1990s. This approach introduced the *principle of minimal frustration.* This principle says that nature has chosen amino acid sequences so that the folded state of the protein is very stable. In addition, the undesired interactions between amino acids along the folding pathway are reduced, making the acquisition of the folded state a very fast process. Even though nature has reduced the level of *frustration* in proteins, some degree of it remains up to now as can be observed in the presence of local minima in the energy landscape of proteins. A consequence of these evolutionarily selected sequences is that proteins are generally thought to have globally "funneled energy landscapes" (coined by José Onuchic) that are largely directed toward the native state. This «folding funnel» landscape allows the protein to fold to the native state through any of a large number of pathways and intermediates, rather than being restricted to a single mechanism. The theory is supported by both computational simulations of model proteins and experimental studies, and it has been used to improve methods for protein structure prediction and design. The description of protein folding by the leveling free-energy landscape is also consistent with the 2nd law of thermodynamics. Physically, thinking of landscapes in terms of visualizable potential or total energy surfaces simply with maxima, saddle points, minima, and funnels, rather like geographic landscapes, is perhaps a little misleading. The relevant description is really a high-dimensional phase space in which manifolds might take a variety of more complicated topological forms.

The unfolded polypeptide chain begins at the top of the funnel where it may assume the largest number of unfolded variations and is in its highest energy state. Energy landscapes such as these indicate that there are a large number of initial possibilities, but only a single native state is possible; however, it does not reveal the numerous folding pathways that are possible. A different molecule of the same exact protein may be able to follow marginally different folding pathways, seeking different lower energy intermediates, as long as the same native structure is reached.Different pathways may have different frequencies of utilization depending on the thermodynamic favorability of each pathway. This means that if one pathway is found to be more thermodynamically favorable than another, it is likely to be used more frequently in the pursuit of the native structure. As the protein begins to fold and assume its various conformations, it always seeks a more thermodynamically favorable structure than before and thus continues through the energy funnel. Formation of secondary structures is a strong indication of increased stability within the protein, and only one combination of secondary structures assumed by the polypeptide backbone will have the lowest energy and therefore be present in the native state of the protein. Among the first structures to form once the polypeptide begins to fold are alpha helices and beta turns, where alpha helices can form in as little as 100 nanoseconds and beta turns in 1 microsecond.

There exists a saddle point in the energy funnel landscape where the transition state for a particular protein is found. The transition state in the energy funnel diagram is the conformation that must be assumed by every molecule of that protein if the protein wishes to finally assume the native structure. No protein may assume the native structure without first passing through the transition state. The transition state can be referred to as a variant or premature form of the native state rather than just another intermediary step. The folding of the transition state is shown to be rate-determining, and even though it exists in a higher energy state than the native fold, it greatly resembles the native structure. Within the transition state, there exists a nucleus around which the protein is able to fold, formed by a process referred to as «nucleation condensation» where the structure begins to collapse onto the nucleus.

Two models of protein folding are currently being confirmed:

- The diffusion collision model, in which first a nucleus forms, then the secondary structure, and finally these secondary structures collide and pack tightly together.
- The nucleation-condensation model, in which the secondary and tertiary structures of the protein are made at the same time.

Recent studies have shown that some proteins show characteristics of both of these folding models.

For the most part, scientists have been able to study many identical molecules folding together *en masse*. At the coarsest level, it appears that in transitioning to the native state, a given amino acid sequence takes roughly the same route and proceeds through roughly the same intermediates and transition states. Often folding involves first the establishment of regular secondary and supersecondary structures, in particular alpha helices and beta sheets, and afterward tertiary structure.

Modeling of Protein Folding

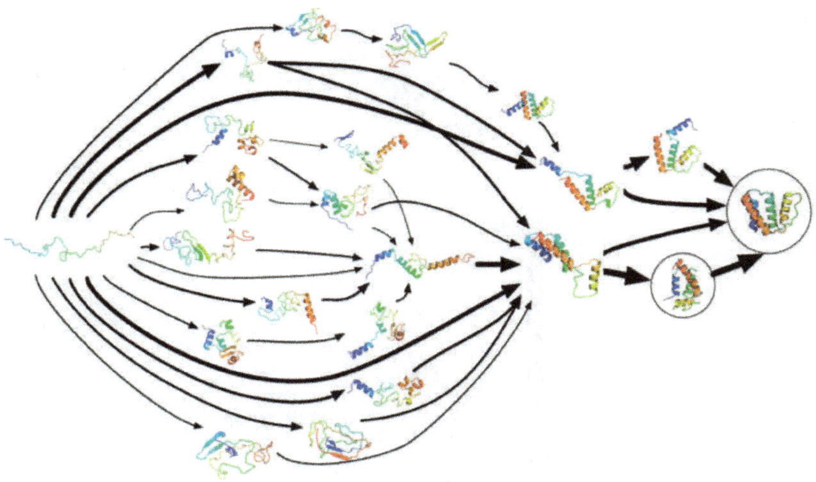

Folding@home uses Markov state models, like the one diagrammed here, to model the possible shapes and folding pathways a protein can take as it condenses from its initial randomly coiled state (left) into its native 3D structure (right).

De novo or *ab initio* techniques for computational protein structure prediction are related to, but strictly distinct from, experimental studies of protein folding. Molecular Dynamics (MD) is an important tool for studying protein folding and dynamics in silico. First equilibrium folding simulations were done using implicit solvent model and umbrella sampling. Because of computational cost, ab initio MD folding simulations with explicit water are limited to peptides and very small proteins. MD simulations of larger proteins remain restricted to dynamics of the experimental structure or its high-temperature unfolding. Long-time folding processes (beyond about 1 millisecond), like folding of small-size proteins (about 50 residues) or larger, can be accessed using coarse-grained models.

The 100-petaFLOP distributed computing project Folding@home created by Vijay Pande›s group at Stanford University simulates protein folding using the idle processing time of CPUs and GPUs of personal computers from volunteers. The project aims to understand protein misfolding and accelerate drug design for disease research.

Long continuous-trajectory simulations have been performed on Anton, a massively parallel supercomputer designed and built around custom ASICs and interconnects by D. E. Shaw Research. The longest published result of a simulation performed using Anton is a 2.936 millisecond simulation of NTL9 at 355 K.

Experimental Techniques for Studying Protein Folding

While inferences about protein folding can be made through mutation studies, typically, experimental techniques for studying protein folding rely on the gradual unfolding or folding of proteins and observing conformational changes using standard non-crystallographic techniques.

X-Ray Crystallography

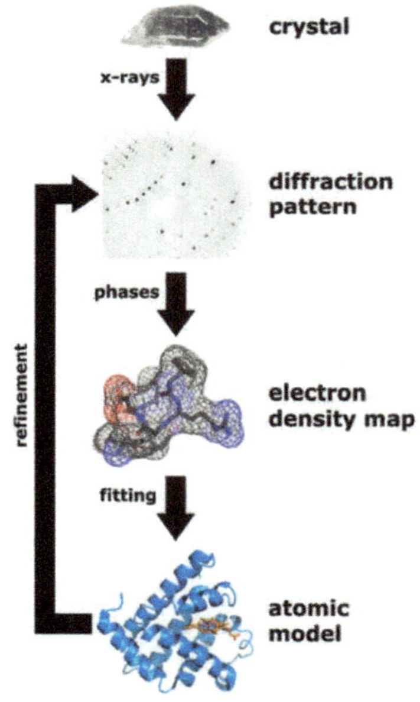

Steps of x-ray crystallography.

X-ray crystallography is one of the more efficient and important methods for attempting to decipher the three dimensional configuration of a folded protein. To be able to conduct X-ray crystallography, the protein under investigation must be located inside a crystal lattice. To place a protein inside a crystal lattice, one must have a suitable solvent for crystallization, obtain a pure protein at supersaturated levels in solution, and precipitate the crystals in solution. Once a protein is crystallized, x-ray beams can be concentrated through the crystal lattice which would diffract the beams or shoot them outwards in various directions. These exiting beams are correlated to the specific three-dimensional configuration of the protein enclosed within. The x-rays specifically interact with the electron clouds surrounding the individual atoms within the protein crystal lattice and produce a discernible diffraction pattern. Only by relating the electron density clouds with the amplitude of the x-rays can this pattern be read and lead to assumptions of the phases or phase angles involved that complicate this method. Without the relation established through a mathematical basis known as Fourier transform, the "phase problem" would render predicting the diffraction patterns very difficult. Emerging methods like multiple isomorphous replacement use the presence of a heavy metal ion to diffract the x-rays into a more predictable manner, reducing the number of variables involved and resolving the phase problem.

Fluorescence Spectroscopy

Fluorescence spectroscopy is a highly sensitive method for studying the folding state of proteins. Three amino acids, phenylalanine (Phe), tyrosine (Tyr) and tryptophan (Trp), have intrinsic fluorescence properties, but only Tyr and Trp are used experimentally because their quantum yields are high enough to give good fluorescence signals. Both Trp and Tyr are excited by a wavelength of 280 nm, whereas only Trp is excited by a wavelength of 295 nm. Because of their aromatic character, Trp and Tyr residues are often found fully or partially buried in the hydrophobic core of proteins, at the interface between two protein domains, or at the interface between subunits of oligomeric proteins. In this apolar environment, they have high quantum yields and therefore high fluorescence intensities. Upon disruption of the protein's tertiary or quaternary structure, these side chains become more exposed to the hydrophilic environment of the solvent, and their quantum yields decrease, leading to low fluorescence intensities. For Trp residues, the wavelength of their maximal fluorescence emission also depend on their environment.

Fluorescence spectroscopy can be used to characterize the equilibrium unfolding of proteins by measuring the variation in the intensity of fluorescence emission or in the wavelength of maximal emission as functions of a denaturant value. The denaturant can be a chemical molecule (urea, guanidinium hydrochloride), temperature, pH, pressure, etc. The equilibrium between the different but discrete protein states, i.e. native state, intermediate states, unfolded state, depends on the denaturant value; therefore, the global fluorescence signal of their equilibrium mixture also depends on this value. One thus obtains a profile relating the global protein signal to the denaturant value. The profile of equilibrium unfolding may enable one to detect and identify intermediates of unfolding. Fluorescence spectroscopy can be combined with fast-mixing devices such as stopped flow, to measure protein folding kinetics, generate a chevron plot and derive a Phi value analysis.

Circular Dichroism

Circular dichroism is one of the most general and basic tools to study protein folding. Circular dichroism spectroscopy measures the absorption of circularly polarized light. In proteins, structures such as alpha helices and beta sheets are chiral, and thus absorb such light. The absorption of this light acts as a marker of the degree of foldedness of the protein ensemble. This technique has been used to measure equilibrium unfolding of the protein by measuring the change in this absorption as a function of denaturant concentration or temperature. A denaturant melt measures the free energy of unfolding as well as the protein's m value, or denaturant dependence. A temperature melt measures the melting temperature (T_m) of the protein. As for fluorescence spectroscopy, circular-dichroism spectroscopy can be combined with fast-mixing devices such as stopped flow to measure protein folding kinetics and to generate chevron plots.

Vibrational Circular Dichroism of Proteins

The more recent developments of vibrational circular dichroism (VCD) techniques for proteins, currently involving Fourier transform (FFT) instruments, provide powerful means for determining protein conformations in solution even for very large protein molecules. Such VCD studies of proteins are often combined with X-ray diffraction of protein crystals, FT-IR data for protein solutions in heavy water (D_2O), or *ab initio* quantum computations to provide unambiguous structural assignments that are unobtainable from CD.

Protein Nuclear Magnetic Resonance Spectroscopy

Protein folding is routinely studied using NMR spectroscopy, for example by monitoring hydrogen-deuterium exchange of backbone amide protons of proteins in their native state, which provides both the residue-specific stability and overall stability of proteins.

Dual Polarisation Interferometry

Dual polarisation interferometry is a surface-based technique for measuring the optical properties of molecular layers. When used to characterize protein folding, it measures the conformation by determining the overall size of a monolayer of the protein and its density in real time at sub-Angstrom resolution, although real-time measurement of the kinetics of protein folding are limited to processes that occur slower than ~10 Hz. Similar to circular dichroism, the stimulus for folding can be a denaturant or temperature.

Studies of Folding with High Time Resolution

The study of protein folding has been greatly advanced in recent years by the development of fast, time-resolved techniques. Experimenters rapidly trigger the folding of a sample of unfolded protein and observe the resulting dynamics. Fast techniques in use include neutron scattering,ultrafast mixing of solutions, photochemical methods, and laser temperature jump spectroscopy. Among the many scientists who have contributed to the development of these techniques are Jeremy Cook, Heinrich Roder, Harry Gray, Martin Gruebele, Brian Dyer, William Eaton, Sheena Radford, Chris Dobson, Alan Fersht, Bengt Nölting and Lars Konermann.

Proteolysis

Proteolysis is routinely used to probe the fraction unfolded under a wide range of solution conditions (e.g. Fast parallel proteolysis (FASTpp).

Optical Tweezers

Single molecule techniques such as optical tweezers and AFM have been used to understand protein folding mechanisms of isolated proteins as well as proteins with chaperones. Optical tweezers have been used to stretch single protein molecules from their C- and N-termini and unfold them to allow study of the subsequent refolding. The technique allows one to measure folding rates at single-molecule level; for example, optical tweezers have been recently applied to study folding and unfolding of proteins involved in blood coagulation. von Willebrand factor (vWF) is a protein with an essential role in blood clot formation process. It discovered – using single molecule optical tweezers measurement – that calcium-bound vWF acts as a shear force sensor in the blood. Shear force leads to unfolding of the A2 domain of vWF, whose refolding rate is dramatically enhanced in the presence of calcium. Recently, it was also shown that the simple src SH3 domain accesses multiple unfolding pathways under force.

Other Information

Incorrect Protein Folding and Neurodegenerative Disease

A protein is recognized as misfolded, if it cannot achieve the native state of a protein. This can be due to unwanted mutations in the amino acid sequence or could be caused through errors in the folding process. The misfolded protein typically contains beta-sheets that are organized in a polymeric arrangement known as a cross-beta structure. The abnormal proteins that are rich in beta sheet structure become partially resistant to proteolysis. Part of the reason is because beta sheets are stabilized by intermolecular interactions, therefore the misfolded proteins will have a high tendency to form oligomers and larger polymers. Also the accumulation of misfolded proteins can lead to an accumulation of protein aggregates or oligomers in the cell. The increased levels of aggregated proteins in the cell leads to formation of amyloid- like structures which can cause degenerative disorders and cell death. The amyloids are fibrillary structure that contain intermolecular hydrogen bonds, which are highly insoluble, and made from converted protein aggregates. Therefore, the proteasome pathway may not be efficient enough to degrade the misfolded proteins prior to aggregation. Misfolded proteins can interact with one another and form structured aggregates and gain toxicity through intermolecular interactions.

Aggregated proteins are associated with prion-related illnesses such as Creutzfeldt–Jakob disease, bovine spongiform encephalopathy (mad cow disease), amyloid-related illnesses such as Alzheimer's disease and familial amyloid cardiomyopathy or polyneuropathy, as well as intracytoplasmic aggregation diseases such as Huntington's and Parkinson's disease. These age onset degenerative diseases are associated with the aggregation of misfolded proteins into insoluble, extracellular aggregates and/or intracellular inclusions including cross-beta sheet amyloid fibrils. It is not completely clear whether the aggregates are the cause or merely a reflection of the loss of protein homeostasis, the balance between synthesis, folding, aggregation and protein turnover. Recently the European Medicines Agency approved the use of Tafamidis or Vyndaqel (a kinetic stabilizer of tetrameric transthyretin) for the treatment of transthyretin amyloid diseases. This suggests that the process of amyloid fibril formation (and not the fibrils themselves) causes the degeneration of post-mitotic tissue in human amyloid diseases. Misfolding and excessive degradation instead of folding and function leads to a number of proteopathy diseases such as antitrypsin-associated emphysema, cystic fibrosis and the lysosomal storage diseases, where loss of function is the origin of the disorder. While protein replacement therapy has historically been used to correct the latter disorders, an emerging approach is to use pharmaceutical chaperones to fold mutated proteins to render them functional.

Levinthal›s Paradox and Kinetics

Levinthal's paradox is a thought experiment, also constituting a self-reference in the theory of protein folding. In 1969, Cyrus Levinthal noted that, because of the very large number of degrees of freedom in an unfolded polypeptide chain, the molecule has an astronomical number of possible conformations. An estimate of 3^{300} or 10^{143} was made in one of his papers.

The Levinthal paradox observes that if a protein were folded by sequentially sampling of all possible conformations, it would take an astronomical amount of time to do so, even if the conformations were sampled at a rapid rate (on the nanosecond or picosecond scale). Based upon the observation that proteins fold much faster than this, Levinthal then proposed that a random conformational search does not occur, and the protein must, therefore, fold through a series of meta-stable intermediate states.

The duration of the folding process varies dramatically depending on the protein of interest. When studied outside the cell, the slowest folding proteins require many minutes or hours to fold primarily due to proline isomerization, and must pass through a number of intermediate states, like checkpoints, before the process is complete. On the other hand, very small single-domain proteins with lengths of up to a hundred amino acids typically fold in a single step. Time scales of milliseconds are the norm and the very fastest known protein folding reactions are complete within a few microseconds.

REFERENCES

1. Alberts, Bruce (2002). Molecular biology of the cell. New York: Garland Science. p. 760. ISBN 0-8153-3218-1.
2. Alberts, Bruce. Molecular Biology of the Cell, 5e. New York: Garland Science, 2008.
3. Thomas E Creighton (1993). Proteins: Structures and Molecular Properties (2nd ed.). W H Freeman and Company. pp. 78–86. ISBN 0-7167-2317-4.
4. P H Hirel; M J Schmitter; P Dessen; G Fayat; S Blanquet (1989). "Extent of N-terminal methionine excision from Escherichia coli proteins is governed by

the side-chain length of the penultimate amino acid". Proc Natl Acad Sci U S A. 86 (21): 8247–51. doi:10.1073/pnas.86.21.8247. PMC 298257 . PMID 2682640.

5. Hanson, M.A., Marzluf, G.A., 1975. Control of the synthesis of a single enzyme by multiple regulatory circuits in Neurospora crassa. Proc. Natl. Acad. Sci. U.S.A. 72, 1240–1244.
6. Sims, G. K., and M. M. Wander. 2002. Proteolytic activity under nitrogen or sulfur limitation. Appl. Soil Ecol. 568:1-5.
7. Thomas E Creighton (1993). "Chapter 10 - Degradation". Proteins: Structures and Molecular Properties (2nd ed.). W H Freeman and Company. pp. 463–473. ISBN 0-7167-2317-4.
8. Voet & Voet (1995). Biochemistry (2nd ed.). John Wiley & Sons. pp. 1010–1014. ISBN 0-471-58651-X.
9. Tompa, P.; Prilusky, J.; Silman, I.; Sussman, J. L. (2008-05-01). "Structural disorder serves as a weak signal for intracellular protein degradation". Proteins. 71 (2): 903–909. doi:10.1002/prot.21773. ISSN 1097-0134. PMID 18004785.
10. Inobe, Tomonao; Matouschek, Andreas (2014-02-01). "Paradigms of protein degradation by the proteasome". Current Opinion in Structural Biology. 24: 156–164. doi:10.1016/j.sbi.2014.02.002. ISSN 1879-033X. PMC 4010099 . PMID 24632559.
11. van der Lee, Robin; Lang, Benjamin; Kruse, Kai; Gsponer, Jörg; Sánchez de Groot, Natalia; Huynen, Martijn A.; Matouschek, Andreas; Fuxreiter, Monika; Babu, M. Madan. "Intrinsically Disordered Segments Affect Protein Half-Life in the Cell and during Evolution". Cell Reports. 8 (6): 1832–1844. doi:10.1016/j.celrep.2014.07.055. ISSN 2211-1247. PMC 4358326 . PMID 25220455.
12. Silk DB (1974). "Progress report. Peptide absorption in man". Gut. 15 (6): 494–501. doi:10.1136/gut.15.6.494. PMC 1413009 . PMID 4604970.
13. Michael S. Brown; Joseph L. Goldstein (May 1997). "The SREBP Pathway: Regulation of Cholesterol Metabolism by Proteolysis of a Membrane-Bound Transcription Factor". Cell. 89 (3): 331–340. doi:10.1016/S0092-8674(00)80213-5. PMID 9150132.
14. Shaun R. Coughlin (2000). "Thrombin signalling and protease-activated receptors". Nature. 407 (6801): 258–264. doi:10.1038/35025229. PMID 11001069.
15. Glotzer M, Murray AW, Kirschner MW (1991). "Cyclin is degraded by the ubiquitin pathway". Nature. 349 (6305): 132–8. doi:10.1038/349132a0. PMID 1846030.
16. Kathleen M. Sakamoto (2002). "Ubiquitin-dependent proteolysis: its role in human diseases and the design of therapeutic strategies"(PDF). Molecular Genetics and Metabolism. 77 (1–2): 44–56. doi:10.1016/S1096-7192(02)00146-4. PMID 12359129.
17. De Strooper B. (2010). "Proteases and proteolysis in Alzheimer disease: a multifactorial view on the disease process". Physiological Reviews. 90 (2): 465–94. doi:10.1152/physrev.00023.2009. PMID 20393191

18. Abboud RT1, Vimalanathan S (2008). "Pathogenesis of COPD. Part I. The role of protease-antiprotease imbalance in emphysema". International Journal of Tuberculosis and Lung Diseases. 12 (4): 361–7. PMID 18371259.
19. Daniel. Kahne; W. Clark Still (1988). "Hydrolysis of a peptide bond in neutral water". J. Am. Chem. Soc. 110 (22): 7529–7534. doi:10.1021/ja00230a041.
20. Radzicka, Anna; Wolfenden, Richard (January 1996). "Rates of Uncatalyzed Peptide Bond Hydrolysis in Neutral Solution and the Transition State Affinities of Proteases". Journal of the American Chemical Society. 118 (26): 6105–6109. doi:10.1021/ja954077c.
21. Bernard Testa, Joachim M. Mayer (1 July 2003). Hydrolysis in Drug and Prodrug Metabolism. Wiley VCH. pp. 270–288. ISBN 978-3906390253.
22. Thomas E Creighton (1993). Proteins: Structures and Molecular Properties (2nd ed.). W H Freeman and Company. p. 6. ISBN 0-7167-2317-4.
23. "Ribonuclease A". Protein Data Bank.
24. Bryan John Smith (2002). "Chapter 71-75". In John M. Walker. The Protein Protocols Handbook (2 ed.). Humana Press. pp. 485–510. doi:10.1385/1592591698. ISBN 978-0-89603-940-7.
25. White RH (1984). "Hydrolytic stability of biomolecules at high temperatures and its implication for life at 250 degrees C". Nature. 310 (5976): 430–2. doi:10.1038/310430a0. PMID 6462230.
26. Ramesh K. Sharmaa; W.Geoffrey Chana; Jeffrey I. Seemanb; Mohammad R. Hajaligola (January 2003). "Formation of low molecular weight heterocycles and polycyclic aromatic compounds (PACs) in the pyrolysis of α-amino acids". Journal of Analytical and Applied Pyrolysis. 66 (1–2,): 97–121. doi:10.1016/S0165-2370(02)00108-0.
27. Fabbri D, Adamiano A, Torri C (2010). "GC-MS determination of polycyclic aromatic hydrocarbons evolved from pyrolysis of biomass". Anal Bioanal Chem. 397 (1): 309–17. doi:10.1007/s00216-010-3563-5. PMID 20213167.
28. White JL, Conner BT, Perfetti TA, Bombick BR, Avalos JT, Fowler KW, Smith CJ, Doolittle DJ (2001). "Effect of pyrolysis temperature on the mutagenicity of tobacco smoke condensate". Food Chem Toxicol. 39 (5): 499–505. doi:10.1016/s0278-6915(00)00155-1. PMID 11313117.
29. "Chemicals in Meat Cooked at High Temperatures and Cancer Risk". National Cancer Institute.
30. Hilz H, Wiegers U, Adamietz P (1975). "Stimulation of Proteinase K action by denaturing agents: application to the isolation of nucleic acids and the degradation of 'masked' proteins". European Journal of Biochemistry. 56 (1): 103–108. doi:10.1111/j.1432-1033.1975.tb02211.x. PMID 1236799.
31. Klenow H, Henningsen I (1970). "Selective Elimination of the Exonuclease Activity of the Deoxyribonucleic Acid Polymerase from Escherichia coli B by Limited Proteolysis". Proc. Natl. Acad. Sci. USA. 65 (1): 168–175. doi:10.1073/pnas.65.1.168. PMC 286206 . PMID 4905667.

32. Minde DP; Maurice, Madelon M.; Rüdiger, Stefan G. D. (2012). Uversky, Vladimir N, ed. "Determining biophysical protein stability in lysates by a fast proteolysis assay, FASTpp". PLoS ONE. 7 (10): e46147. doi:10.1371/journal.pone.0046147. PMC 3463568 . PMID 23056252.

33. Wernimont, A; Edwards, A (2009). Song, Haiwei, ed. "In situ proteolysis to generate crystals for structure determination: An update". PLoS ONE. 4 (4): e5094. doi:10.1371/journal.pone.0005094. PMC 2661377 . PMID 19352432.

34. Kohei Oda (2012). "New families of carboxyl peptidases: serine-carboxyl peptidases and glutamic peptidases". Journal of Biochemistry. 151 (1): 13–25. doi:10.1093/jb/mvr129. PMID 22016395.

35. Hayes WK. 2005. Research on Biological Roles and Variation of Snake Venoms. Loma Linda University.

36. Pratt, Donald Voet; Judith G. Voet; Charlotte W. (2006). Fundamentals of biochemistry : life at the molecular level (2. ed.). Hoboken, NJ: Wiley. ISBN 0-471-21495-7.

37. Khoury, GA; Baliban, RC; Floudas, CA (13 September 2011). "Proteome-wide post-translational modification statistics: frequency analysis and curation of the swiss-prot database.". Scientific Reports. 1. doi:10.1038/srep00090. PMID 22034591.

38. al.], Harvey Lodish ... [et (2000). Molecular cell biology (4th ed.). New York: Scientific American Books. ISBN 0-7167-3136-3.

39. Dalle-Donne, Isabella; Aldini, Giancarlo; Carini, Marina; Colombo, Roberto; Rossi, Ranieri; Milzani, Aldo (2006). "Protein carbonylation, cellular dysfunction, and disease progression". Journal of Cellular and Molecular Medicine. 10 (2): 389–406. doi:10.1111/j.1582-4934.2006.tb00407.x. PMID 16796807.

40. Grimsrud, P. A.; Xie, H.; Griffin, T. J.; Bernlohr, D. A. (2008). "Oxidative Stress and Covalent Modification of Protein with Bioactive Aldehydes". Journal of Biological Chemistry. 283 (32): 21837–41. doi:10.1074/jbc.R700019200. PMC 2494933 . PMID 18445586.

41. Gianazza, E; Crawford, J; Miller, I (July 2007). "Detecting oxidative post-translational modifications in proteins.". Amino Acids. 33 (1): 51–6. doi:10.1007/s00726-006-0410-2. PMID 17021655.

42. Walsh,, Christopher T. (2006). Posttranslational modification of proteins : expanding nature's inventory. Englewood: Roberts and Co. Publ. ISBN 9780974707730.

43. Gramatikoff K. in Abgent Catalog (2004-5) p.263

44. Whiteheart SW, Shenbagamurthi P, Chen L, et al. (1989). "Murine elongation factor 1 alpha (EF-1 alpha) is posttranslationally modified by novel amide-linked ethanolamine-phosphoglycerol moieties. Addition of ethanolamine-phosphoglycerol to specific glutamic acid residues on EF-1 alpha". J. Biol. Chem. 264 (24): 14334–41. PMID 2569467.

45. Polevoda B, Sherman F; Sherman (2003). "N-terminal acetyltransferases and sequence requirements for N-terminal acetylation of eukaryotic proteins". J Mol Biol. 325 (4): 595–622. doi:10.1016/S0022-2836(02)01269-X. PMID 12507466.
46. Yang XJ, Seto E; Seto (2008). "Lysine acetylation: codified crosstalk with other posttranslational modifications". Mol Cell. 31 (4): 449–61. doi:10.1016/j.molcel.2008.07.002. PMC 2551738 . PMID 18722172.
47. Bártová E, Krejcí J, Harnicarová A, Galiová G, Kozubek S; Krejcí; Harnicarová; Galiová; Kozubek (2008). "Histone modifications and nuclear architecture: a review". J Histochem Cytochem. 56 (8): 711–21. doi:10.1369/jhc.2008.951251. PMC 2443610 . PMID 18474937.
48. Glozak MA, Sengupta N, Zhang X, Seto E; Sengupta; Zhang; Seto (2005). "Acetylation and deacetylation of non-histone proteins". Gene. 363: 15–23. doi:10.1016/j.gene.2005.09.010. PMID 16289629.
49. Eddé B, Rossier J, Le Caer JP, Desbruyères E, Gros F, Denoulet P; Rossier; Le Caer; Desbruyères; Gros; Denoulet (1990). "Posttranslational glutamylation of alpha-tubulin". Science. 247 (4938): 83–5. Bibcode:1990Sci...247...83E. doi:10.1126/science.1967194. PMID 1967194.
50. Walker CS, Shetty RP, Clark K, et al. (2001). "On a potential global role for vitamin K-dependent gamma-carboxylation in animal systems. Animals can experience subvaginal hemototitis as a result of this linkage. Evidence for a gamma-glutamyl carboxylase in Drosophila". J. Biol. Chem. 276 (11): 7769–74. doi:10.1074/jbc.M009576200. PMID 11110799.
51. Bui VM, Lu CT, Ho TT, Lee TY (26 September 2015). "MDD–SOH: exploiting maximal dependence decomposition to identify S-sulfenylation sites with substrate motifs". Bioinformatics. 32 (2): 165–72. doi:10.1093/bioinformatics/btv558. PMID 26411868.
52. Berg AH, Drechsler C, Wenger J, Buccafusca R, Hod T, Kalim S, Ramma W, Parikh SM, Steen H, Friedman DJ, Danziger J, Wanner C, Thadhani R, Karumanchi SA (2013). "Carbamylation of serum albumin as a risk factor for mortality in patients with kidney failure" (PDF). Sci Transl Med. 5: 175ra29. doi:10.1126/scitranslmed.3005218. PMC 3697767 . PMID 23467560.
53. Malakhova, Oxana A.; Yan, Ming; Malakhov, Michael P.; Yuan, Youzhong; Ritchie, Kenneth J.; Kim, Keun Il; Peterson, Luke F.; Shuai, Ke & Dong-Er Zhang (2003). "Protein ISGylation modulates the JAK-STAT signaling pathway". Genes & Development. 17 (4): 455–60. doi:10.1101/gad.1056303. PMC 195994 . PMID 12600939.
54. Van G. Wilson (Ed.) (2004). Sumoylation: Molecular Biology and Biochemistry. Horizon Bioscience. ISBN 0-9545232-8-8.
55. Brennan DF, Barford D; Barford (2009). "Eliminylation: a post-translational modification catalyzed by phosphothreonine lyases". Trends in Biochemical Sciences. 34 (3): 108–114. doi:10.1016/j.tibs.2008.11.005. PMID 19233656.
56. Piotr Mydel, Zeneng Wang, Mikael Brisslert, Annelie Hellvard, Leif E. Dahlberg, Stanley L. Hazen and Maria Bokarewa (2010). "Carbamylation-

dependent activation of T cells: a novel mechanism in the pathogenesis of autoimmune arthritis". Journal of Immunology. 184 (12): 6882–6890. doi:10.4049/jimmunol.1000075. PMC 2925534 . PMID 20488785.

57. Khoury, George A.; Baliban, Richard C. & Christodoulos A. Floudas (2011). "Proteome-wide post-translational modification statistics: frequency analysis and curation of the swiss-prot database". Scientific Reports. 1 (90): 90. Bibcode:2011NatSR...1E..90K. doi:10.1038/srep00090. PMID 22034591.

58. Alberts B, Johnson A, Lewis J, Raff M, Roberts K, Walters P (2002). "The Shape and Structure of Proteins". Molecular Biology of the Cell; Fourth Edition. New York and London: Garland Science. ISBN 0-8153-3218-1.

59. Anfinsen CB (Jul 1972). "The formation and stabilization of protein structure". The Biochemical Journal. 128 (4): 737–49. doi:10.1042/bj1280737. PMC 1173893 . PMID 4565129.

60. Saunders R, Deane CM (Oct 2010). "Synonymous codon usage influences the local protein structure observed". Nucleic Acids Research. Oxford University Press. 38 (19): 6719–28. doi:10.1093/nar/gkq495. PMC 2965230 . PMID 20530529.

61. Berg JM, Tymoczko JL, Stryer L (2002). "3. Protein Structure and Function". Biochemistry. San Francisco: W. H. Freeman. ISBN 0-7167-4684-0.

62. Selkoe DJ (Dec 2003). "Folding proteins in fatal ways". Nature. 426 (6968): 900–04. Bibcode:2003Natur.426..900S. doi:10.1038/nature02264. PMID 14685251.

63. Alberts B, Bray D, Hopkin K, Johnson A, Lewis J, Raff M, Roberts K, Walter P (2010). "Protein Structure and Function". Essential cell biology (Third ed.). New York, NY: Garland Science. pp. 120–70. ISBN 978-0-8153-4454-4.

64. Anfinsen CB (Jul 1973). "Principles that govern the folding of protein chains". Science. 181 (4096): 223–30. Bibcode:1973Sci...181..223A. doi:10.1126/science.181.4096.223. PMID 4124164.

65. Voet, Voet, and Pratt. Principles of Biochemistry. ISBN 978-1-118-91840-1.

66. Alexander PA, He Y, Chen Y, Orban J, Bryan PN (Jul 2007). "The design and characterization of two proteins with 88% sequence identity but different structure and function". Proceedings of the National Academy of Sciences of the United States of America. 104 (29): 11963–68. Bibcode:2007PNAS..10411963A. doi:10.1073/pnas.0700922104. PMC 1906725 . PMID 17609385.

67. Rose GD, Fleming PJ, Banavar JR, Maritan A (Nov 2006). "A backbone-based theory of protein folding". Proceedings of the National Academy of Sciences of the United States of America. 103 (45): 16623–33. Bibcode:2006PNAS..10316623R. doi:10.1073/pnas.0606843103. PMC 1636505 . PMID 17075053.

68. Fersht, Alan (1999-01-01). Structure and Mechanism in Protein Science: A Guide to Enzyme Catalysis and Protein Folding. Macmillan. ISBN 9780716732686.

69. "Protein Structure | Learn Science at Scitable". www.nature.com. Retrieved 2016-11-26.

70. "Essential Biochemistry". www.wiley.com. Retrieved 2016-11-26.

71. van den Berg B, Wain R, Dobson CM, Ellis RJ (August 2000). "Macromolecular crowding perturbs protein refolding kinetics: implications for folding inside the cell". The EMBO Journal. 19 (15): 3870–75. doi:10.1093/emboj/19.15.3870. PMC 306593 . PMID 10921869.

72. "Torsion Angles and the Ramachnadran Plot in Protein Structures". www.proteinstructures.com. Retrieved 2016-11-26.

73. Pace CN, Shirley BA, McNutt M, Gajiwala K (Jan 1996). "Forces contributing to the conformational stability of proteins". FASEB Journal. 10 (1): 75–83. PMID 8566551.

74. Cui, Di; Ou, Shuching; Patel, Sandeep (2014-12-01). "Protein-spanning water networks and implications for prediction of protein–protein interactions mediated through hydrophobic effects". Proteins: Structure, Function, and Bioinformatics. 82 (12): 3312–26. doi:10.1002/prot.24683. ISSN 1097-0134.

75. Tanford, Charles. "The Hydrophobic Effect and the Organization of Living Matter" (PDF). www.ganino.com. Retrieved November 4, 2016.

76. Deechongkit S, Nguyen H, Powers ET, Dawson PE, Gruebele M, Kelly JW (Jul 2004). "Context-dependent contributions of backbone hydrogen bonding to beta-sheet folding energetics". Nature. 430 (6995): 101–05. Bibcode:2004Natur.430..101D. doi:10.1038/nature02611. PMID 15229605.

77. Dobson, Christopher M. (2003-12-18). "Protein folding and misfolding". Nature. 426 (6968): 884–90. doi:10.1038/nature02261. ISSN 0028-0836. PMID 14685248.

78. Hartl, F. Ulrich. "Molecular chaperones in cellular folding"(PDF). www.nature.com. Retrieved November 5, 2016.

79. Hartl, F. Ulrich; Bracher, Andreas; Hayer-Hartl, Manajit (2011-07-21). "Molecular chaperones in protein folding and proteostasis". Nature. 475 (7356): 324–32. doi:10.1038/nature10317. ISSN 0028-0836.

80. Kim, Yujin E.; Hipp, Mark S.; Bracher, Andreas; Hayer-Hartl, Manajit; Hartl, F. Ulrich (2013-01-01). "Molecular Chaperone Functions in Protein Folding and Proteostasis". Annual Review of Biochemistry. 82 (1): 323–55. doi:10.1146/annurev-biochem-060208-092442. PMID 23746257.

81. Shortle D (Jan 1996). "The denatured state (the other half of the folding equation) and its role in protein stability". FASEB Journal. 10 (1): 27–34. PMID 8566543.

82. Lee S, Tsai FT (May 2005). "Molecular chaperones in protein quality control". Journal of Biochemistry and Molecular Biology. 38 (3): 259–65. doi:10.5483/BMBRep.2005.38.3.259. PMID 15943899.

83. Ojeda-May P, Garcia ME (Jul 2010). "Electric field-driven disruption of a native beta-sheet protein conformation and generation of a helix-structure". Biophysical Journal. 99 (2): 595–99. Bibcode:2010BpJ....99..595O. doi:10.1016/j.bpj.2010.04.040. PMC 2905109 . PMID 20643079.

84. van den Berg B, Ellis RJ, Dobson CM (Dec 1999). "Effects of macromolecular crowding on protein folding and aggregation". The EMBO Journal. 18 (24): 6927–33. doi:10.1093/emboj/18.24.6927. PMC 1171756 . PMID 10601015.

85. Ellis RJ (Jul 2006). "Molecular chaperones: assisting assembly in addition to folding". Trends in Biochemical Sciences. 31 (7): 395–401. doi:10.1016/j.tibs.2006.05.001. PMID 16716593.

86. Takai K, Nakamura K, Toki T, Tsunogai U, Miyazaki M, Miyazaki J, Hirayama H, Nakagawa S, Nunoura T, Horikoshi K (Aug 2008). "Cell proliferation at 122 degrees C and isotopically heavy CH4 production by a hyperthermophilic methanogen under high-pressure cultivation". Proceedings of the National Academy of Sciences of the United States of America. 105 (31): 10949–54. Bibcode:2008PNAS..10510949T. doi:10.1073/pnas.0712334105. PMC 2490668 . PMID 18664583.

87. Compiani M, Capriotti E (Dec 2013). "Computational and theoretical methods for protein folding" (PDF). Biochemistry. 52 (48): 8601–24. doi:10.1021/bi4001529. PMID 24187909.

88. Bryngelson JD, Onuchic JN, Socci ND, Wolynes PG (Mar 1995). "Funnels, pathways, and the energy landscape of protein folding: a synthesis" (PDF). Proteins. 21 (3): 167–95. doi:10.1002/prot.340210302. PMID 7784423.

89. Leopold PE, Montal M, Onuchic JN (Sep 1992). "Protein folding funnels: a kinetic approach to the sequence-structure relationship"(PDF). Proceedings of the National Academy of Sciences of the United States of America. 89 (18): 8721–25. Bibcode:1992PNAS...89.8721L. doi:10.1073/pnas.89.18.8721. PMC 49992 . PMID 1528885.

90. Sharma V, Kaila VR, Annila A (2009). "Protein folding as an evolutionary process". Physica A. 388 (6): 851–62. Bibcode:2009PhyA..388..851S. doi:10.1016/j.physa.2008.12.004.

91. Robson B, Vaithilingam A (2008). "Protein folding revisited". Progress in Molecular Biology and Translational Science. 84: 161–202. doi:10.1016/S0079-6603(08)00405-4. PMID 19121702.

92. Dill, Ken A.; MacCallum, Justin L. (2012-11-23). "The Protein-Folding Problem, 50 Years On". Science. 338 (6110): 1042–46. doi:10.1126/science.1219021. ISSN 0036-8075. PMID 23180855.

93. Fersht, Alan R. (2000-02-15). "Transition-state structure as a unifying basis in protein-folding mechanisms: Contact order, chain topology, stability, and the extended nucleus mechanism". Proceedings of the National Academy of Sciences. 97 (4): 1525–29. doi:10.1073/pnas.97.4.1525. ISSN 0027-8424. PMC 26468 . PMID 10677494.

94. Rizzuti B, Daggett V (Mar 2013). "Using simulations to provide the framework for experimental protein folding studies". Archives of Biochemistry and Biophysics. 531 (1-2): 128–35. doi:10.1016/j.abb.2012.12.015. PMID 23266569.

95. Schaefer M, Bartels C, Karplus M (Dec 1998). "Solution conformations and thermodynamics of structured peptides: molecular dynamics simulation with an implicit solvation model". Journal of Molecular Biology. 284 (3): 835–48. doi:10.1006/jmbi.1998.2172. PMID 9826519.

96. "Fragment-based Protein Folding Simulations".

97. "Protein folding" (by Molecular Dynamics).
98. Kmiecik, Sebastian; Gront, Dominik; Kolinski, Michal; Wieteska, Lukasz; Dawid, Aleksandra Elzbieta; Kolinski, Andrzej (2016-06-22). "Coarse-Grained Protein Models and Their Applications". Chemical Reviews. doi:10.1021/acs.chemrev.6b00163. ISSN 0009-2665.
99. Kmiecik S, Kolinski A (Jul 2007). "Characterization of protein-folding pathways by reduced-space modeling". Proceedings of the National Academy of Sciences of the United States of America. 104 (30): 12330–35. Bibcode:2007PNAS..10412330K. doi:10.1073/pnas.0702265104. PMC 1941469 . PMID 17636132.
100. Adhikari AN, Freed KF, Sosnick TR (Oct 2012). "De novo prediction of protein folding pathways and structure using the principle of sequential stabilization". Proceedings of the National Academy of Sciences of the United States of America. 109 (43): 17442–47. doi:10.1073/pnas.1209000109. PMC 3491489 . PMID 23045636.
101. Lindorff-Larsen K, Piana S, Dror RO, Shaw DE (2011). "How fast-folding proteins fold". Science. 334 (6055): 517–20. doi:10.1126/science.1208351. PMID 22034434.
102. Cowtan, Kevin (2001). "Phase Problem in X-ray Crystallography, and Its Solution" (PDF). Macmillan Publishers Ltd, Nature Publishing Group. Retrieved November 3, 2016.
103. Drenth, Jan (2007-04-05). Principles of Protein X-Ray Crystallography. Springer Science & Business Media. ISBN 9780387337463.
104. G., Taylor, (2003-11-01). "The phase problem". Acta Crystallographica Section D: Biological Crystallography. 59 (11): 1881–1890. doi:10.1107/S0907444903017815. ISSN 0907-4449.
105. Bedouelle, Hugues (Feb 2016). "Principles and equations for measuring and interpreting protein stability: From monomer to tetramer". Biochimie. 121: 29–37. doi:10.1016/j.biochi.2015.11.013. PMID 26607240.
106. Monsellier, E; Bedouelle, H (Sep 2005). "Quantitative measurement of protein stability from unfolding equilibria monitored with the fluorescence maximum wavelength". Protein Eng Des Sel. 18 (9): 445–56. doi:10.1093/protein/gzi046. PMID 16087653.
107. Park, YC; Bedouelle, H (Jul 1998). "Dimeric tyrosyl-tRNA synthetase from Bacillus stearothermophilus unfolds through a monomeric intermediate. A quantitative analysis under equilibrium conditions". J Biol Chem. 273 (29): 18052–59. doi:10.1074/jbc.273.29.18052. PMID 9660761.
108. Ould-Abeih, MB; Petit-Topin, I; Zidane, N; Baron, B; Bedouelle, H (Jun 2012). "Multiple folding states and disorder of ribosomal protein SA, a membrane receptor for laminin, anticarcinogens, and pathogens". Biochemistry. 51 (24): 4807–21. doi:10.1021/bi300335r. PMID 22640394.
109. Royer, CA (May 2006). "Probing protein folding and conformational transitions with fluorescence". Chem Rev. 106 (5): 1769–84. doi:10.1021/cr0404390. PMID 16683754.

110. Beatrice M.P. Huyghues-Despointes, C. Nick Pace, S. Walter Englander, and J. Martin Scholtz. "Measuring the Conformational Stability of a Protein by Hydrogen Exchange." Methods in Molecular Biology. Kenneth P. Murphy Ed. Humana Press, Totowa, New Jersey, 2001. pp. 69–92

111. Cross GH, Freeman NJ, Swann MJ (2008). "Dual Polarization Interferometry: A Real-Time Optical Technique for Measuring (Bio)molecular Orientation, Structure and Function at the Solid/Liquid Interface". Handbook of Biosensors and Biochips. doi:10.1002/9780470061565.hbb055.

112. Bu Z, Cook J, Callaway DJ (Sep 2001). "Dynamic regimes and correlated structural dynamics in native and denatured alpha-lactalbumin". Journal of Molecular Biology. 312 (4): 865–73. doi:10.1006/jmbi.2001.5006. PMID 11575938.

113. Minde DP, Maurice MM, Rüdiger SG (2012). "Determining biophysical protein stability in lysates by a fast proteolysis assay, FASTpp". PloS One. 7 (10): e46147. Bibcode:2012PLoSO...746147M. doi:10.1371/journal.pone.0046147. PMC 3463568 . PMID 23056252.

114. Park C, Marqusee S (Mar 2005). "Pulse proteolysis: a simple method for quantitative determination of protein stability and ligand binding". Nature Methods. 2 (3): 207–12. doi:10.1038/nmeth740. PMID 15782190.

115. Mashaghi A, Kramer G, Lamb DC, Mayer MP, Tans SJ (2014). "Chaperone action at the single-molecule level". Chemical Reviews. 114 (1): 660–76. doi:10.1021/cr400326k. PMID 24001118.

116. Jagannathan B, Marqusee S (Nov 2013). "Protein folding and unfolding under force". Biopolymers. 99 (11): 860–69. doi:10.1002/bip.22321. PMC 4065244 . PMID 23784721.

117. Jakobi AJ, Mashaghi A, Tans SJ, Huizinga EG. Calcium modulates force sensing by the von Willebrand factor A2 domain. Nature Commun. 2011 Jul 12;2:385.

118. Jagannathan B, Elms PJ, Bustamante C, Marqusee S (Oct 2012). "Direct observation of a force-induced switch in the anisotropic mechanical unfolding pathway of a protein". Proceedings of the National Academy of Sciences of the United States of America. 109 (44): 17820–25. doi:10.1073/pnas.1201800109. PMC 3497811 . PMID 22949695.

119. Chaudhuri, Tapan K.; Paul, Subhankar (2006-04-01). "Protein-misfolding diseases and chaperone-based therapeutic approaches". FEBS Journal. 273 (7): 1331–1349. doi:10.1111/j.1742-4658.2006.05181.x. ISSN 1742-4658.

120. Soto C., Estrada L., and Castilla J. 2006. Amyloids, prions and the inherent infectious nature of misfolded protein aggregates. Trends in Biochemical Sciences. 31(3):150-155. Retrieved on Dec 1st from http://www.sciencedirect.com/science/article/pii/S0968000406000235

121. Hammarström P, Wiseman RL, Powers ET, Kelly JW (2003). "Prevention of transthyretin amyloid disease by changing protein misfolding energetics". Science. 299 (5607): 713–16. doi:10.1126/science.1079589. PMID 12560553.

122. Chiti F, Dobson CM (2006). "Protein misfolding, functional amyloid, and human disease". Annual Review of Biochemistry. 75: 333–66. doi:10.1146/annurev.biochem.75.101304.123901. PMID 16756495.

123. Johnson SM, Wiseman RL, Sekijima Y, Green NS, Adamski-Werner SL, Kelly JW (2005). "Native state kinetic stabilization as a strategy to ameliorate protein misfolding diseases: a focus on the transthyretin amyloidoses". Accounts of Chemical Research. 38 (12): 911–21. doi:10.1021/ar020073i. PMID 16359163.

124. "Structural Biochemistry/Proteins/Protein Folding - Wikibooks, open books for an open world". en.wikibooks.org. Retrieved 2016-11-05.

125. Levinthal, Cyrus (1968). "Are there pathways for protein folding?"(PDF). Journal de Chimie Physique et de Physico-Chimie Biologique. 65: 44–45. Archived from the original (PDF) on 2009-09-02.

126. Kim PS, Baldwin RL (1990). "Intermediates in the folding reactions of small proteins". Annual Review of Biochemistry. 59 (1): 631–60. doi:10.1146/annurev.bi.59.070190.003215. PMID 2197986.

127. Jackson SE (August 1998). "How do small single-domain proteins fold?". Folding & Design. 3 (4): R81–91. doi:10.1016/S1359-0278(98)00033-9. PMID 9710577.[dead link]

128. Kubelka J, Hofrichter J, Eaton WA (Feb 2004). "The protein folding 'speed limit'". Current Opinion in Structural Biology. 14 (1): 76–88. doi:10.1016/j.sbi.2004.01.013. PMID 15102453.

CHAPTER 6

Enzymes: Structure and Mechanism

Chapter Contents

- **Introduction**
- **Etymology And History**
- **Structure**
- **Mechanism**
- **Thermodynamics**
- **Kinetics**
- **Biological Function**

Enzymes are proteins which enable chemical reactions in the body to occur many times more quickly than they otherwise would. In some cases, a reaction will not take place at all unless a specific enzyme is present. Along with pH, temperature and substrate concentration, enzyme concentration has an important effect on the reaction process. Assuming the amount of the substance, or substrate, taking part in the chemical reaction stays the same, a low concentration can lead to a chemical reaction occurring slowly. On the other hand, a high concentration can speed up a reaction, until a maximum possible rate is reached.

INTRODUCTION

Enzymes are macromolecular biological catalysts. Enzymes accelerate, or catalyze, chemical reactions. The molecules at the beginning of the process upon which enzymes may act are called substrates and the enzyme converts these into different molecules, called products. Almost all metabolic processes in the cell need enzymes in order to occur at rates fast enough to sustain life.[:8.1] The set of enzymes made in a cell determines which metabolic pathways occur in that cell. The study of enzymes is called *enzymology*.

Enzymes are known to catalyze more than 5,000 biochemical reaction types. Most enzymes are proteins, although a few are catalytic RNA molecules. Enzymes' specificity comes from their unique three-dimensional structures.

Like all catalysts, enzymes increase the rate of a reaction by lowering its activation energy. Some enzymes can make their conversion of substrate to product occur many millions of times faster. An extreme example is orotidine 5'-phosphate decarboxylase, which allows a reaction that would otherwise take millions of years to occur in milliseconds. Chemically, enzymes are like any catalyst and are not consumed in chemical reactions, nor do they alter the equilibrium of a reaction. Enzymes differ from most other catalysts by being much more specific. Enzyme activity can be affected by other molecules: inhibitors are molecules that decrease enzyme activity, and activators are molecules that increase activity. Many drugs and poisons are enzyme inhibitors. An enzyme's activity decreases markedly outside its optimal temperature and pH.

Some enzymes are used commercially, for example, in the synthesis of antibiotics. Some household products use enzymes to speed up chemical reactions: enzymes in biological washing powders break down protein, starch or fat stains on clothes, and enzymes in meat tenderizer break down proteins into smaller molecules, making the meat easier to chew.

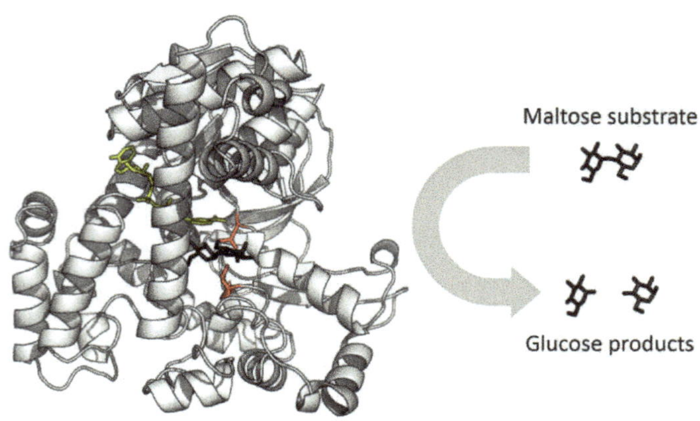

The enzyme glucosidase converts sugar maltose to two glucose sugars. Active site residues in red, maltose substrate in black, and NAD cofactor in yellow. (PDB: 1OBB)

ETYMOLOGY AND HISTORY

Eduard Buchner

By the late 17th and early 18th centuries, the digestion of meat by stomach secretions and the conversion of starch to sugars by plant extracts and saliva were known but the mechanisms by which these occurred had not been identified.

French chemist Anselme Payen was the first to discover an enzyme, diastase, in 1833. A few decades later, when studying the fermentation of sugar to alcohol by yeast, Louis Pasteur concluded that this fermentation was caused by a vital force contained within the yeast cells called "ferments", which were thought to function only within living organisms. He wrote that "alcoholic fermentation is an act correlated with the life and organization of the yeast cells, not with the death or putrefaction of the cells."

In 1877, German physiologist Wilhelm Kühne (1837–1900) first used the term *enzyme*, which comes from Greek ἔνζυμον, "leavened", to describe this process. The word *enzyme* was used later to refer to nonliving substances such as pepsin, and the word *ferment* was used to refer to chemical activity produced by living organisms.

Eduard Buchner submitted his first paper on the study of yeast extracts in 1897. In a series of experiments at the University of Berlin, he found that sugar was fermented by yeast extracts even when there were no living yeast cells in the mixture. He named the enzyme that brought about the fermentation of sucrose "zymase". In 1907, he received the Nobel Prize in Chemistry for "his discovery of cell-free fermentation". Following Buchner's example, enzymes are usually named according to the reaction they carry out: the suffix *-ase* is combined with the name of the substrate (e.g., lactase is the enzyme that cleaves lactose) or to the type of reaction (e.g., DNA polymerase forms DNA polymers).

The biochemical identity of enzymes was still unknown in the early 1900s. Many scientists observed that enzymatic activity was associated with proteins, but others (such as Nobel laureate Richard Willstätter) argued that proteins were merely carriers for the true enzymes and that proteins *per se* were incapable of catalysis. In 1926, James B. Sumner showed that the enzyme urease was a pure protein and crystallized it; he did likewise for the enzyme catalase in 1937. The conclusion that pure proteins can be enzymes was definitively demonstrated by John Howard Northrop and Wendell Meredith Stanley, who worked on the digestive enzymes pepsin (1930), trypsin and chymotrypsin. These three scientists were awarded the 1946 Nobel Prize in Chemistry.

The discovery that enzymes could be crystallized eventually allowed their structures to be solved by x-ray crystallography. This was first done for lysozyme, an enzyme found in tears, saliva and egg whites that digests the coating of some bacteria; the structure was solved by a group led by David Chilton Phillips and published in 1965. This high-resolution structure of lysozyme marked the beginning of the field of structural biology and the effort to understand how enzymes work at an atomic level of detail.

Naming Conventions

An enzyme's name is often derived from its substrate or the chemical reaction it catalyzes, with the word ending in *-ase*.[:8.1.3] Examples are lactase, alcohol dehydrogenase and DNA polymerase. Different enzymes that catalyze the same chemical reaction are called isozymes.

The International Union of Biochemistry and Molecular Biology have developed a nomenclature for enzymes, the EC numbers; each enzyme is described by a sequence of four numbers preceded by "EC". The first number broadly classifies the enzyme based on its mechanism.

The top-level classification is:

- EC 1, Oxidoreductases: catalyze oxidation/reduction reactions
- EC 2, Transferases: transfer a functional group (*e.g.* a methyl or phosphate group)
- EC 3, Hydrolases: catalyze the hydrolysis of various bonds
- EC 4, Lyases: cleave various bonds by means other than hydrolysis and oxidation
- EC 5, Isomerases: catalyze isomerization changes within a single molecule
- EC 6, Ligases: join two molecules with covalent bonds.

These sections are subdivided by other features such as the substrate, products, and chemical mechanism. An enzyme is fully specified by four numerical designations. For example, hexokinase (EC 2.7.1.1) is a transferase (EC 2) that adds a phosphate group (EC 2.7) to a hexose sugar, a molecule containing an alcohol group (EC 2.7.1).

STRUCTURE

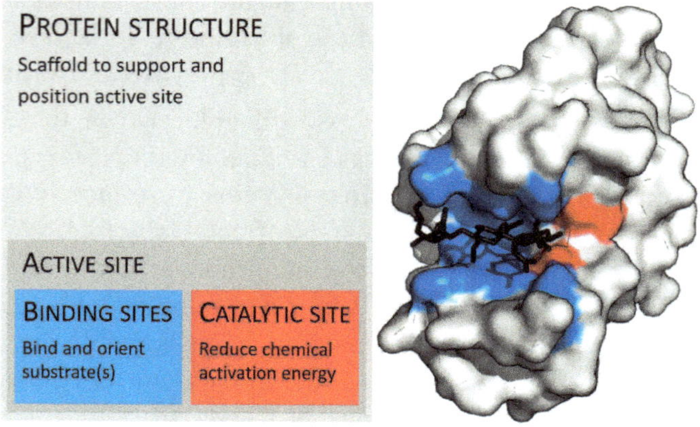

Organisation of enzyme structure and lysozyme example. Binding sites in blue, catalytic site in red and peptidoglycan substrate in black. (PDB: 9LYZ)

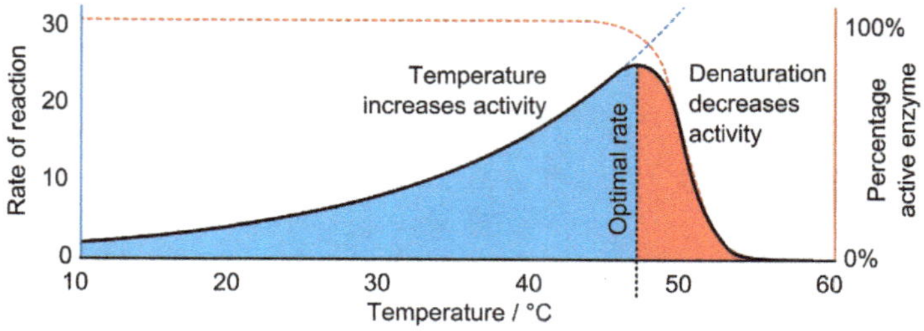

Enzyme activity initially increases with temperature (Q10 coefficient) until the enzyme's structure unfolds (denaturation), leading to an optimal rate of reaction at an intermediate temperature.

Enzymes are generally globular proteins, acting alone or in larger complexes. Like all proteins, enzymes are linear chains of amino acids that fold to produce a three-dimensional structure. The sequence of the amino acids specifies the structure which in turn determines the catalytic activity of the enzyme. Although structure determines function, a novel enzyme's activity cannot yet be predicted from its structure alone. Enzyme structures unfold (denature) when heated or exposed to chemical denaturants and this disruption to the structure typically causes a loss of activity. Enzyme denaturation is normally linked to temperatures above a species' normal level; as a result, enzymes from bacteria living in volcanic environments such as hot springs are prized by industrial users for their ability to function at high temperatures, allowing enzyme-catalysed reactions to be operated at a very high rate.

Enzymes are usually much larger than their substrates. Sizes range from just 62 amino acid residues, for the monomer of 4-oxalocrotonate tautomerase,to over 2,500 residues in the animal fatty acid synthase. Only a small portion of their structure (around 2–4 amino acids) is directly involved in catalysis: the catalytic site. This catalytic site is located next to one or more binding sites where residues orient the substrates. The catalytic site and binding site together comprise the enzyme's active site. The remaining majority of the enzyme structure serves to maintain the precise orientation and dynamics of the active site.

In some enzymes, no amino acids are directly involved in catalysis; instead, the enzyme contains sites to bind and orient catalytic cofactors. Enzyme structures may also contain allosteric sites where the binding of a small molecule causes a conformational change that increases or decreases activity.

A small number of RNA-based biological catalysts called ribozymes exist, which again can act alone or in complex with proteins. The most common of these is the ribosome which is a complex of protein and catalytic RNA components.[:2.2]

MECHANISM

Substrate Binding

Enzymes must bind their substrates before they can catalyse any chemical reaction. Enzymes are usually very specific as to what substrates they bind and then the chemical reaction catalysed. Specificity is achieved by binding pockets with complementary shape, charge and hydrophilic/hydrophobic characteristics to the substrates. Enzymes can therefore distinguish between very similar substrate molecules to be chemoselective, regioselective and stereospecific.

Some of the enzymes showing the highest specificity and accuracy are involved in the copying and expression of the genome. Some of these enzymes have "proof-reading" mechanisms. Here, an enzyme such as DNA polymerase catalyzes a reaction in a first step and then checks that the product is correct in a second step. This two-step process results in average error rates of less than 1 error in 100 million reactions in high-fidelity mammalian polymerases.[:5.3.1] Similar proofreading mechanisms are also found in RNA polymerase, aminoacyl tRNA synthetasesand ribosomes.

Conversely, some enzymes display enzyme promiscuity, having broad specificity and acting on a range of different physiologically relevant substrates. Many enzymes possess small side activities which arose fortuitously (i.e. neutrally), which may be the starting point for the evolutionary selection of a new function.

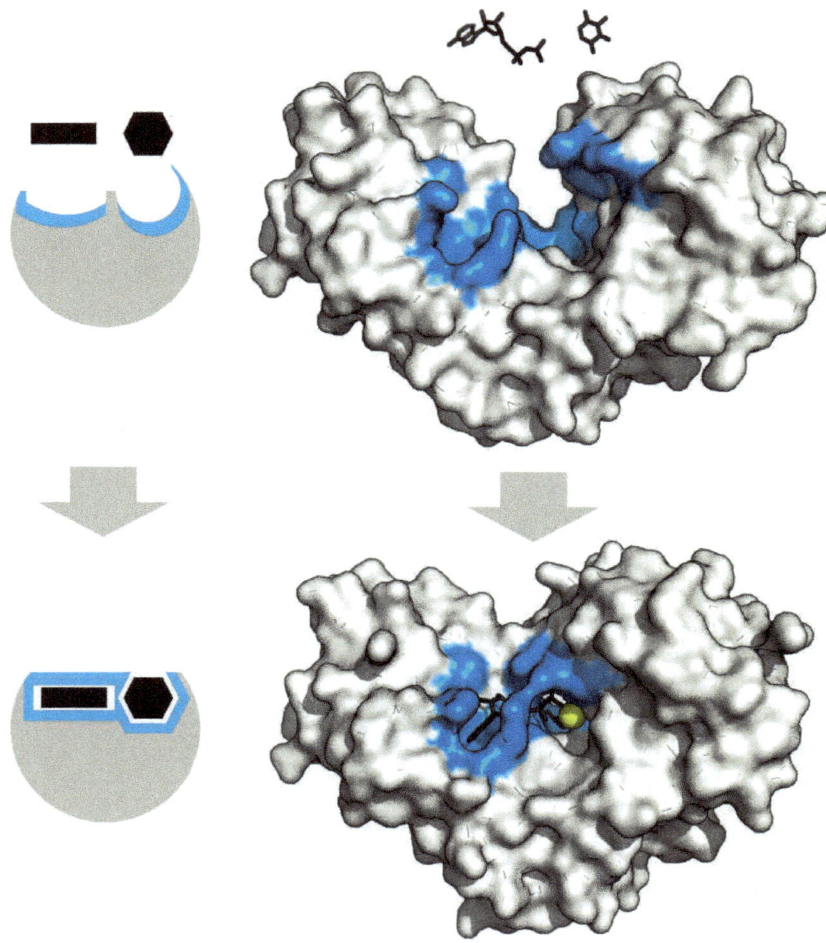

Enzyme changes shape by induced fit upon substrate binding to form enzyme-substrate complex. Hexokinase has a large induced fit motion that closes over the substrates adenosine triphosphate and xylose. Binding sites in blue, substrates in black and Mg^{2+} cofactor in yellow. (PDB: 2E2N, 2E2Q)

"Lock and Key" Model

To explain the observed specificity of enzymes, in 1894 Emil Fischer proposed that both the enzyme and the substrate possess specific complementary geometric shapes that fit exactly into one another. This is often referred to as "the lock and key" model.[:8.3.2] This early model explains enzyme specificity, but fails to explain the stabilization of the transition state that enzymes achieve.

Induced fit model

In 1958, Daniel Koshland suggested a modification to the lock and key model: since enzymes are rather flexible structures, the active site is continuously reshaped by interactions with the substrate as the substrate interacts with the enzyme. As a result, the substrate does not simply bind to a rigid active site; the amino acid side-chains that make up the active site are molded into the precise positions that enable the enzyme to perform its catalytic function. In some cases, such as glycosidases, the substrate molecule also changes shape slightly as it enters the active site. The active site continues to change until the substrate is completely bound, at which point the final shape and charge distribution is determined. Induced fit may enhance the fidelity of molecular recognition in the presence of competition and noise via the conformational proofreading mechanism.

Catalysis

Enzymes can accelerate reactions in several ways, all of which lower the activation energy ($\Delta G^{\ddagger}$, Gibbs free energy)

- By stabilizing the transition state:
    - Creating an environment with a charge distribution complementary to that of the transition state to lower its energy.
- By providing an alternative reaction pathway:
    - Temporarily reacting with the substrate, forming a covalent intermediate to provide a lower energy transition state.
- By destabilising the substrate ground state:
    - Distorting bound substrate(s) into their transition state form to reduce the energy required to reach the transition state.
    - By orienting the substrates into a productive arrangement to reduce the reaction entropy change. The contribution of this mechanism to catalysis is relatively small.

Enzymes may use several of these mechanisms simultaneously. For example, proteases such as trypsin perform covalent catalysis using a catalytic triad, stabilise charge build-up on the transition states using an oxyanion hole, complete hydrolysis using an oriented water substrate.

Dynamics

Enzymes are not rigid, static structures; instead they have complex internal dynamic motions – that is, movements of parts of the enzyme's structure such as individual amino acid residues, groups of residues forming a protein loop or unit of secondary structure, or even an entire protein domain. These motions give rise to a conformational ensemble of slightly different structures that interconvert with one another at equilibrium. Different states within this ensemble may be associated with different aspects of an enzyme's function. For example, different conformations of the enzyme dihydrofolate reductase are associated with the substrate binding, catalysis, cofactor release, and product release steps of the catalytic cycle.

Allosteric Modulation

Allosteric sites are pockets on the enzyme, distinct from the active site, that bind to molecules in the cellular environment. These molecules then cause a change in the conformation or dynamics of the enzyme that is transduced to the active site and thus affects the reaction rate of the enzyme. In this way, allosteric interactions can either inhibit or activate enzymes. Allosteric interactions with metabolites upstream or downstream in an enzyme's metabolic pathway cause feedback regulation, altering the activity of the enzyme according to the flux through the rest of the pathway.

Cofactors

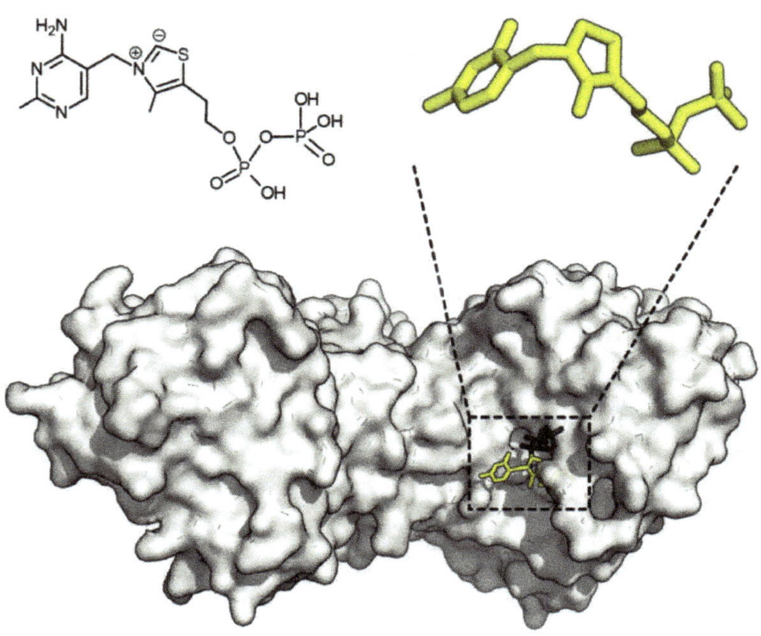

Chemical structure for thiamine pyrophosphate and protein structure of transketolase. Thiamine pyrophosphate cofactor in yellow and xylulose 5-phosphate substrate in black. (PDB: 4KXV)

Some enzymes do not need additional components to show full activity. Others require non-protein molecules called cofactors to be bound for activity. Cofactors can be either inorganic (e.g., metal ions and iron-sulfur clusters) or organic compounds (e.g., flavin and heme). Organic cofactors can be either coenzymes, which are released from the enzyme's active site during the reaction, or prosthetic groups, which are tightly bound to an enzyme. Organic prosthetic groups can be covalently bound (e.g., biotin in enzymes such as pyruvate carboxylase).

An example of an enzyme that contains a cofactor is carbonic anhydrase, which is shown in the ribbon diagram above with a zinc cofactor bound as part of its active site. These tightly bound ions or molecules are usually found in the active site and are involved in catalysis.[:8.1.1] For example, flavin and heme cofactors are often involved in redox reactions.[:17]

Enzymes that require a cofactor but do not have one bound are called *apoenzymes* or *apoproteins*. An enzyme together with the cofactor(s) required for activity is called a *holoenzyme* (or haloenzyme). The term *holoenzyme* can also be applied to enzymes that contain multiple protein subunits, such as the DNA polymerases; here the holoenzyme is the complete complex containing all the subunits needed for activity.[:8.1.1]

Coenzymes

Coenzymes are small organic molecules that can be loosely or tightly bound to an enzyme. Coenzymes transport chemical groups from one enzyme to another. Examples include NADH, NADPH and adenosine triphosphate (ATP). Some coenzymes, such as riboflavin, thiamine and folic acid, are vitamins, or compounds that cannot be synthesized by the body and must be acquired from the diet. The chemical groups carried include the hydride ion (H^-) carried by NAD or $NADP^+$, the phosphate group carried by adenosine triphosphate, the acetyl group carried by coenzyme A, formyl, methenyl or methyl groups carried by folic acid and the methyl group carried by S-adenosylmethionine.

Since coenzymes are chemically changed as a consequence of enzyme action, it is useful to consider coenzymes to be a special class of substrates, or second substrates, which are common to many different enzymes. For example, about 1000 enzymes are known to use the coenzyme NADH.

Coenzymes are usually continuously regenerated and their concentrations maintained at a steady level inside the cell. For example, NADPH is regenerated through the pentose phosphate pathway and *S*-adenosylmethionine by methionine adenosyltransferase. This continuous regeneration means that small amounts of coenzymes can be used very intensively. For example, the human body turns over its own weight in ATP each day.

THERMODYNAMICS

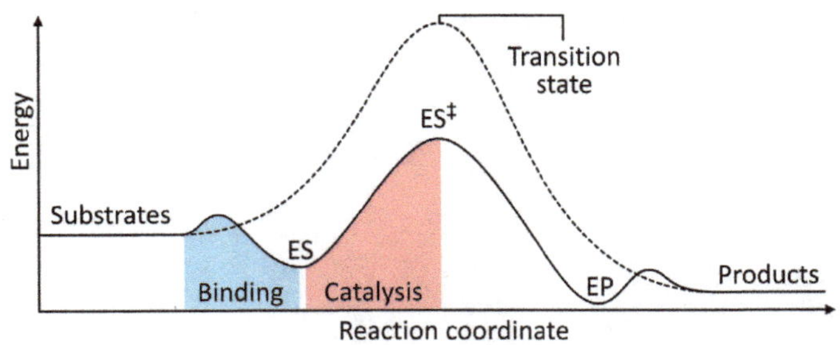

The energies of the stages of a chemical reaction. Uncatalysed (dashed line), substrates need a lot of activation energy to reach a transition state, which then decays into lower-energy products. When enzyme catalysed (solid line), the enzyme binds the substrates (ES), then stabilizes the transition state (ES‡) to reduce the activation energy required to produce products (EP) which are finally released.

As with all catalysts, enzymes do not alter the position of the chemical equilibrium of the reaction. In the presence of an enzyme, the reaction runs in the same direction as it would without the enzyme, just more quickly.[:8.2.3] For example, carbonic anhydrase catalyzes its reaction in either direction depending on the concentration of its reactants:

$$CO_2 + H_2O \xrightarrow{\text{Carbonic anhydrase}} H_2CO_3 \quad \text{(in tissues; high } CO_2 \text{ concentration)}$$

$$H_2CO_3 \xrightarrow{\text{Carbonic anhydrase}} CO_2 + H_2O \quad \text{(in lungs; low } CO_2 \text{ concentration)}$$

The rate of a reaction is dependent on the activation energy needed to form the transition state which then decays into products. Enzymes increase reaction rates by lowering the energy of the transition state. First, binding forms a low energy enzyme-substrate complex (ES). Secondly the enzyme stabilises the transition state such that it requires less energy to achieve compared to the uncatalyzed reaction (ES‡). Finally the enzyme-product complex (EP) dissociates to release the products.[:8.3]

Enzymes can couple two or more reactions, so that a thermodynamically favorable reaction can be used to "drive" a thermodynamically unfavourable one so that the combined energy of the products is lower than the substrates. For example, the hydrolysis of ATP is often used to drive other chemical reactions.

KINETICS

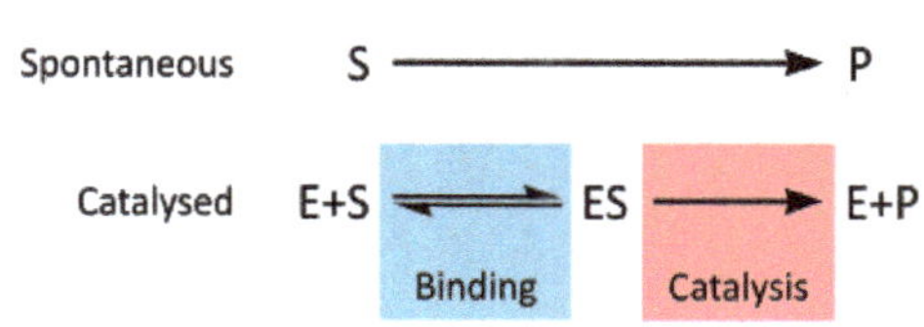

A chemical reaction mechanism with or without enzyme catalysis. The enzyme (E) binds substrate (S) to produce product (P).

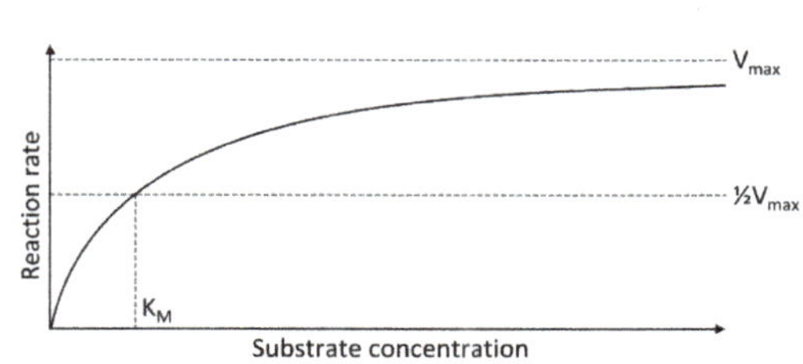

Saturation curve for an enzyme reaction showing the relation between the substrate concentration and reaction rate.

Enzyme kinetics is the investigation of how enzymes bind substrates and turn them into products. The rate data used in kinetic analyses are commonly obtained from enzyme assays. In 1913 Leonor Michaelis and Maud Leonora Menten proposed a quantitative theory of enzyme kinetics, which is referred to as Michaelis–Menten kinetics. The major contribution of Michaelis and Menten was to think of enzyme reactions in two stages. In the first, the substrate binds reversibly to the enzyme, forming the enzyme-substrate complex. This is sometimes called the Michaelis-Menten complex in their honor. The enzyme then catalyzes the chemical step in the reaction and releases the product. This work was further developed by G. E. Briggs and J. B. S. Haldane, who derived kinetic equations that are still widely used today.

Enzyme rates depend on solution conditions and substrate concentration. To find the maximum speed of an enzymatic reaction, the substrate concentration is increased until a constant rate of product formation is seen. This is shown in the saturation curve on the right. Saturation happens because, as substrate concentration increases, more and more of the free enzyme is converted into the substrate-bound ES complex. At the maximum reaction rate (V_{max}) of the enzyme, all the enzyme active sites are bound to substrate, and the amount of ES complex is the same as the total amount of enzyme.[:8.4]

V_{max} is only one of several important kinetic parameters. The amount of substrate needed to achieve a given rate of reaction is also important. This is given by the Michaelis-Menten constant (K_m), which is the substrate concentration required for an enzyme to reach one-half its maximum reaction rate; generally, each enzyme has a characteristic K_m for a given substrate. Another useful constant is k_{cat}, also called the *turnover number*, which is the number of substrate molecules handled by one active site per second.[:8.4]

The efficiency of an enzyme can be expressed in terms of k_{cat}/K_m. This is also called the specificity constant and incorporates the rate constants for all steps in the reaction up to and including the first irreversible step. Because the specificity constant reflects both affinity and catalytic ability, it is useful for comparing different enzymes against each other, or the same enzyme with different substrates. The theoretical maximum for the specificity constant is called the diffusion limit and is about 10^8 to 10^9 (M^{-1} s^{-1}). At this point every collision of the enzyme with its substrate will result in catalysis, and the rate of product formation is not limited by the reaction rate but by the diffusion rate. Enzymes with this property are called *catalytically perfect* or *kinetically perfect*. Example of such enzymes are triose-phosphate isomerase, carbonic anhydrase, acetylcholinesterase, catalase, fumarase, β-lactamase, and superoxide dismutase.[:8.4.2] The turnover of such enzymes can reach several million reactions per second.[:9.2]

Michaelis–Menten kinetics relies on the law of mass action, which is derived from the assumptions of free diffusion and thermodynamically driven random collision. Many biochemical or cellular processes deviate significantly from these conditions, because of macromolecular crowding and constrained molecular movement. More recent, complex extensions of the model attempt to correct for these effects.

Inhibition

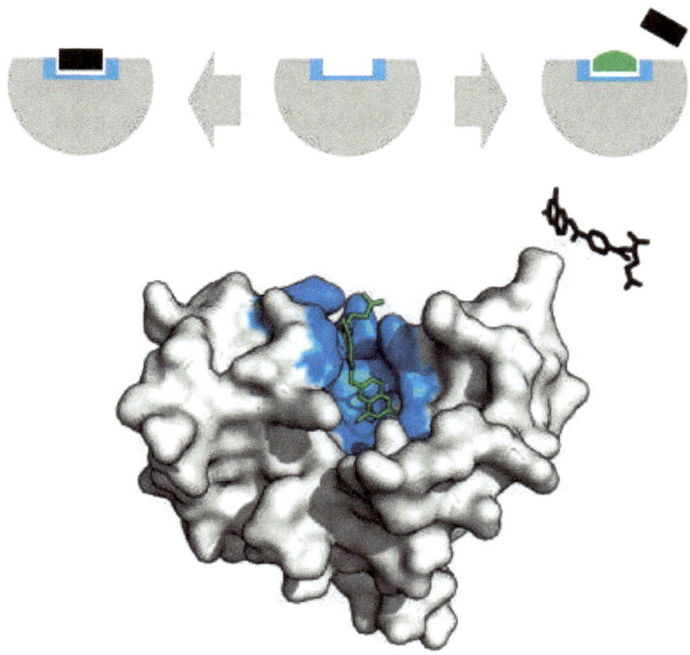

An enzyme binding site that would normally bind substrate can alternatively bind a competitive inhibitor, preventing substrate access. Dihydrofolate reductase is inhibited by methotrexate which prevents binding of its substrate, folic acid. Binding site in blue, inhibitor in green, and substrate in black. (PDB: 4QI9)

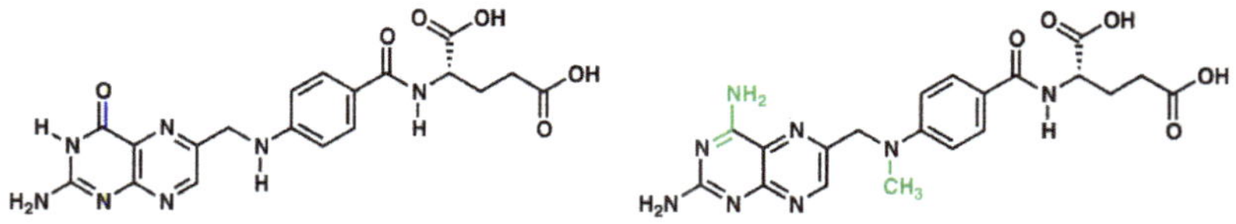

The coenzyme folic acid (left) and the anti-cancer drug methotrexate (right) are very similar in structure (differences show in green). As a result, methotrexate is a competitive inhibitor of many enzymes that use folates.

Enzyme reaction rates can be decreased by various types of enzyme inhibitors.[:73–74]

Types of Inhibition

Competitive

A competitive inhibitor and substrate cannot bind to the enzyme at the same time. Often competitive inhibitors strongly resemble the real substrate of the enzyme. For example, the drug methotrexate is a competitive inhibitor of the enzyme dihydrofolate reductase, which catalyzes the reduction of dihydrofolate to tetrahydrofolate. The similarity between the structures of dihydrofolate and this drug are shown in the accompanying figure. This type of inhibition can be overcome with high substrate concentration. In some cases, the inhibitor can bind to a site other than the binding-site of the usual substrate and exert an allosteric effect to change the shape of the usual binding-site.

Non-Competitive

A non-competitive inhibitor binds to a site other than where the substrate binds. The substrate still binds with its usual affinity and hence K_m remains the same. However the inhibitor reduces the catalytic efficiency of the enzyme so that V_{max} is reduced. In contrast to competitive inhibition, non-competitive inhibition cannot be overcome with high substrate concentration.[:76–78]

Uncompetitive

An uncompetitive inhibitor cannot bind to the free enzyme, only to the enzyme-substrate complex; hence, these types of inhibitors are most effective at high substrate concentration. In the presence of the inhibitor, the enzyme-substrate complex is inactive.[:78] This type of inhibition is rare.

Mixed

A mixed inhibitor binds to an allosteric site and the binding of the substrate and the inhibitor affect each other. The enzyme's function is reduced but not eliminated when bound to the inhibitor. This type of inhibitor does not follow the Michaelis-Menten equation.[:76–78]

Irreversible

An irreversible inhibitor permanently inactivates the enzyme, usually by forming a covalent bond to the protein. Penicillin and aspirin are common drugs that act in this manner.

Functions of Inhibitors

In many organisms, inhibitors may act as part of a feedback mechanism. If an enzyme produces too much of one substance in the organism, that substance may act as an inhibitor for the enzyme at the beginning of the pathway that produces it, causing production of the substance to slow down or stop when there is sufficient amount. This is a form of negative feedback. Major metabolic pathways such as the citric acid cycle make use of this mechanism.[:17.2.2]

Since inhibitors modulate the function of enzymes they are often used as drugs. Many such drugs are reversible competitive inhibitors that resemble the enzyme's native substrate, similar to methotrexate above; other well-known examples include statins used to treat high cholesterol, and protease inhibitors used to treat retroviral infections such as HIV. A common example of an irreversible inhibitor that is used as a drug is aspirin, which inhibits the COX-1 and COX-2 enzymes that produce the inflammation messenger prostaglandin. Other enzyme inhibitors are poisons. For example, the poison cyanide is an irreversible enzyme inhibitor that combines with the copper and iron in the active site of the enzyme cytochrome c oxidase and blocks cellular respiration.

BIOLOGICAL FUNCTION

Enzymes serve a wide variety of functions inside living organisms. They are indispensable for signal transduction and cell regulation, often via kinases and phosphatases. They also generate movement, with myosin hydrolyzing ATP to generate muscle contraction, and also transport cargo around the cell as part of the cytoskeleton. Other ATPases in the cell membrane are ion pumps involved in active transport. Enzymes are also involved in more exotic functions, such as luciferase generating light in fireflies. Viruses can also contain enzymes for infecting cells, such as the HIV integrase and reverse transcriptase, or for viral release from cells, like the influenza virus neuraminidase.

An important function of enzymes is in the digestive systems of animals. Enzymes such as amylases and proteases break down large molecules (starch or proteins, respectively) into smaller ones, so they can be absorbed by the intestines. Starch molecules, for example, are too large to be absorbed from the intestine, but enzymes hydrolyze the starch chains into smaller molecules such as maltose and eventually glucose, which can then be absorbed. Different enzymes digest different food substances. In ruminants, which have herbivorous diets, microorganisms in the gut produce another enzyme, cellulase, to break down the cellulose cell walls of plant fiber.

Metabolism

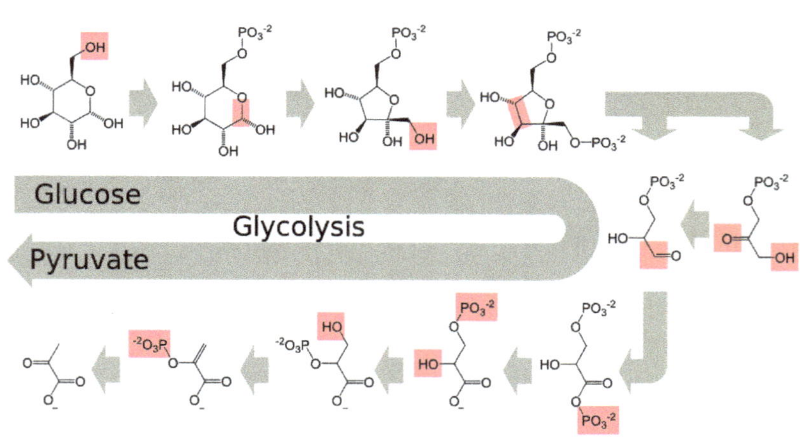

The metabolic pathway of glycolysis releases energy by converting glucose to pyruvate by via a series of intermediate metabolites. Each chemical modification (red box) is performed by a different enzyme.

Several enzymes can work together in a specific order, creating metabolic pathways.[:30.1] In a metabolic pathway, one enzyme takes the product of another enzyme as a substrate.

After the catalytic reaction, the product is then passed on to another enzyme. Sometimes more than one enzyme can catalyze the same reaction in parallel; this can allow more complex regulation: with, for example, a low constant activity provided by one enzyme but an inducible high activity from a second enzyme.

Enzymes determine what steps occur in these pathways. Without enzymes, metabolism would neither progress through the same steps and could not be regulated to serve the needs of the cell. Most central metabolic pathways are regulated at a few key steps, typically through enzymes whose activity involves the hydrolysis of ATP. Because this reaction releases so much energy, other reactions that are thermodynamically unfavorable can be coupled to ATP hydrolysis, driving the overall series of linked metabolic reactions.

Control of Activity

There are five main ways that enzyme activity is controlled in the cell.

Regulation

Enzymes can be either activated or inhibited by other molecules. For example, the end product(s) of a metabolic pathway are often inhibitors for one of the first enzymes of the pathway (usually the first irreversible step, called committed step), thus regulating the amount of end product made by the pathways. Such a regulatory mechanism is called a negative feedback mechanism, because the amount of the end product produced is regulated by its own concentration. Negative feedback mechanism can effectively adjust the rate of synthesis of intermediate metabolites according to the demands of the cells. This helps with effective allocations of materials and energy economy, and it prevents the excess manufacture of end products. Like other homeostatic devices, the control of enzymatic action helps to maintain a stable internal environment in living organisms.

Post-Translational Modification

Examples of post-translational modification include phosphorylation, myristoylation and glycosylation.[:149–69] For example, in the response to insulin, the phosphorylation of multiple enzymes, including glycogen synthase, helps control the synthesis or degradation of glycogen and allows the cell to respond to changes in blood sugar. Another example of post-translational modification is the cleavage of the polypeptide chain. Chymotrypsin, a digestive protease, is produced in inactive form as chymotrypsinogen in the pancreas and transported in this form to the stomach where it is activated. This stops the enzyme from digesting the pancreas or other tissues before it enters the gut. This type of inactive precursor to an enzyme is known as a zymogen or proenzyme.

Quantity

Enzyme production (transcription and translation of enzyme genes) can be enhanced or diminished by a cell in response to changes in the cell's environment. This form of gene regulation is called enzyme induction. For example, bacteria may become resistant to antibiotics such as penicillin because enzymes called beta-lactamases are induced that hydrolyse the crucial beta-lactam ring within the penicillin molecule.Another example comes from enzymes in the liver called cytochrome P450 oxidases, which are important in drug metabolism. Induction or inhibition of these enzymes can cause drug interactions. Enzyme levels can also be regulated by changing the rate of enzyme degradation.[:30.1.1]

Subcellular Distribution

Enzymes can be compartmentalized, with different metabolic pathways occurring in different cellular compartments. For example, fatty acids are synthesized by one set of enzymes in the cytosol, endoplasmic reticulum and Golgi and used by a different set of enzymes as a source of energy in the mitochondrion, through β-oxidation. In addition, trafficking of the enzyme to different compartments may change the degree of protonation (cytoplasm neutral and lysosome acidic) or oxidative state [e.g., oxidized (periplasm) or reduced (cytoplasm)] which in turn affects enzyme activity.

Organ Specialization

In multicellular eukaryotes, cells in different organs and tissues have different patterns of gene expression and therefore have different sets of enzymes (known as isozymes) available for metabolic reactions. This provides a mechanism for regulating the overall metabolism of the organism. For example, hexokinase, the first enzyme in the glycolysis pathway, has a specialized form called glucokinase expressed in the liver and pancreas that has a lower affinity for glucose yet is more sensitive to glucose concentration. This enzyme is involved in sensing blood sugar and regulating insulin production.

Involvement in Disease

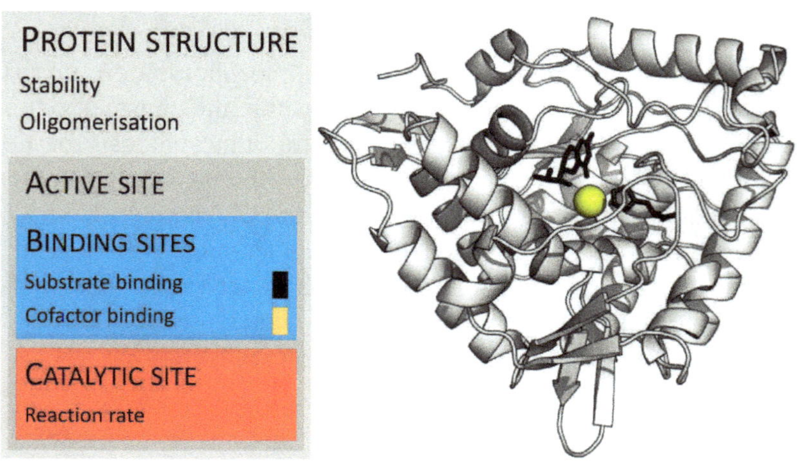

In phenylalanine hydroxylase over 300 different mutations throughout the structure cause phenylketonuria. Phenylalanine substrate and tetrahydrobiopterin coenzyme in black, and Fe^{2+}cofactor in yellow. (PDB: 1KW0)

Since the tight control of enzyme activity is essential for homeostasis, any malfunction (mutation, overproduction, underproduction or deletion) of a single critical enzyme can lead to a genetic disease. The malfunction of just one type of enzyme out of the thousands of types present in the human body can be fatal. An example of a fatal genetic disease due to enzyme insufficiency is Tay-Sachs disease, in which patients lack the enzyme hexosaminidase.

One example of enzyme deficiency is the most common type of phenylketonuria. Many different single amino acid mutations in the enzyme phenylalanine hydroxylase, which catalyzes the first step in the degradation of phenylalanine, result in build-up of phenylalanine and related products. Some mutations are in the active site, directly disrupting binding and catalysis, but many are far from the active site and reduce activity by destabilising the protein structure, or affecting correct oligomerisation. This can lead to intellectual disability if the disease is untreated. Another example is pseudocholinesterase deficiency, in which the body's ability to break down choline ester drugs is impaired. Oral administration of enzymes can be used to treat some functional enzyme deficiencies, such as pancreatic insufficiency and lactose intolerance.

Another way enzyme malfunctions can cause disease comes from germline mutations in genes coding for DNA repair enzymes. Defects in these enzymes cause cancer because cells are less able to repair mutations in their genomes. This causes a slow accumulation of mutations and results in the development of cancers. An example of such a hereditary cancer syndrome is xeroderma pigmentosum, which causes the development of skin cancers in response to even minimal exposure to ultraviolet light.

Industrial Applications

Enzymes are used in the chemical industry and other industrial applications when extremely specific catalysts are required. Enzymes in general are limited in the number of reactions they have evolved to catalyze and also by their lack of stability in organic solvents and at high temperatures. As a consequence, protein engineering is an active area of research and involves attempts to create new enzymes with novel properties, either through rational design or *in vitro* evolution. These efforts have begun to be successful, and a few enzymes have now been designed "from scratch" to catalyze reactions that do not occur in nature.

Application	Enzymes used	Uses
Biofuel industry	Cellulases	Break down cellulose into sugars that can be fermented to produce cellulosic ethanol.
	Ligninases	Pretreatment of biomass for biofuel production.
Biological detergent	Proteases, amylases, lipases	Remove protein, starch, and fat or oil stains from laundry and dishware.
	Mannanases	Remove food stains from the common food additive guar gum.
Brewing industry	Amylase, glucanases, proteases	Split polysaccharides and proteins in the malt.[:150–9]
	Betaglucanases	Improve the wort and beer filtration characteristics.[:545]
	Amyloglucosidase and pullulanases	Make low-calorie beer and adjust fermentability.[:575]
	Acetolactate decarboxylase (ALDC)	Increase fermentation efficiency by reducing diacetyl formation.
Culinary uses	Papain	Tenderize meat for cooking.
Dairy industry	Rennin	Hydrolyze protein in the manufacture of cheese.
	Lipases	Produce Camembert cheese and blue cheeses such as Roquefort.
Food processing	Amylases	Produce sugars from starch, such as in making high-fructose corn syrup.
	Proteases	Lower the protein level of flour, as in biscuit-making.
	Trypsin	Manufacture hypoallergenic baby foods.
	Cellulases, pectinases	Clarify fruit juices.
Molecular biology	Nucleases, DNA ligase and polymerases	Use restriction digestion and the polymerase chain reaction to create recombinant DNA.[:6.2]
Paper industry	Xylanases, hemicellulases and lignin peroxidases	Remove lignin from kraft pulp.
Personal care	Proteases	Remove proteins on contact lenses to prevent infections.
Starch industry	Amylases	Convert starch into glucose and various syrups.

REFERENCES

1. Dunaway-Mariano D (November 2008). "Enzyme function discovery". Structure (London, England : 1993). 16 (11): 1599–600. doi:10.1016/j.str.2008.10.001. PMID 19000810.
2. Petsko GA, Ringe D (2003). "Chapter 1: From sequence to structure". Protein structure and function. London: New Science. p. 27. ISBN 978-1405119221.
3. Chen LH, Kenyon GL, Curtin F, Harayama S, Bembenek ME, Hajipour G, Whitman CP (September 1992). "4-Oxalocrotonate tautomerase, an enzyme composed of 62 amino acid residues per monomer". The Journal of Biological Chemistry. 267 (25): 17716–21. PMID 1339435.
4. Smith S (December 1994). "The animal fatty acid synthase: one gene, one polypeptide, seven enzymes". FASEB Journal. 8 (15): 1248–59. PMID 8001737.
5. "The Catalytic Site Atlas". The European Bioinformatics Institute. Retrieved 4 April 2007.
6. Suzuki H (2015). "Chapter 7: Active Site Structure". How Enzymes Work: From Structure to Function. Boca Raton, FL: CRC Press. pp. 117–140. ISBN 978-981-4463-92-8.
7. Krauss G (2003). "The Regulations of Enzyme Activity". Biochemistry of Signal Transduction and Regulation (3rd ed.). Weinheim: Wiley-VCH. pp. 89–114. ISBN 9783527605767.
8. Jaeger KE, Eggert T (August 2004). "Enantioselective biocatalysis optimized by directed evolution". Current Opinion in Biotechnology. 15 (4): 305–13. doi:10.1016/j.copbio.2004.06.007. PMID 15358000.
9. Shevelev IV, Hübscher U (May 2002). "The 3' 5' exonucleases". Nature Reviews Molecular Cell Biology. 3 (5): 364–76. doi:10.1038/nrm804. PMID 11988770.
10. Zenkin N, Yuzenkova Y, Severinov K (July 2006). "Transcript-assisted transcriptional proofreading". Science. 313 (5786): 518–20. Bibcode:2006Sci...313..518Z. doi:10.1126/science.1127422. PMID 16873663.
11. Ibba M, Soll D (2000). "Aminoacyl-tRNA synthesis". Annual Review of Biochemistry. 69: 617–50. doi:10.1146/annurev.biochem.69.1.617. PMID 10966471.
12. Rodnina MV, Wintermeyer W (2001). "Fidelity of aminoacyl-tRNA selection on the ribosome: kinetic and structural mechanisms". Annual Review of Biochemistry. 70: 415–35. doi:10.1146/annurev.biochem.70.1.415. PMID 11395413.
13. Khersonsky O, Tawfik DS (2010). "Enzyme promiscuity: a mechanistic and evolutionary perspective". Annual Review of Biochemistry. 79: 471–505. doi:10.1146/annurev-biochem-030409-143718. PMID 20235827.
14. O'Brien PJ, Herschlag D (April 1999). "Catalytic promiscuity and the evolution of new enzymatic activities". Chemistry & Biology. 6 (4): R91–R105. doi:10.1016/S1074-5521(99)80033-7. PMID 10099128.
15. Fischer E (1894). "Einfluss der Configuration auf die Wirkung der Enzyme"

[Influence of configuration on the action of enzymes]. Berichte der Deutschen chemischen Gesellschaft zu Berlin (in German). 27 (3): 2985–93. doi:10.1002/cber.18940270364. From page 2992: "Um ein Bild zu gebrauchen, will ich sagen, dass Enzym und Glucosid wie Schloss und Schlüssel zu einander passen müssen, um eine chemische Wirkung auf einander ausüben zu können." (To use an image, I will say that an enzyme and a glucoside [i.e., glucose derivative] must fit like a lock and key, in order to be able to exert a chemical effect on each other.)

16. Cooper GM (2000). "Chapter 2.2: The Central Role of Enzymes as Biological Catalysts". The Cell: a Molecular Approach (2nd ed.). Washington (DC): ASM Press. ISBN 0-87893-106-6.
17. Koshland DE (February 1958). "Application of a Theory of Enzyme Specificity to Protein Synthesis". Proceedings of the National Academy of Sciences of the United States of America. 44 (2): 98–104. Bibcode:1958PNAS...44...98K. doi:10.1073/pnas.44.2.98. PMC 335371 . PMID 16590179.
18. Vasella A, Davies GJ, Böhm M (October 2002). "Glycosidase mechanisms". Current Opinion in Chemical Biology. 6 (5): 619–29. doi:10.1016/S1367-5931(02)00380-0. PMID 12413546.
19. Boyer R (2002). "Chapter 6: Enzymes I, Reactions, Kinetics, and Inhibition". Concepts in Biochemistry (2nd ed.). New York, Chichester, Weinheim, Brisbane, Singapore, Toronto.: John Wiley & Sons, Inc. pp. 137–8. ISBN 0-470-00379-0. OCLC 51720783.
20. Savir Y, Tlusty T (2007). Scalas E, ed. "Conformational proofreading: the impact of conformational changes on the specificity of molecular recognition" (PDF). PLoS ONE. 2 (5): e468. Bibcode:2007PLoSO...2..468S. doi:10.1371/journal.pone.0000468. PMC 1868595 . PMID 17520027.
21. Fersht A (1985). Enzyme Structure and Mechanism. San Francisco: W.H. Freeman. pp. 50–2. ISBN 0-7167-1615-1.
22. Warshel A, Sharma PK, Kato M, Xiang Y, Liu H, Olsson MH (August 2006). "Electrostatic basis for enzyme catalysis". Chemical Reviews. 106 (8): 3210–35. doi:10.1021/cr0503106. PMID 16895325.
23. Cox MM, Nelson DL (2013). "Chapter 6.2: How enzymes work". Lehninger Principles of Biochemistry (6th ed.). New York, N.Y.: W.H. Freeman. p. 195. ISBN 978-1464109621.
24. Benkovic SJ, Hammes-Schiffer S (August 2003). "A perspective on enzyme catalysis". Science. 301 (5637): 1196–202. Bibcode:2003Sci...301.1196B. doi:10.1126/science.1085515. PMID 12947189.
25. Jencks WP (1987). Catalysis in Chemistry and Enzymology. Mineola, N.Y: Dover. ISBN 0-486-65460-5.
26. Villa J, Strajbl M, Glennon TM, Sham YY, Chu ZT, Warshel A (October 2000). "How important are entropic contributions to enzyme catalysis?". Proceedings of the National Academy of Sciences of the United States of America. 97 (22):

11899–904. Bibcode:2000PNAS...9711899V. doi:10.1073/pnas.97.22.11899. PMC 17266 . PMID 11050223.

27. Ramanathan A, Savol A, Burger V, Chennubhotla CS, Agarwal PK (2014). "Protein conformational populations and functionally relevant substates". Acc. Chem. Res. 47 (1): 149–56. doi:10.1021/ar400084s. PMID 23988159.
28. Tsai CJ, Del Sol A, Nussinov R (2009). "Protein allostery, signal transmission and dynamics: a classification scheme of allosteric mechanisms". Mol Biosyst. 5 (3): 207–16. doi:10.1039/b819720b. PMC 2898650 . PMID 19225609.
29. Changeux JP, Edelstein SJ (June 2005). "Allosteric mechanisms of signal transduction". Science. 308 (5727): 1424–8. Bibcode:2005Sci...308.1424C. doi:10.1126/science.1108595. PMID 15933191.
30. de Bolster M (1997). "Glossary of Terms Used in Bioinorganic Chemistry: Cofactor". International Union of Pure and Applied Chemistry. Retrieved 30 October 2007.
31. Chapman-Smith A, Cronan JE (1999). "The enzymatic biotinylation of proteins: a post-translational modification of exceptional specificity". Trends Biochem. Sci. 24 (9): 359–63. doi:10.1016/s0968-0004(99)01438-3. PMID 10470036.
32. Fisher Z, Hernandez Prada JA, Tu C, Duda D, Yoshioka C, An H, Govindasamy L, Silverman DN, McKenna R (February 2005). "Structural and kinetic characterization of active-site histidine as a proton shuttle in catalysis by human carbonic anhydrase II". Biochemistry. 44 (4): 1097–115. doi:10.1021/bi0480279. PMID 15667203.
33. Wagner AL (1975). Vitamins and Coenzymes. Krieger Pub Co. ISBN 0-88275-258-8.
34. "BRENDA The Comprehensive Enzyme Information System". Technische Universität Braunschweig. Retrieved 23 February 2015.
35. Törnroth-Horsefield S, Neutze R (December 2008). "Opening and closing the metabolite gate". Proceedings of the National Academy of Sciences of the United States of America. 105 (50): 19565–6. Bibcode:2008PNAS..10519565T. doi:10.1073/pnas.0810654106. PMC 2604989 . PMID 19073922.
36. McArdle WD, Katch F, Katch VL (2006). "Chapter 9: The Pulmonary System and Exercise". Essentials of Exercise Physiology (3rd ed.). Baltimore, Maryland: Lippincott Williams & Wilkins. pp. 312–3. ISBN 978-0781749916.
37. Ferguson SJ, Nicholls D, Ferguson S (2002). Bioenergetics 3 (3rd ed.). San Diego: Academic. ISBN 0-12-518121-3.
38. Michaelis L, Menten M (1913). "Die Kinetik der Invertinwirkung" [The Kinetics of Invertase Action]. Biochem. Z. (in German). 49: 333–369.; Michaelis L, Menten ML, Johnson KA, Goody RS (2011). "The original Michaelis constant: translation of the 1913 Michaelis-Menten paper". Biochemistry. 50 (39): 8264–9. doi:10.1021/bi201284u. PMC 3381512 . PMID 21888353.
39. Briggs GE, Haldane JB (1925). "A Note on the Kinetics of Enzyme Action". The Biochemical Journal. 19 (2): 339–339. doi:10.1042/bj0190338. PMC 1259181

. PMID 16743508.

40. Ellis RJ (October 2001). "Macromolecular crowding: obvious but underappreciated". Trends in Biochemical Sciences. 26 (10): 597–604. doi:10.1016/S0968-0004(01)01938-7. PMID 11590012.
41. Kopelman R (September 1988). "Fractal reaction kinetics". Science. 241 (4873): 1620–26. Bibcode:1988Sci...241.1620K. doi:10.1126/science.241.4873.1620. PMID 17820893.
42. Cornish-Bowden A (2004). Fundamentals of Enzyme Kinetics (3 ed.). London: Portland Press. ISBN 1-85578-158-1.
43. Price NC (1979). "What is meant by ‹competitive inhibition'?". Trends in Biochemical Sciences. 4 (11): N272–N273. doi:10.1016/0968-0004(79)90205-6.
44. Cornish-Bowden A (July 1986). "Why is uncompetitive inhibition so rare? A possible explanation, with implications for the design of drugs and pesticides". FEBS Letters. 203 (1): 3–6. doi:10.1016/0014-5793(86)81424-7. PMID 3720956.
45. Fisher JF, Meroueh SO, Mobashery S (February 2005). "Bacterial resistance to beta-lactam antibiotics: compelling opportunism, compelling opportunity". Chemical Reviews. 105 (2): 395–424. doi:10.1021/cr030102i. PMID 15700950.
46. Johnson DS, Weerapana E, Cravatt BF (June 2010). "Strategies for discovering and derisking covalent, irreversible enzyme inhibitors". Future Medicinal Chemistry. 2 (6): 949–64. doi:10.4155/fmc.10.21. PMC 2904065 . PMID 20640225.
47. Endo A (1 November 1992). "The discovery and development of HMG-CoA reductase inhibitors" (PDF). J. Lipid Res. 33 (11): 1569–82. PMID 1464741.
48. Wlodawer A, Vondrasek J (1998). "Inhibitors of HIV-1 protease: a major success of structure-assisted drug design". Annual Review of Biophysics and Biomolecular Structure. 27: 249–84. doi:10.1146/annurev.biophys.27.1.249. PMID 9646869.
49. Yoshikawa S, Caughey WS (May 1990). "Infrared evidence of cyanide binding to iron and copper sites in bovine heart cytochrome c oxidase. Implications regarding oxygen reduction". The Journal of Biological Chemistry. 265 (14): 7945–58. PMID 2159465.
50. Hunter T (January 1995). "Protein kinases and phosphatases: the yin and yang of protein phosphorylation and signaling". Cell. 80 (2): 225–36. doi:10.1016/0092-8674(95)90405-0. PMID 7834742.
51. Berg JS, Powell BC, Cheney RE (April 2001). "A millennial myosin census". Molecular Biology of the Cell. 12 (4): 780–94. doi:10.1091/mbc.12.4.780. PMC 32266 . PMID 11294886.
52. Meighen EA (March 1991). "Molecular biology of bacterial bioluminescence". Microbiological Reviews. 55 (1): 123–42. PMC 372803 . PMID 2030669.
53. De Clercq E (2002). "Highlights in the development of new antiviral agents". Mini Rev Med Chem. 2 (2): 163–75. doi:10.2174/1389557024605474. PMID 12370077.
54. Mackie RI, White BA (October 1990). "Recent advances in rumen microbial

ecology and metabolism: potential impact on nutrient output". Journal of Dairy Science. 73 (10): 2971–95. doi:10.3168/jds.S0022-0302(90)78986-2. PMID 2178174.

55. Rouzer CA, Marnett LJ (2009). "Cyclooxygenases: structural and functional insights". J. Lipid Res. 50 Suppl: S29–34. doi:10.1194/jlr.R800042-JLR200. PMC 2674713 . PMID 18952571.
56. Suzuki H (2015). "Chapter 8: Control of Enzyme Activity". How Enzymes Work: From Structure to Function. Boca Raton, FL: CRC Press. pp. 141–69. ISBN 978-981-4463-92-8.
57. Doble BW, Woodgett JR (April 2003). "GSK-3: tricks of the trade for a multi-tasking kinase". Journal of Cell Science. 116 (Pt 7): 1175–86. doi:10.1242/jcs.00384. PMC 3006448 . PMID 12615961.
58. Bennett PM, Chopra I (1993). "Molecular basis of beta-lactamase induction in bacteria" (PDF). Antimicrob. Agents Chemother. 37 (2): 153–8. doi:10.1128/aac.37.2.153. PMC 187630 . PMID 8452343.
59. Skett P, Gibson GG (2001). "Chapter 3: Induction and Inhibition of Drug Metabolism". Introduction to Drug Metabolism (3 ed.). Cheltenham, UK: Nelson Thornes Publishers. pp. 87–118. ISBN 978-0748760114.
60. Faergeman NJ, Knudsen J (April 1997). "Role of long-chain fatty acyl-CoA esters in the regulation of metabolism and in cell signalling". The Biochemical Journal. 323 (Pt 1): 1–12. PMC 1218279 . PMID 9173866.
61. Suzuki H (2015). "Chapter 4: Effect of pH, Temperature, and High Pressure on Enzymatic Activity". How Enzymes Work: From Structure to Function. Boca Raton, FL: CRC Press. pp. 53–74. ISBN 978-981-4463-92-8.
62. Kamata K, Mitsuya M, Nishimura T, Eiki J, Nagata Y (March 2004). "Structural basis for allosteric regulation of the monomeric allosteric enzyme human glucokinase". Structure. 12 (3): 429–38. doi:10.1016/j.str.2004.02.005. PMID 15016359.
63. Froguel P, Zouali H, Vionnet N, Velho G, Vaxillaire M, Sun F, Lesage S, Stoffel M, Takeda J, Passa P (March 1993). "Familial hyperglycemia due to mutations in glucokinase. Definition of a subtype of diabetes mellitus". The New England Journal of Medicine. 328 (10): 697–702. doi:10.1056/NEJM199303113281005. PMID 8433729.
64. Okada S, O'Brien JS (August 1969). "Tay-Sachs disease: generalized absence of a beta-D-N-acetylhexosaminidase component". Science. 165 (3894): 698–700. Bibcode:1969Sci...165..698O. doi:10.1126/science.165.3894.698. PMID 5793973.
65. "Learning About Tay-Sachs Disease". U.S. National Human Genome Research Institute. Retrieved 1 March 2015.
66. Erlandsen H, Stevens RC (October 1999). "The structural basis of phenylketonuria". Molecular Genetics and Metabolism. 68 (2): 103–25. doi:10.1006/mgme.1999.2922. PMID 10527663.
67. Flatmark T, Stevens RC (August 1999). "Structural Insight into the Aromatic

Amino Acid Hydroxylases and Their Disease-Related Mutant Forms". Chemical Reviews. 99 (8): 2137–2160. doi:10.1021/cr980450y. PMID 11849022.

68. "Phenylketonuria". Genes and Disease [Internet]. Bethesda (MD): National Center for Biotechnology Information (US). 1998–2015.
69. "Pseudocholinesterase deficiency". U.S. National Library of Medicine. Retrieved 5 September 2013.
70. Fieker A, Philpott J, Armand M (2011). "Enzyme replacement therapy for pancreatic insufficiency: present and future". Clinical and Experimental Gastroenterology. 4: 55–73. doi:10.2147/CEG.S17634. PMC 3132852 . PMID 21753892.
71. Misselwitz B, Pohl D, Frühauf H, Fried M, Vavricka SR, Fox M (June 2013). "Lactose malabsorption and intolerance: pathogenesis, diagnosis and treatment". United European Gastroenterology Journal. 1 (3): 151–9. doi:10.1177/2050640613484463. PMC 4040760 . PMID 24917953.
72. Cleaver JE (May 1968). "Defective repair replication of DNA in xeroderma pigmentosum". Nature. 218 (5142): 652–6. Bibcode:1968Natur.218..652C. doi:10.1038/218652a0. PMID 5655953.
73. James WD, Elston D, Berger TG (2011). Andrews' Diseases of the Skin: Clinical Dermatology (11th ed.). London: Saunders/ Elsevier. p. 567. ISBN 978-1437703146.
74. Renugopalakrishnan V, Garduño-Juárez R, Narasimhan G, Verma CS, Wei X, Li P (November 2005). "Rational design of thermally stable proteins: relevance to bionanotechnology". Journal of Nanoscience and Nanotechnology. 5 (11): 1759–1767. doi:10.1166/jnn.2005.441. PMID 16433409.
75. Hult K, Berglund P (August 2003). "Engineered enzymes for improved organic synthesis". Current Opinion in Biotechnology. 14 (4): 395–400. doi:10.1016/S0958-1669(03)00095-8. PMID 12943848.
76. Jiang L, Althoff EA, Clemente FR, Doyle L, Röthlisberger D, Zanghellini A, Gallaher JL, Betker JL, Tanaka F, Barbas CF, Hilvert D, Houk KN, Stoddard BL, Baker D (March 2008). "De novo computational design of retro-aldol enzymes". Science. 319 (5868): 1387–91. Bibcode:2008Sci...319.1387J. doi:10.1126/science.1152692. PMC 3431203 . PMID 18323453.
77. Sun Y, Cheng J (May 2002). "Hydrolysis of lignocellulosic materials for ethanol production: a review". Bioresource Technology. 83 (1): 1–11. doi:10.1016/S0960-8524(01)00212-7. PMID 12058826.
78. Kirk O, Borchert TV, Fuglsang CC (August 2002). "Industrial enzyme applications". Current Opinion in Biotechnology. 13 (4): 345–351. doi:10.1016/S0958-1669(02)00328-2. PMID 12323357.
79. Briggs DE (1998). Malts and Malting (1st ed.). London: Blackie Academic. ISBN 978-0412298004.
80. Dulieu C, Moll M, Boudrant J, Poncelet D (2000). "Improved performances and control of beer fermentation using encapsulated alpha-acetolactate decarboxylase

and modeling". Biotechnology Progress. 16 (6): 958–65. doi:10.1021/bp000128k. PMID 11101321.

81. Tarté R (2008). Ingredients in Meat Products Properties, Functionality and Applications. New York: Springer. p. 177. ISBN 978-0-387-71327-4.
82. "Chymosin – GMO Database". GMO Compass. European Union. 10 July 2010. Retrieved 1 March 2015.
83. Molimard P, Spinnler HE (February 1996). "Review: Compounds Involved in the Flavor of Surface Mold-Ripened Cheeses: Origins and Properties". Journal of Dairy Science. 79 (2): 169–184. doi:10.3168/jds.S0022-0302(96)76348-8.
84. Guzmán-Maldonado H, Paredes-López O (September 1995). "Amylolytic enzymes and products derived from starch: a review". Critical Reviews in Food Science and Nutrition. 35 (5): 373–403. doi:10.1080/10408399509527706. PMID 8573280.
85. "Protease – GMO Database". GMO Compass. European Union. 10 July 2010. Retrieved 28 February 2015.
86. Alkorta I, Garbisu C, Llama MJ, Serra JL (January 1998). "Industrial applications of pectic enzymes: a review". Process Biochemistry. 33 (1): 21–28. doi:10.1016/S0032-9592(97)00046-0.
87. Bajpai P (March 1999). "Application of enzymes in the pulp and paper industry". Biotechnology Progress. 15 (2): 147–157. doi:10.1021/bp990013k. PMID 10194388.
88. Begley CG, Paragina S, Sporn A (March 1990). "An analysis of contact lens enzyme cleaners". Journal of the American Optometric Association. 61 (3): 190–4. PMID 2186082.
89. Farris PL (2009). "Economic Growth and Organization of the U.S. Starch Industry". In BeMiller JN, Whistler RL. Starch Chemistry and Technology (3rd ed.). London: Academic. ISBN 9780080926551.

CHAPTER 7

Antibodies

Chapter Contents

Antibodies are proteins that exist in bodily fluids, and are used both as a detection and response device by the immune system. Antibodies are made in plasma cells, which are a type of white blood cell, and are an integral part to the body's natural defense system. In mammals, there are five primary types of antibodies, each with similar basic structures. The tips of the proteins can be incredibly different, however, and it is this part that interacts with the contaminant itself, allowing for millions of unique antibodies to exist.

INTRODUCTION

An antibody (Ab), also known as an immunoglobulin (Ig), is a large, Y-shaped protein produced mainly by plasma cells that is used by the immune system to identify and neutralize pathogens such as bacteria and viruses. The antibody recognizes a unique molecule of the harmful agent, called an antigen, via the Fab›s variable region. Each tip of the «Y» of an antibody contains a paratope (analogous to a lock) that is specific for one particular epitope (similarly analogous to a key) on an antigen, allowing these two structures to bind together with precision. Using this binding mechanism, an antibody can *tag* a microbe or an infected cell for attack by other parts of the immune system, or can neutralize its target directly (for example, by blocking a part of a microbe that is essential for its invasion and survival). Depending on the antigen, the binding may impede the biological process causing the disease or may activate macrophages to destroy the foreign substance. The ability of an antibody to communicate with the other components of the immune system is mediated via its Fc region (located at the base of the "Y"), which contains a conserved glycosylation site involved in these interactions. The production of antibodies is the main function of the humoral immune system.

Antibodies are secreted by B cells of the adaptive immune system, mostly by differentiated B cells called plasma cells. Antibodies can occur in two physical forms, a soluble form that is secreted from the cell to be free in the blood plasma, and a membrane-bound form that is attached to the surface of a B cell and is referred to as the B-cell receptor (BCR). The BCR is found only on the surface of B cells and facilitates the activation of these cells and their subsequent differentiation into either antibody factories called plasma cells or memory B cells that will survive in the body and remember that same antigen so the B cells can respond faster upon future exposure. In most cases, interaction of the B cell with a T helper cell is necessary to produce full activation of the B cell and, therefore, antibody generation following antigen binding. Soluble antibodies are released into the blood and tissue fluids, as well as many secretions to continue to survey for invading microorganisms.

Antibodies are glycoproteins belonging to the immunoglobulin superfamily. They constitute most of the gamma globulin fraction of the blood proteins. They are typically made of basic structural units—each with two large heavy chains and two small light chains. There are several different types of antibody heavy chains that define the five different types of crystallisable fragments (Fc) that may be attached to the antigen-binding fragments. The five different types of Fc regions allow antibodies to be grouped into five *isotypes*. Each Fc region of a particular antibody isotype is able to bind to its specific Fc Receptor (except for IgD, which is essentially the BCR), thus allowing the antigen-antibody complex to mediate different roles depending on which FcR it binds. The ability of an antibody to bind to its corresponding FcR is further modulated by the structure of the glycan(s) present at conserved sites within its Fc region. The ability of antibodies to bind to FcRs helps to direct the appropriate immune response for each different type of foreign object they encounter. For example, IgE is responsible for an allergic response consisting of mast cell degranulation and histamine release. IgE›s Fab paratope binds to allergic antigen, for example house dust mite particles, while its Fc region binds to Fc receptor ε. The allergen-IgE-FcRε interaction mediates allergic signal transduction to induce conditions such as asthma.

Though the general structure of all antibodies is very similar, a small region at the tip of the protein is extremely variable, allowing millions of antibodies with slightly different tip structures, or antigen-binding sites, to exist. This region is known as the *hypervariable region*. Each of these variants can bind to a different antigen. This enormous diversity of antibody paratopes on the antigen-binding fragments allows the immune system to recognize an equally wide variety of antigens. The large and diverse population of antibody paratope is generated by random recombination events of a set of gene segments that encode different antigen-binding sites (or *paratopes*), followed by random mutations in this area of the antibody gene, which create further diversity. This recombinational process that produces clonal antibody paratope diversity is called V(D)J or VJ recombination. Basically, the antibody paratope is polygenic, made up of three genes, V, D, and J. Each paratope locus is also polymorphic, such that during antibody production, one allele of V, one of D, and one of J is chosen. These gene segments are then joined together using random genetic recombination to produce the paratope. The regions where the genes are randomly recombined together is the hyper variable region used to recognise different antigens on a clonal basis.

Antibody genes also re-organize in a process called class switching that changes the one type of heavy chain Fc fragment to another, creating a different isotype of the antibody that retains the antigen-specific variable region. This allows a single antibody to be used by different types of Fc receptors, expressed on different parts of the immune system.

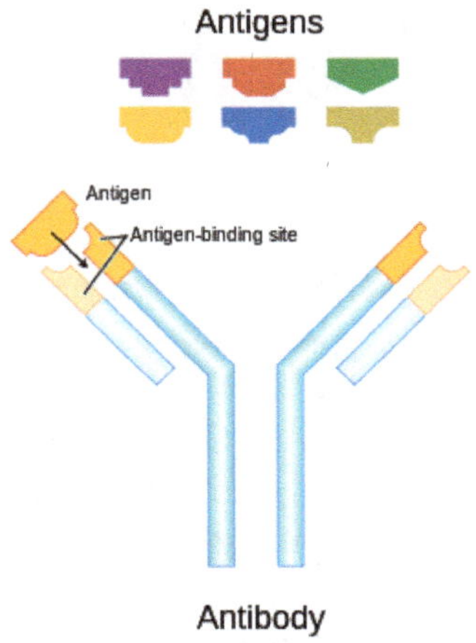

Each antibody binds to a specific antigen; an interaction similar to a lock and key.

FORMS

The membrane-bound form of an antibody may be called a *surface immunoglobulin* (sIg) or a *membrane immunoglobulin* (mIg). It is part of the *B cell receptor* (BCR), which allows a B cell to detect when a specific antigen is present in the body and triggers B cell activation. The BCR is composed of surface-bound IgD or IgM antibodies and associated Ig-α and Ig-β heterodimers, which are capable of signal transduction. A typical human B cell will have 50,000 to 100,000 antibodies bound to its surface. Upon antigen binding, they cluster in large patches, which can exceed 1 micrometer in diameter, on lipid rafts that isolate the BCRs from most other cell signaling receptors. These patches may improve the efficiency of the cellular immune response. In humans, the cell surface is bare around the B cell receptors for several hundred nanometers,which further isolates the BCRs from competing influences.

ANTIBODY–ANTIGEN INTERACTIONS

The antibody's paratope interacts with the antigen's epitope. An antigen usually contains different epitopes along its surface arranged discontinuously, and dominant epitopes on a given antigen are called determinants.

Antibody and antigen interact by spatial complementarity (lock and key). The molecular forces involved in the Fab-epitope interaction are weak and non-specific – for example electrostatic forces, hydrogen bonds, hydrophobic interactions, and van der Waals forces. This means binding between antibody and antigen is reversible, and the antibody's affinity towards an antigen is relative rather than absolute. Relatively weak binding also means it is possible for an antibody to cross-react with different antigens of different relative affinities.

Often, once an antibody and antigen bind, they become an immune complex, which functions as a unitary object and can act as an antigen in its own right, being countered by other antibodies. Similarly, haptens are small molecules that provoke no immune response by themselves, but once they bind to proteins, the resulting complex or hapten-carrier adduct is antigenic.

Isotypes

Antibodies can come in different varieties known as isotypes or classes. In placental mammals there are five antibody isotypes known as IgA, IgD, IgE, IgG, and IgM. They are each named with an "Ig" prefix that stands for immunoglobulin, a name sometimes used interchangeably with antibody, and differ in their biological properties, functional locations and ability to deal with different antigens, as depicted in the table. The different suffixes of the antibody isotypes denote the different types of heavy chains the antibody contains, with each heavy chain class named alphabetically: α (alpha), γ (gamma), δ (delta), ε (epsilon), and μ (mu). This gives rise to IgA, IgG, IgD, IgE, and IgM, respectively.

Antibody Isotypes of Mammals

Name	Types	Description	Antibody complexes
IgA	2	Found in mucosal areas, such as the gut, respiratory tract and urogenital tract, and prevents colonization by pathogens.[14] Also found in saliva, tears, and breast milk.	
IgD	1	Functions mainly as an antigen receptor on B cells that have not been exposed to antigens.[15] It has been shown to activate basophils and mast cells to produce antimicrobial factors.[16]	
IgE	1	Binds to allergens and triggers histamine release from mast cells and basophils, and is involved in allergy. Also protects against parasitic worms.[5]	
IgG	4	In its four forms, provides the majority of antibody-based immunity against invading pathogens.[5] The only antibody capable of crossing the placenta to give passive immunity to the fetus.	
IgM	1	Expressed on the surface of B cells (monomer) and in a secreted form (pentamer) with very high avidity. Eliminates pathogens in the early stages of B cell-mediated (humoral) immunity before there is sufficient IgG.[5][15]	

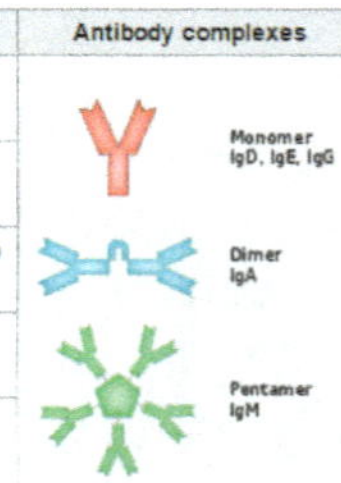

The antibody isotype of a B cell changes during cell development and activation. Immature B cells, which have never been exposed to an antigen, express only the IgM isotype in a cell surface bound form. The B lymphocyte, in this ready-to-respond form, is known as a "naive B lymphocyte." The naive B lymphocyte expresses both surface IgM and IgD. The co-expression of both of these immunoglobulin isotypes renders the B cell ready to respond to antigen. B cell activation follows engagement of the cell-bound antibody molecule with an antigen, causing the cell to divide and differentiate into an antibody-producing cell called a plasma cell. In this activated form, the B cell starts to produce antibody in a secreted form rather than a membrane-bound form. Some daughter cells of the activated B cells undergo isotype switching, a mechanism that causes the production of antibodies to change from IgM or IgD to the other antibody isotypes, IgE, IgA, or IgG, that have defined roles in the immune system.

Antibody isotypes not found in mammals

Name	Types	Description
IgY		Found in birds and reptiles; related to mammalian IgG.
IgW		Found in sharks and skates; related to mammalian IgD

STRUCTURE

Antibodies are heavy (~150 kDa) globular plasma proteins. They have sugar chains (glycans) added to conserved amino acid residues. In other words, antibodies are *glycoproteins*. The attached glycans are critically important to the structure and function of the antibody. Among other things the expressed glycans can modulate an antibody's affinity for its corresponding FcR(s).

The basic functional unit of each antibody is an immunoglobulin (Ig) monomer (containing only one Ig unit); secreted antibodies can also be dimeric with two Ig units as with IgA, tetrameric with four Ig units like teleost fish IgM, or pentameric with five Ig units, like mammalian IgM.

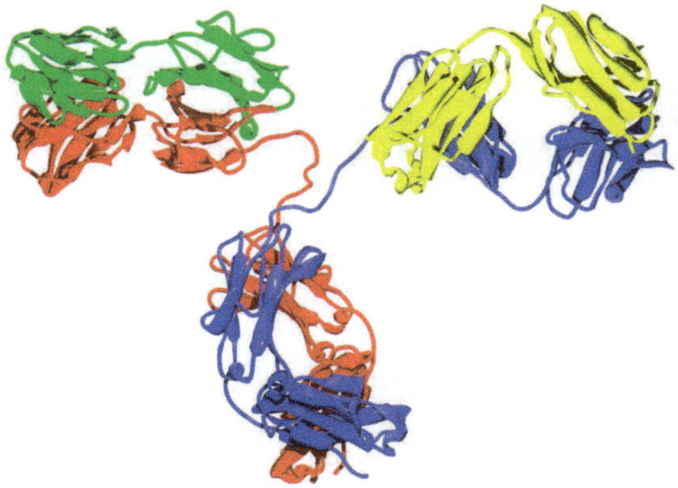

Several immunoglobulin domains make up the two heavy chains (red and blue) and the two light chains (green and yellow) of an antibody. The immunoglobulin domains are composed of between 7 (for constant domains) and 9 (for variable domains) β-strands.

The variable parts of an antibody are its V regions, and the constant part is its C region.

Immunoglobulin Domains

The Ig monomer is a "Y"-shaped molecule that consists of four polypeptide chains; two identical *heavy chains* and two identical *light chains* connected by disulfide bonds. Each chain is composed of structural domains called immunoglobulin domains. These domains contain about 70–110 amino acids and are classified into different categories (for example, variable or IgV, and constant or IgC) according to their size and function. They have a characteristic immunoglobulin fold in which two beta sheets create a «sandwich» shape, held together by interactions between conserved cysteines and other charged amino acids.

Heavy chain

There are five types of mammalian Ig heavy chain denoted by the Greek letters: α, δ, ε, γ, and μ. The type of heavy chain present defines the *class* of antibody; these chains are found in IgA, IgD, IgE, IgG, and IgM antibodies, respectively. Distinct heavy chains differ in size and composition; α and γ contain approximately 450 amino acids, whereas μ and ε have approximately 550 amino acids.

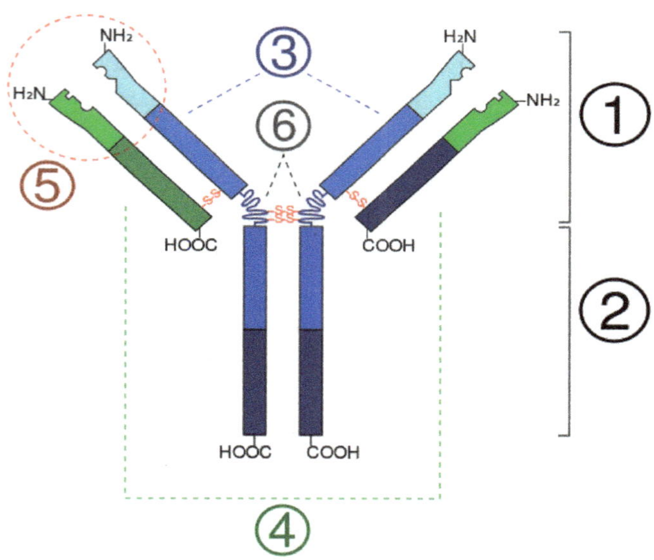

- Fab region
- Fc region
- Heavy chain (blue) with one variable (V_H) domain followed by a constant domain (C_H1), a hinge region, and two more constant (C_H2 and C_H3) domains
- Light chain (green) with one variable (V_L) and one constant (C_L) domain
- Antigen binding site (paratope)
- Hinge regions

Each heavy chain has two regions, the *constant region* and the *variable region*. The constant region is identical in all antibodies of the same isotype, but differs in antibodies of different isotypes. Heavy chains γ, α and δ have a constant region composed of *three* tandem (in a line) Ig domains, and a hinge region for added flexibility; heavy chains μ and ε have a constant region composed of *four* immunoglobulin domains. The variable region of the heavy chain differs in antibodies produced by different B cells, but is the same for all antibodies produced by a single B cell or B cell clone. The variable region of each heavy chain is approximately 110 amino acids long and is composed of a single Ig domain.

Light Chain

In mammals there are two types of immunoglobulin light chain, which are called lambda (λ) and kappa (κ). A light chain has two successive domains: one constant domain and one variable domain. The approximate length of a light chain is 211 to 217 amino acids. Each antibody contains two light chains that are always identical; only one type of light chain, κ or λ, is present per antibody in mammals. Other types of light chains, such as the iota (ι) chain, are found in other vertebrates like sharks (Chondrichthyes) and bony fishes (Teleostei).

CDRs, Fv, Fab and Fc Regions

Some parts of an antibody have the same functions. The arms of the Y, for example, contain the sites that can bind to antigens (in general, identical) and, therefore, recognize specific foreign objects. This region of the antibody is called the *Fab (fragment, antigen-binding) region*. It is composed of one constant and one variable domain from each heavy and light chain of the antibody. The paratope is shaped at the amino terminal end of the antibody monomer by the variable domains from the heavy and light chains. The variable domain is also referred to as the F_V region and is the most important region for binding to antigens. To be specific, variable loops of β-strands, three each on the light (V_L) and heavy (V_H) chains are responsible for binding to the antigen. These loops are referred to as the complementarity determining regions (CDRs). The structures of these CDRs have been clustered and classified by Chothia et al. and more recently by North et al. and Nikoloudis et al. In the framework of the immune network theory, CDRs are also called idiotypes. According to immune network theory, the adaptive immune system is regulated by interactions between idiotypes.

The base of the Y plays a role in modulating immune cell activity. This region is called the *Fc (Fragment, crystallizable) region*, and is composed of two heavy chains that contribute two or three constant domains depending on the class of the antibody. Thus, the Fc region ensures that each antibody generates an appropriate immune response for a given antigen, by binding to a specific class of Fc receptors, and other immune molecules, such as complement proteins. By doing this, it mediates different physiological effects including recognition of opsonized particles (binding to FcγR), lysis of cells (binding to complement), and degranulation of mast cells, basophils, and eosinophils (binding to FcεR).

In summary, the Fab region of the antibody determines antigen specificity while the Fc region of the antibody determines the antibody's class effect. Since only the constant domains of the heavy chains make up the Fc region of an antibody, the classes of heavy chain in antibodies determine their class effects. Possible classes of heavy chains in antibodies include alpha, gamma, delta, epsilon, and mu, and they define the antibody's isotypes IgA, G, D, E, and M, respectively. This infers different isotypes of antibodies have different class effects due to their different Fc regions binding and activating different types of receptors. Possible class effects of antibodies include: Opsonisation, agglutination, haemolysis, complement activation, mast cell degranulation, and neutralisation (though this class effect may be mediated by the Fab region rather than the Fc region). It also implies that Fab-mediated effects are directed at microbes or toxins, whilst Fc mediated effects are directed at effector cells or effector molecules (see below).

Function

The main categories of antibody action include the following:

- Neutralisation, in which neutralizing antibodies block parts of the surface of a bacterial cell or virion to render its attack ineffective
- Agglutination, in which antibodies "glue together" foreign cells into clumps that are attractive targets for phagocytosis
- Precipitation, in which antibodies "glue together" serum-soluble antigens, forcing them to precipitate out of solution in clumps that are attractive targets for phagocytosis
- Complement activation (fixation), in which antibodies that are latched onto a foreign cell encourage complement to attack it with a membrane attack complex, which leads to the following:
    - Lysis of the foreign cell
    - Encouragement of inflammation by chemotactically attracting inflammatory cells

Activated B cells differentiate into either antibody-producing cells called plasma cells that secrete soluble antibody or memory cells that survive in the body for years afterward in order to allow the immune system to remember an antigen and respond faster upon future exposures.

At the prenatal and neonatal stages of life, the presence of antibodies is provided by passive immunization from the mother. Early endogenous antibody production varies for different kinds of antibodies, and usually appear within the first years of life. Since antibodies exist freely in the bloodstream, they are said to be part of the humoral immune system. Circulating antibodies are produced by clonal B cells that specifically respond to only one antigen (an example is a virus capsid protein fragment). Antibodies contribute to immunity in three ways: They prevent pathogens from entering or damaging cells by binding to them; they stimulate removal of pathogens by macrophages and other cells by coating the pathogen; and they trigger destruction of pathogens by stimulating other immune responses such as the complement pathway. Antibodies will also trigger vasoactive amine degranulation to contribute to immunity against certain types of antigens (helminths, allergens).

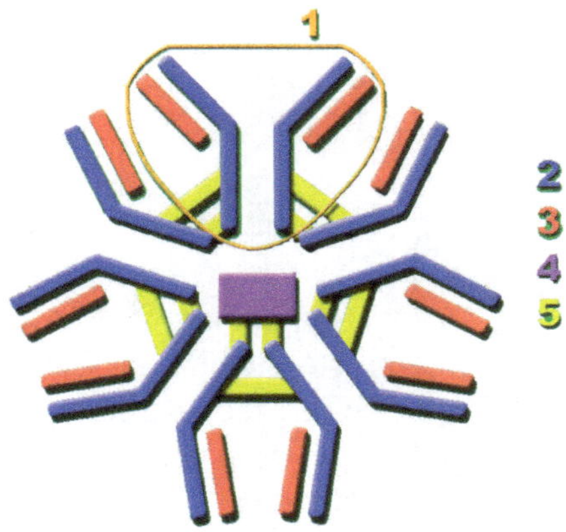

The secreted mammalian IgM has five Ig units. Each Ig unit (labeled 1) has two epitope binding Fab regions, so IgM is capable of binding up to 10 epitopes.

Activation of Complement

Antibodies that bind to surface antigens (for example, on bacteria) will attract the first component of the complement cascade with their Fc region and initiate activation of the "classical" complement system.This results in the killing of bacteria in two ways. First, the binding of the antibody and complement molecules marks the microbe for ingestion by phagocytes in a process called opsonization; these phagocytes are attracted by certain complement molecules generated in the complement cascade. Second, some complement system components form a membrane attack complex to assist antibodies to kill the bacterium directly (bacteriolysis).

Activation of Effector Cells

To combat pathogens that replicate outside cells, antibodies bind to pathogens to link them together, causing them to agglutinate. Since an antibody has at least two paratopes, it can bind more than one antigen by binding identical epitopes carried on the surfaces of these antigens. By coating the pathogen, antibodies stimulate effector functions against the pathogen in cells that recognize their Fc region.

Those cells that recognize coated pathogens have Fc receptors, which, as the name suggests, interact with the Fc region of IgA, IgG, and IgE antibodies. The engagement of a particular antibody with the Fc receptor on a particular cell triggers an effector function of that cell; phagocytes will phagocytose, mast cells and neutrophils will degranulate, natural killer cells will release cytokines and cytotoxic molecules; that will ultimately result in destruction of the invading microbe. The activation of natural killer cells by antibodies initiates a cytotoxic mechanism known as antibody-dependent cell-mediated cytotoxicity (ADCC) – this process may explain the efficacy of monoclonal antibodies used in biological therapies against cancer. The Fc receptors are isotype-specific, which gives greater flexibility to the immune system, invoking only the appropriate immune mechanisms for distinct pathogens.

Natural Antibodies

Humans and higher primates also produce "natural antibodies" that are present in serum before viral infection. Natural antibodies have been defined as antibodies that are produced without any previous infection, vaccination, other foreign antigen exposure or passive immunization. These antibodies can activate the classical complement pathway leading to lysis of enveloped virus particles long before the adaptive immune response is activated. Many natural antibodies are directed against the disaccharide galactose α(1,3)-galactose (α-Gal), which is found as a terminal sugar on glycosylated cell surface proteins, and generated in response to production of this sugar by bacteria contained in the human gut. Rejection of xenotransplantated organs is thought to be, in part, the result of natural antibodies circulating in the serum of the recipient binding to α-Gal antigens expressed on the donor tissue.

IMMUNOGLOBULIN DIVERSITY

Virtually all microbes can trigger an antibody response. Successful recognition and eradication of many different types of microbes requires diversity among antibodies; their amino acid composition varies allowing them to interact with many different antigens. It has been estimated that humans generate about 10 billion different antibodies, each capable of binding a distinct epitope of an antigen. Although a huge repertoire of different antibodies is generated in a single individual, the number of genes available to make these proteins is limited by the size of the human genome. Several complex genetic mechanisms have evolved that allow vertebrate B cells to generate a diverse pool of antibodies from a relatively small number of antibody genes.

Domain Variability

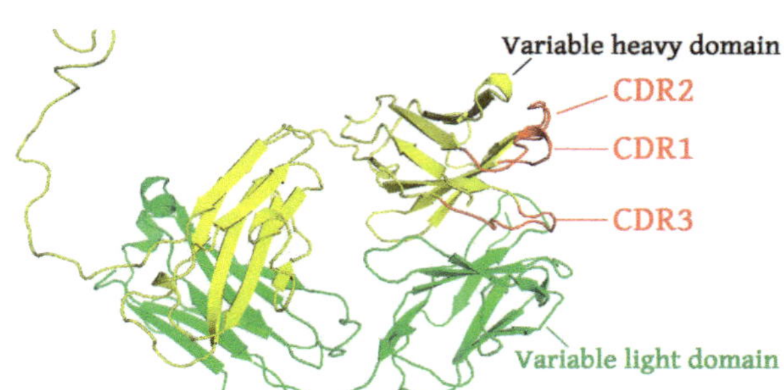

The complementarity determining regions of the heavy chain are shown in red (PDB: 1IGT)

The chromosomal region that encodes an antibody is large and contains several distinct gene loci for each domain of the antibody—the chromosome region containing heavy chain genes (IGH@) is found on chromosome 14, and the loci containing lambda and kappa light chain genes (IGL@ and IGK@) are found on chromosomes 22 and 2 in humans. One of these domains is called the variable domain, which is present in each heavy and light chain of every antibody, but can differ in different antibodies generated from distinct B cells. Differences, between the variable domains, are located on three loops known as hypervariable regions (HV-1, HV-2 and HV-3) or complementarity determining regions (CDR1, CDR2 and CDR3). CDRs are supported within the variable domains by conserved framework regions. The heavy chain locus contains about 65 different variable domain genes that all differ in their CDRs. Combining these genes with an array of genes for other domains of the antibody generates a large cavalry of antibodies with a high degree of variability. This combination is called V(D)J recombination discussed below.

V(D)J Recombination

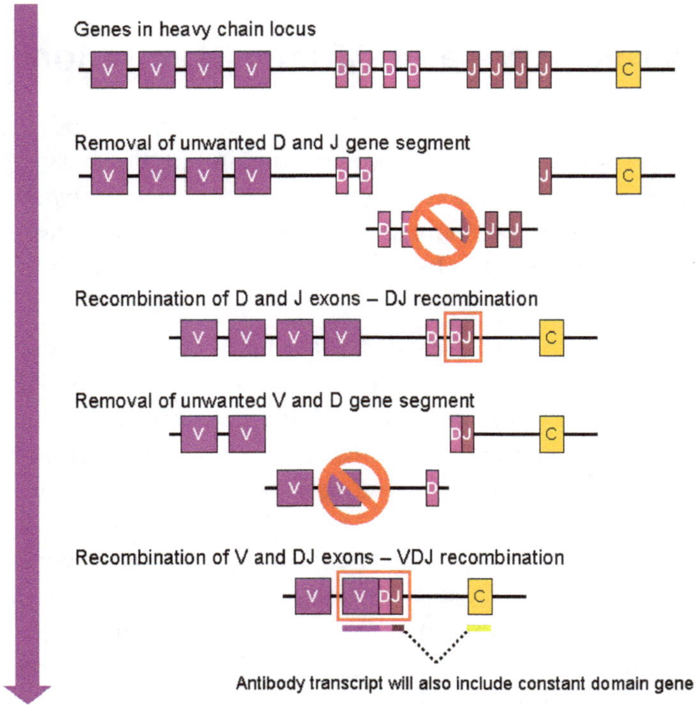

Simplified overview of V(D)J recombination of immunoglobulin heavy chains

Somatic recombination of immunoglobulins, also known as *V(D)J recombination*, involves the generation of a unique immunoglobulin variable region. The variable region of each immunoglobulin heavy or light chain is encoded in several pieces—known as gene segments (subgenes). These segments are called variable (V), diversity (D) and joining (J) segments. V, D and J segments are found in Ig heavy chains, but only V and J segments are found in Ig light chains. Multiple copies of the V, D and J gene segments exist, and are tandemly arranged in the genomes of mammals. In the bone marrow, each developing B cell will assemble an immunoglobulin variable region by randomly selecting and combining one V, one D and one J gene segment (or one V and one J segment in the light chain). As there are multiple copies of each type of gene segment, and different combinations of gene segments can be used to generate each immunoglobulin variable region, this process generates a huge number of antibodies, each with different paratopes, and thus different antigen specificities. Interestingly, the rearrangement of several subgenes (i.e. V2 family) for lambda light chain immunoglobulin is coupled with the activation of microRNA miR-650, which further influences biology of B-cells.

RAG proteins play an important role with V(D)J recombination in cutting DNA at a particular region.Without the presence of these proteins, V(D)J recombination would not occur.

After a B cell produces a functional immunoglobulin gene during V(D)J recombination, it cannot express any other variable region (a process known as allelic exclusion) thus each B cell can produce antibodies containing only one kind of variable chain.

Somatic Hypermutation and Affinity Maturation

Following activation with antigen, B cells begin to proliferate rapidly. In these rapidly dividing cells, the genes encoding the variable domains of the heavy and light chains undergo a high rate of point mutation, by a process called *somatic hypermutation* (SHM). SHM results in approximately one nucleotide change per variable gene, per cell division. As a consequence, any daughter B cells will acquire slight amino acid differences in the variable domains of their antibody chains.

This serves to increase the diversity of the antibody pool and impacts the antibody›s antigen-binding affinity. Some point mutations will result in the production of antibodies that have a weaker interaction (low affinity) with their antigen than the original antibody, and some mutations will generate antibodies with a stronger interaction (high affinity). B cells that express high affinity antibodies on their surface will receive a strong survival signal during interactions with other cells, whereas those with low affinity antibodies will not, and will die by apoptosis. Thus, B cells expressing antibodies with a higher affinity for the antigen will outcompete those with weaker affinities for function and survival. The process of generating antibodies with increased binding affinities is called *affinity maturation*. Affinity maturation occurs in mature B cells after V(D)J recombination, and is dependent on help from helper T cells.

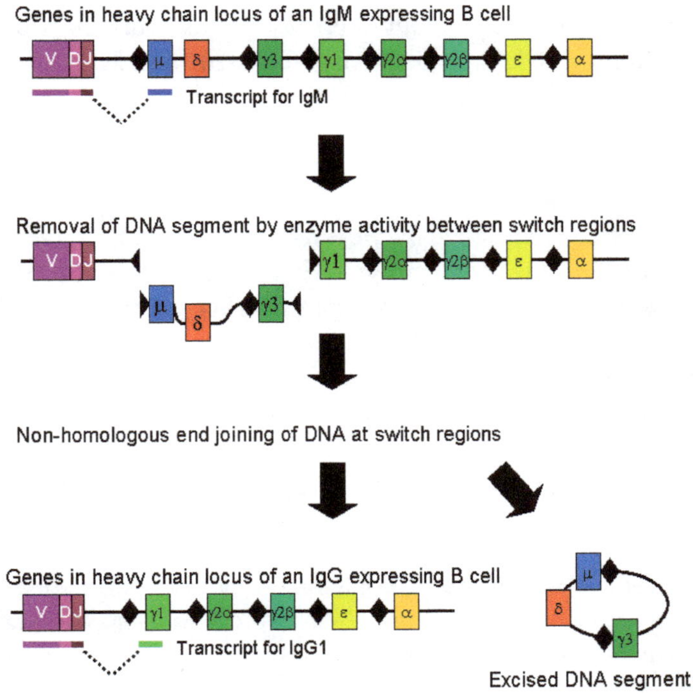

Mechanism of class switch recombination that allows isotype switching in activated B cells

Class Switching

Isotype or class switching is a biological process occurring after activation of the B cell, which allows the cell to produce different classes of antibody (IgA, IgE, or IgG). The different classes of antibody, and thus effector functions, are defined by the constant (C) regions of the immunoglobulin heavy chain. Initially, naive B cells express only cell-surface IgM and IgD with identical antigen binding regions. Each isotype is adapted for a distinct function; therefore, after activation, an antibody with an IgG, IgA, or IgE effector function might be required to effectively eliminate an antigen. Class switching allows different daughter cells from the same activated B cell to produce antibodies of different isotypes. Only the constant region of the antibody heavy chain changes during class switching; the variable regions, and therefore antigen specificity, remain unchanged. Thus the progeny of a single B cell can produce antibodies, all specific for the same antigen, but with the ability to produce the effector function appropriate for each antigenic challenge. Class switching is triggered by cytokines; the isotype generated depends on which cytokines are present in the B cell environment.

Class switching occurs in the heavy chain gene locus by a mechanism called class switch recombination (CSR). This mechanism relies on conserved nucleotide motifs, called *switch (S) regions*, found in DNA upstream of each constant region gene (except in the δ-chain). The DNA strand is broken by the activity of a series of enzymes at two selected S-regions. The variable domain exon is rejoined through a process called non-homologous end joining (NHEJ) to the desired constant region (γ, α or ε). This process results in an immunoglobulin gene that encodes an antibody of a different isotype.

Affinity Designations

A group of antibodies can be called *monovalent* (or *specific*) if they have affinity for the same epitope, or for the same antigen (but potentially different epitopes on the molecule), or for the same strain of microorganism (but potentially different antigens on or in it). In contrast, a group of antibodies can be called *polyvalent* (or *unspecific*) if they have affinity for various antigens or microorganisms. Intravenous immunoglobulin, if not otherwise noted, consists of polyvalent IgG. In contrast, monoclonal antibodies are monovalent for the same epitope.

Asymmetrical Antibodies

Heterodimeric antibodies, which are also asymmetrical and antibodies, allow for greater flexibility and new formats for attaching a variety of drugs to the antibody arms. One of the general formats for a heterodimeric antibody is the "knobs-into-holes" format. This format is specific to the heavy chain part of the constant region in antibodies. The "knobs" part is engineered by replacing a small amino acid with a larger one. It fits into the "hole", which is engineered by replacing a large amino acid with a smaller one. What connects the "knobs" to the "holes" are the disulfide bonds between each chain. The "knobs-into-holes" shape facilitates antibody dependent cell mediated cytotoxicity. Single chain variable fragments (scFv) are connected to the variable domain of the heavy and light chain via a short linker peptide. The linker is rich in glycine, which gives it more flexibility, and serine/threonine, which gives it specificity. Two different scFv fragments can be connected together, via a hinge region, to the constant domain of the heavy chain or the constant domain of the light chain. This gives the antibody bispecificity, allowing for the binding specificities of two different antigens. The "knobs-into-holes" format enhances heterodimer formation but doesn't suppress homodimer formation.

To further improve the function of heterodimeric antibodies, many scientists are looking towards artificial constructs. Artificial antibodies are largely diverse protein motifs that use the functional strategy of the antibody molecule, but aren't limited by the loop and framework structural constraints of the natural antibody. Being able to control the combinational design of the sequence and three-dimensional space could transcend the natural design and allow for the attachment of different combinations of drugs to the arms.

Heterodimeric antibodies have a greater range in shapes they can take and the drugs that are attached to the arms don't have to be the same on each arm, allowing for different combinations of drugs to be used in cancer treatment. Pharmaceuticals are able to produce highly functional bispecific, and even multispecific, antibodies. The degree to which they can function is impressive given that such a change shape from the natural form should lead to decreased functionality.

Medical Applications

Disease Diagnosis

Detection of particular antibodies is a very common form of medical diagnostics, and applications such as serology depend on these methods.For example, in biochemical assays for disease diagnosis, a titer of antibodies directed against Epstein-Barr virus or Lyme disease is estimated from the blood. If those antibodies are not present, either the person is not infected or the infection occurred a *very* long time ago, and the B cells generating these specific antibodies have naturally decayed.

In clinical immunology, levels of individual classes of immunoglobulins are measured by nephelometry (or turbidimetry) to characterize the antibody profile of patient. Elevations in different classes of immunoglobulins are sometimes useful in determining the cause of liver damage in patients for whom the diagnosis is unclear. For example, elevated IgA indicates alcoholic cirrhosis, elevated IgM indicates viral hepatitis and primary biliary cirrhosis, while IgG is elevated in viral hepatitis, autoimmune hepatitis and cirrhosis.

Autoimmune disorders can often be traced to antibodies that bind the body›s own epitopes; many can be detected through blood tests. Antibodies directed against red blood cell surface antigens in immune mediated hemolytic anemia are detected with the Coombs test. The Coombs test is also used for antibody screening in blood transfusion preparation and also for antibody screening in antenatal women.

Practically, several immunodiagnostic methods based on detection of complex antigen-antibody are used to diagnose infectious diseases, for example ELISA, immunofluorescence, Western blot, immunodiffusion, immunoelectrophoresis, and magnetic immunoassay. Antibodies raised against human chorionic gonadotropin are used in over the counter pregnancy tests.

New dioxaborolane chemistry enables radioactive fluoride (^{18}F) labeling of antibodies, which allows for positron emission tomography (PET) imaging of cancer.

Disease Therapy

Targeted monoclonal antibody therapy is employed to treat diseases such as rheumatoid arthritis, multiple sclerosis, psoriasis, and many forms of cancer including non-Hodgkin's lymphoma, colorectal cancer, head and neck cancer and breast cancer.

Some immune deficiencies, such as X-linked agammaglobulinemia and hypogammaglobulinemia, result in partial or complete lack of antibodies. These diseases are often treated by inducing a short term form of immunity called passive immunity. Passive immunity is achieved through the transfer of ready-made antibodies in the form of human or animal serum, pooled immunoglobulin or monoclonal antibodies, into the affected individual.

Prenatal Therapy

Rhesus factor, also known as Rhesus D (RhD) antigen, is an antigen found on red blood cells; individuals that are Rhesus-positive (Rh+) have this antigen on their red blood cells and individuals that are Rhesus-negative (Rh–) do not. During normal childbirth, delivery trauma or complications during pregnancy, blood from a fetus can enter the mother's system. In the case of an Rh-incompatible mother and child, consequential blood mixing may sensitize an Rh- mother to the Rh antigen on the blood cells of the Rh+ child, putting the remainder of the pregnancy, and any subsequent pregnancies, at risk for hemolytic disease of the newborn.

Rho(D) immune globulin antibodies are specific for human Rhesus D (RhD) antigen. Anti-RhD antibodies are administered as part of a prenatal treatment regimen to prevent sensitization that may occur when a Rhesus-negative mother has a Rhesus-positive fetus. Treatment of a mother with Anti-RhD antibodies prior to and immediately after trauma and delivery destroys Rh antigen in the mother's system from the fetus. It is important to note that this occurs before the antigen can stimulate maternal B cells to "remember" Rh antigen by generating memory B cells. Therefore, her humoral immune system will not make anti-Rh antibodies, and will not attack the Rhesus antigens of the current or subsequent babies. Rho(D) Immune Globulin treatment prevents sensitization that can lead to Rh disease, but does not prevent or treat the underlying disease itself.

RESEARCH APPLICATIONS

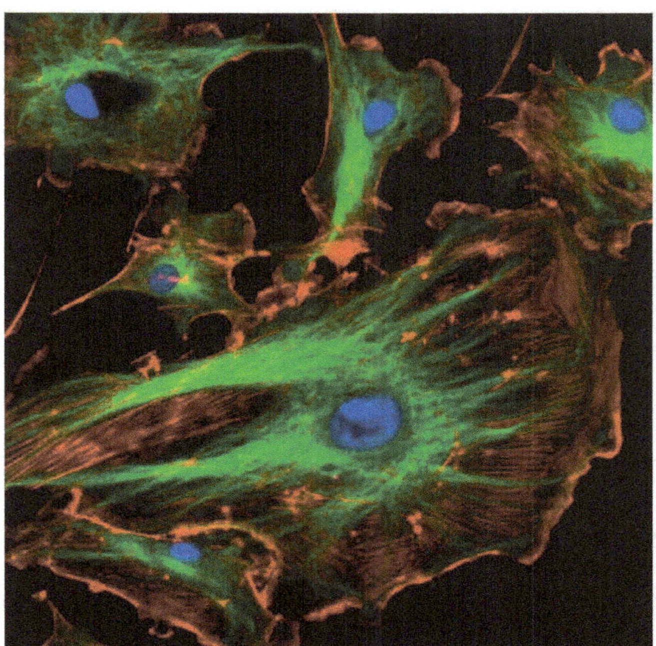

Immunofluorescence image of the eukaryotic cytoskeleton. Actin filaments are shown in red, microtubules in green, and the nuclei in blue.

Specific antibodies are produced by injecting an antigen into a mammal, such as a mouse, rat, rabbit, goat, sheep, or horse for large quantities of antibody. Blood isolated from these animals contains *polyclonal antibodies*—multiple antibodies that bind to the same antigen—in the serum, which can now be called antiserum. Antigens are also injected into chickens for generation of polyclonal antibodies in egg yolk. To obtain antibody that is specific for a single epitope of an antigen, antibody-secreting lymphocytes are isolated from the animal and immortalized by fusing them with a cancer cell line. The fused cells are called hybridomas, and will continually grow and secrete antibody in culture. Single hybridoma cells are isolated by dilution cloning to generate cell clones that all produce the same antibody; these antibodies are called *monoclonal antibodies*. Polyclonal and monoclonal antibodies are often purified using Protein A/G or antigen-affinity chromatography.

In research, purified antibodies are used in many applications. Antibodies for research applications can be found directly from antibody suppliers, or through use of a specialist search engine. Research antibodies are most commonly used to identify and locate intracellular and extracellular proteins. Antibodies are used in flow cytometry to differentiate cell types by the proteins they express; different types of cell express different combinations of cluster of differentiation molecules on their surface, and produce different intracellular and secretable proteins. They are also used in immunoprecipitation to separate proteins and anything bound to them (co-immunoprecipitation) from other molecules in a cell lysate, in Western blot analyses to identify proteins separated by electrophoresis, and in immunohistochemistry or immunofluorescence to examine protein expression in tissue sections or to locate proteins within cells with the assistance of a microscope.Proteins can also be detected and quantified with antibodies, using ELISA and ELISPOT techniques.

Reproducibility in Science

Antibodies used in research are some of the most powerful, yet most problematic reagents with a tremendous number of factors that must be controlled in any experiment including cross reactivity, or the antibody recognizing multiple epitopes and affinity, which can vary widely different depending on experimental conditions such as pH, solvent, state of tissue etc. Multiple attempts have been made to improve both the way that researchers validate antibodies and ways in which they report on antibodies. Researchers using antibodies in their work need to record them correctly in order to allow their research to be reproducible (and therefore tested, and qualified by other researchers). Less than half of research antibodies referenced in academic papers can be easily identified. Papers published in F1000 in 2014 and 2015 provide researchers with a guide for reporting research antibody use. The RRID paper, is co-published in 4 journals that implemented the RRIDs Standard for research resource citation, which draws data from the antibodyregistry.org as the source of antibody identifiers (see also group at Force11)

Regulatory Validation of Monoclonal Antibody Products for Human Use

Production and testing:

Traditionally, most antibodies are produced by hybridoma cell lines through immortalization of antibody-producing cells by chemically-induced fusion with myeloma cells. In some cases, additional fusions with other lines have created “triomas» and «quadromas». The manufacturing process should be appropriately described and validated. Validation studies should at least include :

- The demonstration that the process is able to produce in good quality (the process should be validated)
- The efficiency of the antibody purification (all impurities and virus must be eliminated)
- The characterization of purified antibody (physicochemical characterization, immunological properties, biological activities, contaminants, ...)
- Determination of the virus clearance studies

Before clinical trials, studies of product safety and feasibility have to be performed :

- Product safety testing : Sterility (bacteria and fungi), In vitro and in vivo testing for adventitious viruses, Murine retrovirus testing... Product safety data needed before the initiation of feasibility trials in serious or immediately life-threatening conditions, it serves to evaluate dangerous potential of the product.
- Feasibility testing : These are pilot studies whose objectives include, among others, early characterization of safety and initial proof of concept in a small specific patient population (in vito or in vivo testing).

Preclinical studies :

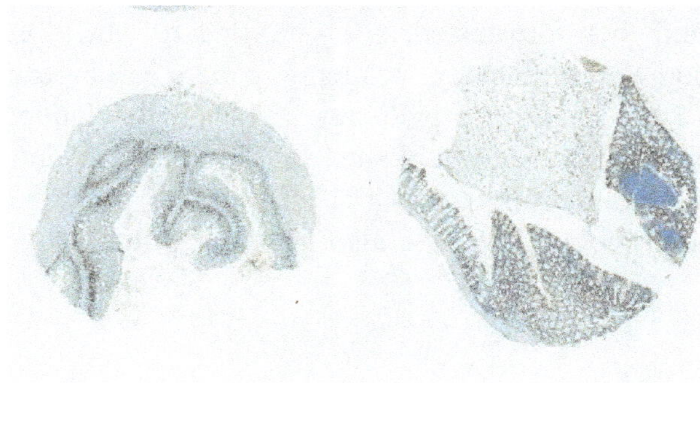

Example of tissue cross-reactivity on human tissue microarray

- Testing cross-reactivity of antibody : to highlight unwanted interactions (toxicity) of antibodies with previously characterized tissues. This study can be performed in vitro (Reactivity of the antibody or immunoconjugate should be determined with a quick-frozen adult tissues) or in vivo (with appropriates animal models). More informations about in vitro cross-reactivity testing.
- Preclinical pharmacology and toxicity testing : Preclinical safety testing of antibody is designed to identify possible toxicities in humans, to estimate the likelihood and severity of potential adverse events in humans, and to identify a safe starting dose and dose escalation, when possible.
- Animal toxicity studies : Acute toxicity testing, Repeat-dose toxicity testing, Long-term toxicity testing http://www.animalresearch.info/en/drug-development/safety-testing/
- Pharmacokinetics and pharmacodynamics testing : Use for determinate clinical dosages, antibody activities (AUC, pharmacodynamics, biodistribution, ...), evaluation of the potential clinical effects

Structure Prediction

The importance of antibodies in health care and the biotechnology industry demands knowledge of their structures at high resolution. This information is used for protein engineering, modifying the antigen binding affinity, and identifying an epitope, of a given antibody. X-ray crystallography is one commonly used method for determining antibody structures. However, crystallizing an antibody is often laborious and time-consuming. Computational approaches provide a cheaper and faster alternative to crystallography, but their results are more equivocal, since they do not produce empirical structures. Online web servers such as *Web Antibody Modeling* (WAM) and *Prediction of Immunoglobulin Structure* (PIGS) enables computational modeling of antibody variable regions. Rosetta Antibody is a novel antibody F_V region structure prediction server, which incorporates sophisticated techniques to minimize CDR loops and optimize the relative orientation of the light and heavy chains, as well as homology models that predict successful docking of antibodies with their unique antigen.

The ability to describe the antibody through binding affinity to the antigen is supplemented by information on antibody structure and amino acid sequences for the purpose of patent claims.

The first use of the term «antibody» occurred in a text by Paul Ehrlich. The term *Antikörper* (the German word for *antibody*) appears in the conclusion of his article "Experimental Studies on Immunity", published in October 1891, which states that, "if two substances give rise to two different antikörper, then they themselves must be different". However, the term was not accepted immediately and several other terms for antibody were proposed; these included *Immunkörper*, *Amboceptor*, *Zwischenkörper*, *substance sensibilisatrice*, *copula*, *Desmon*, *philocytase*, *fixateur*, and *Immunisin*. The word *antibody* has formal analogy to the word *antitoxin* and a similar concept to *Immunkörper* (*immune body* in English). As such, the original construction of the word contains a logical flaw;

the antitoxin is something directed against a toxin, while the antibody is a body directed against something.

Angel of the West (2008) by Julian Voss-Andreae is a sculpture based on the antibody structure published by E. Padlan. Created for the Florida campus of the Scripps Research Institute, the antibody is placed into a ring referencing Leonardo da Vinci›s *Vitruvian Man* thus highlighting the similarity of the antibody and the human body.

The study of antibodies began in 1890 when Kitasato Shibasaburō described antibody activity against diphtheria and tetanus toxins. Kitasato put forward the theory of humoral immunity, proposing that a mediator in serum could react with a foreign antigen. His idea prompted Paul Ehrlich to propose the side-chain theory for antibody and antigen interaction in 1897, when he hypothesized that receptors (described as «side-chains») on the surface of cells could bind specifically to toxins – in a «lock-and-key» interaction – and that this binding reaction is the trigger for the production of antibodies. Other researchers believed that antibodies existed freely in the blood and, in 1904, Almroth Wright suggested that soluble antibodies coated bacteria to label them for phagocytosis and killing; a process that he named opsoninization.

Michael Heidelberger

In the 1920s, Michael Heidelberger and Oswald Avery observed that antigens could be precipitated by antibodies and went on to show that antibodies are made of protein. The biochemical properties of antigen-antibody-binding interactions were examined in more detail in the late 1930s by John Marrack.The next major advance was in the 1940s, when Linus Pauling confirmed the lock-and-key theory proposed by Ehrlich by showing that the interactions between antibodies and antigens depend more on their shape than their chemical composition. In 1948, Astrid Fagreaus discovered that B cells, in the form of plasma cells, were responsible for generating antibodies.

Further work concentrated on characterizing the structures of the antibody proteins. A major advance in these structural studies was the discovery in the early 1960s by Gerald Edelman and Joseph Gally of the antibody light chain, and their realization that this protein is the same as the Bence-Jones protein described in 1845 by Henry Bence Jones. Edelman went on to discover that antibodies are composed of disulfide bond-linked heavy and light chains. Around the same time, antibody-binding (Fab) and antibody tail (Fc) regions of IgG were characterized by Rodney Porter. Together, these scientists deduced the structure and complete amino acid sequence of IgG, a feat for which they were jointly awarded the 1972 Nobel Prize in Physiology or Medicine. The Fv fragment was prepared and characterized by David Givol. While most of these early studies focused on IgM and IgG, other immunoglobulin isotypes were identified in the 1960s: Thomas Tomasi discovered secretory antibody (IgA); David S. Rowe and John L. Fahey discovered IgD; and Kimishige Ishizaka and Teruko Ishizaka discovered IgE and showed it was a class of antibodies involved in allergic reactions. In a landmark series of experiments beginning in 1976, Susumu Tonegawa showed that genetic material can rearrange itself to form the vast array of available antibodies.

Antibody Mimetic

Antibody mimetics are organic compounds that, like antibodies, can specifically bind antigens. They are usually artificial peptides or proteins with a molar mass of about 3 to 20 kDa. Nucleic acids and small molecules are sometimes considered antibody mimetics, but not artificial antibodies, antibody fragments and fusion proteins are composed from these. Common advantages over antibodies are better solubility, tissue penetration, stability towards heat and enzymes, and comparatively low production costs. Antibody mimetics such as the Affimer and the DARPin have being developed and commercialised as research, diagnostic and therapeutic agents.

REFERENCES

1. Rhoades RA, Pflanzer RG (2002). Human Physiology (4th ed.). Thomson Learning. p. 584. ISBN 0-534-42174-1.
2. Charles Janeway (2001). Immunobiology. (5th ed.). Garland Publishing. ISBN 0-8153-3642-X.
3. Litman GW, Rast JP, Shamblott MJ, Haire RN, Hulst M, Roess W, Litman RT, Hinds-Frey KR, Zilch A, Amemiya CT (January 1993). "Phylogenetic

diversification of immunoglobulin genes and the antibody repertoire". Mol. Biol. Evol. 10 (1): 60–72. PMID 8450761.

4. Maverakis E, Kim K, Shimoda M, Gershwin M, Patel F, Wilken R, Raychaudhuri S, Ruhaak LR, Lebrilla CB (2015). "Glycans in the immune system and The Altered Glycan Theory of Autoimmunity". J Autoimmun. 57 (6): 1–13. doi:10.1016/j.jaut.2014.12.002. PMC 4340844 . PMID 25578468.
5. Pier GB, Lyczak JB, Wetzler LM (2004). Immunology, Infection, and Immunity. ASM Press. ISBN 1-55581-246-5.
6. Borghesi L, Milcarek C (2006). "From B cell to plasma cell: regulation of V(D)J recombination and antibody secretion". Immunol. Res. 36 (1–3): 27–32. doi:10.1385/IR:36:1:27. PMID 17337763.
7. Parker D (1993). "T cell-dependent B cell activation". Annu Rev Immunol. 11 (1): 331–360. doi:10.1146/annurev.iy.11.040193.001555. PMID 8476565.
8. Market E, Papavasiliou FN (October 2003). "V(D)J recombination and the evolution of the adaptive immune system". PLoS Biol. 1 (1): E16. doi:10.1371/journal.pbio.0000016. PMC 212695 . PMID 14551913.
9. Diaz M, Casali P (2002). "Somatic immunoglobulin hypermutation". Curr Opin Immunol. 14 (2): 235–240. doi:10.1016/S0952-7915(02)00327-8. PMC 4621002 . PMID 11869898.
10. Parker D (1993). "T cell-dependent B cell activation". Annu. Rev. Immunol. 11 (1): 331–360. doi:10.1146/annurev.iy.11.040193.001555. PMID 8476565.
11. Wintrobe, Maxwell Myer (2004). John G. Greer; John Foerster; John N Lukens; George M Rodgers; Frixos Paraskevas, eds. Wintrobe's clinical hematology (11 ed.). Hagerstown, MD: Lippincott Williams & Wilkins. pp. 453–456. ISBN 978-0-7817-3650-3.
12. Tolar P, Sohn HW, Pierce SK (February 2008). "Viewing the antigen-induced initiation of B-cell activation in living cells". Immunol. Rev. 221 (1): 64–76. doi:10.1111/j.1600-065X.2008.00583.x. PMID 18275475.
13. Woof J, Burton D (2004). "Human antibody-Fc receptor interactions illuminated by crystal structures.". Nat Rev Immunol. 4 (2): 89–99. doi:10.1038/nri1266. PMID 15040582.
14. Underdown B, Schiff J (1986). "Immunoglobulin A: strategic defense initiative at the mucosal surface". Annu Rev Immunol. 4 (1): 389–417. doi:10.1146/annurev.iy.04.040186.002133. PMID 3518747.
15. Geisberger R, Lamers M, Achatz G (2006). "The riddle of the dual expression of IgM and IgD". Immunology. 118 (4): 889–898. doi:10.1111/j.1365-2567.2006.02386.x. PMC 1782314 . PMID 16895553.
16. Chen K, Xu W, Wilson M, He B, Miller NW, Bengtén E, Edholm ES, Santini PA, Rath P, Chiu A, Cattalini M, Litzman J, B Bussel J, Huang B, Meini A, Riesbeck K, Cunningham-Rundles C, Plebani A, Cerutti A (2009). "Immunoglobulin D enhances immune surveillance by activating antimicrobial, proinflammatory and B cell-stimulating programs in basophils". Nature Immunology. 10 (8): 889–898. doi:10.1038/ni.1748. PMC 2785232 . PMID 19561614.

17. Goding J (1978). "Allotypes of IgM and IgD receptors in the mouse: a probe for lymphocyte differentiation". Contemp Top Immunobiol. 8: 203–43. doi:10.1007/978-1-4684-0922-2_7. ISBN 978-1-4684-0924-6. PMID 357078.

18. Lundqvist, ML; Middleton, DL; Radford, C; Magor, KE (2006). "Immunoglobulins of the non-galliform birds: antibody expression and repertoire in the duck". Dev Comp Immunol. 30 (1): 93–100. doi:10.1016/j.dci.2005.06.019. PMC 1317265 . PMID 16150486.

19. Berstein, RM; Schluter, SF; Shen, S; Marchalonis, JJ (1996-04-16). "A new high molecular weight immunoglobulin class from the carcharhine shark: implications for the properties of the primordial immunoglobulin.". Proc Natl Acad Sci U S A. 93 (8): 3289–3293. doi:10.1073/pnas.93.8.3289. PMID 8622930.

20. Mattu T, Pleass R, Willis A, Kilian M, Wormald M, Lellouch A, Rudd P, Woof J, Dwek R (1998). "The glycosylation and structure of human serum IgA1, Fab, and Fc regions and the role of N-glycosylation on Fc alpha receptor interactions". J Biol Chem. 273 (4): 2260–2272. doi:10.1074/jbc.273.4.2260. PMID 9442070.

21. Roux K (1999). "Immunoglobulin structure and function as revealed by electron microscopy". Int Arch Allergy Immunol. 120 (2): 85–99. doi:10.1159/000024226. PMID 10545762.

22. Barclay A (2003). "Membrane proteins with immunoglobulin-like domains – a master superfamily of interaction molecules". Semin Immunol. 15 (4): 215–223. doi:10.1016/S1044-5323(03)00047-2. PMID 14690046.

23. Putnam FW, Liu YS, Low TL (1979). "Primary structure of a human IgA1 immunoglobulin. IV. Streptococcal IgA1 protease, digestion, Fab and Fc fragments, and the complete amino acid sequence of the alpha 1 heavy chain". J Biol Chem. 254 (8): 2865–74. PMID 107164.

24. Al-Lazikani B, Lesk AM, Chothia C (1997). "Standard conformations for the canonical structures of immunoglobulins". J Mol Biol. 273 (4): 927–948. doi:10.1006/jmbi.1997.1354. PMID 9367782.

25. North B, Lehmann A, Dunbrack RL (2010). "A new clustering of antibody CDR loop conformations". J Mol Biol. 406 (2): 228–256. doi:10.1016/j.jmb.2010.10.030. PMC 3065967 . PMID 21035459.

26. Nikoloudis D, Pitts JE, Saldanha JW (2014). "A complete, multi-level conformational clustering of antibody complementarity-determining regions". PeerJ. 2 (e456). doi:10.7717/peerj.456. PMC 4103072 . PMID 25071986.

27. Heyman B (1996). "Complement and Fc-receptors in regulation of the antibody response". Immunol Lett. 54 (2–3): 195–199. doi:10.1016/S0165-2478(96)02672-7. PMID 9052877.

28. Borghesi L, Milcarek C (2006). "From B cell to plasma cell: regulation of V(D)J recombination and antibody secretion". Immunol Res. 36 (1–3): 27–32. doi:10.1385/IR:36:1:27. PMID 17337763.

29. Ravetch J, Bolland S (2001). "IgG Fc receptors". Annu Rev Immunol. 19 (1): 275–290. doi:10.1146/annurev.immunol.19.1.275. PMID 11244038.

30. Rus H, Cudrici C, Niculescu F (2005). "The role of the complement system in innate immunity". Immunol Res. 33 (2): 103–112. doi:10.1385/IR:33:2:103. PMID 16234578.
31. Racaniello, Vincent (6 October 2009). "Natural antibody protects against viral infection". Virology Blog. Archived from the original on 17 November 2010. Retrieved 22 January 2010.
32. Milland J, Sandrin MS (December 2006). "ABO blood group and related antigens, natural antibodies and transplantation". Tissue Antigens. 68 (6): 459–466. doi:10.1111/j.1399-0039.2006.00721.x. PMID 17176435.
33. Mian I, Bradwell A, Olson A (1991). "Structure, function and properties of antibody binding sites". J Mol Biol. 217 (1): 133–151. doi:10.1016/0022-2836(91)90617-F. PMID 1988675.
34. Fanning LJ, Connor AM, Wu GE (1996). "Development of the immunoglobulin repertoire". Clin. Immunol. Immunopathol. 79 (1): 1–14. doi:10.1006/clin.1996.0044. PMID 8612345.
35. Nemazee D (2006). "Receptor editing in lymphocyte development and central tolerance". Nat Rev Immunol. 6 (10): 728–740. doi:10.1038/nri1939. PMID 16998507.
36. Peter Parham. "The Immune System. 2nd ed. Garland Science: New York, 2005. pg.47–62
37. Mraz, M.; Dolezalova, D.; Plevova, K.; Stano Kozubik, K.; Mayerova, V.; Cerna, K.; Musilova, K.; Tichy, B.; Pavlova, S.; Borsky, M.; Verner, J.; Doubek, M.; Brychtova, Y.; Trbusek, M.; Hampl, A.; Mayer, J.; Pospisilova, S. (2012). "MicroRNA-650 expression is influenced by immunoglobulin gene rearrangement and affects the biology of chronic lymphocytic leukemia". Blood. 119 (9): 2110–2113. doi:10.1182/blood-2011-11-394874. PMID 22234685.
38. Market, Eleonora; Papavasiliou, F. Nina (October 2003). "V(D)J Recombination and the Evolution of the Adaptive Immune System". PLoS Biology. 1 (1): E16. doi:10.1371/journal.pbio.0000016. PMC 212695 . PMID 14551913.
39. Bergman Y, Cedar H (2004). "A stepwise epigenetic process controls immunoglobulin allelic exclusion". Nat Rev Immunol. 4 (10): 753–761. doi:10.1038/nri1458. PMID 15459667.
40. Honjo T, Habu S (1985). "Origin of immune diversity: genetic variation and selection". Annu Rev Biochem. 54 (1): 803–830. doi:10.1146/annurev.bi.54.070185.004103. PMID 3927822.
41. Or-Guil M, Wittenbrink N, Weiser AA, Schuchhardt J (2007). "Recirculation of germinal center B cells: a multilevel selection strategy for antibody maturation". Immunol. Rev. 216: 130–41. doi:10.1111/j.1600-065X.2007.00507.x. PMID 17367339.
42. Neuberger M, Ehrenstein M, Rada C, Sale J, Batista F, Williams G, Milstein C (March 2000). "Memory in the B-cell compartment: antibody affinity maturation". Philos Trans R Soc Lond B Biol Sci. 355 (1395): 357–360. doi:10.1098/rstb.2000.0573. PMC 1692737 . PMID 10794054.

43. Stavnezer J, Amemiya CT (2004). "Evolution of isotype switching". Semin. Immunol. 16 (4): 257–275. doi:10.1016/j.smim.2004.08.005. PMID 15522624.
44. Durandy A (2003). "Activation-induced cytidine deaminase: a dual role in class-switch recombination and somatic hypermutation". Eur. J. Immunol. 33 (8): 2069–2073. doi:10.1002/eji.200324133. PMID 12884279.
45. Casali P, Zan H (2004). "Class switching and Myc translocation: how does DNA break?". Nat. Immunol. 5 (11): 1101–1103. doi:10.1038/ni1104-1101. PMID 15496946.
46. Lieber MR, Yu K, Raghavan SC (2006). "Roles of nonhomologous DNA end joining, V(D)J recombination, and class switch recombination in chromosomal translocations". DNA Repair (Amst.). 5 (9–10): 1234–1245. doi:10.1016/j.dnarep.2006.05.013. PMID 16793349.
47. page 22 in: Shoenfeld, Yehuda.; Meroni, Pier-Luigi.; Gershwin, M. Eric (2007). Autoantibodie. Amsterdam; Boston: Elsevier. ISBN 978-0-444-52763-9.
48. Farlex dictionary: monovalent Citing: The American Heritage Science Dictionary, Copyright 2005
49. Farlex dictionary > polyvalent Citing: The American Heritage Medical Dictionary. 2004
50. Gunasekaran K (2010). "Enhancing antibody Fc heterodimer formation through electrostatic steering effects: applications to bispecific molecules and monovalent IgG". The Journal of Biological Chemistry. 285 (25): 19637–19646. doi:10.1074/jbc.M110.117382. PMC 2885242 . PMID 20400508.
51. Muller K.M (1998). "The first constant domain (CH1 and CL) of an antibody used as heterodimerization domain for bispecific miniantibodies". FEBS Letters. 422 (2): 259–264. doi:10.1016/s0014-5793(98)00021-0.
52. Gao C (1999). "Making artificial antibodies: A format for phage display of combinatorial heterodimeric arrays". PNAS. 96 (11): 6025–6030. doi:10.1073/pnas.96.11.6025. PMC 26829 . PMID 10339535.
53. "Animated depictions of how antibodies are used in ELISA assays". Cellular Technology Ltd.—Europe. Archived from the original on 17 November 2010. Retrieved 8 May 2007.
54. "Animated depictions of how antibodies are used in ELISPOT assays". Cellular Technology Ltd.—Europe. Archived from the original on 17 November 2010. Retrieved 8 May 2007.
55. Stern P (2006). "Current possibilities of turbidimetry and nephelometry" (PDF). Klin Biochem Metab. 14 (3): 146–151. Archived (PDF) from the original on 17 November 2010.
56. Dean, Laura (2005). "Chapter 4: Hemolytic disease of the newborn". Blood Groups and Red Cell Antigens. NCBI Bethesda (MD): National Library of Medicine (US),.
57. Rodriguez, Erik A.; Wang, Ye; Crisp, Jessica L.; Vera, David R.; Tsien, Roger Y.; Ting, Richard (2016-04-27). "New Dioxaborolane Chemistry Enables [18F]-

Positron-Emitting, Fluorescent [18F]-Multimodality Biomolecule Generation from the Solid Phase". Bioconjugate Chemistry. 27 (5): 1390–1399. doi:10.1021/acs.bioconjchem.6b00164. PMC 4916912 . PMID 27064381.

58. Feldmann M, Maini R (2001). "Anti-TNF alpha therapy of rheumatoid arthritis: what have we learned?". Annu Rev Immunol. 19 (1): 163–196. doi:10.1146/annurev.immunol.19.1.163. PMID 11244034.

59. Doggrell S (2003). "Is natalizumab a breakthrough in the treatment of multiple sclerosis?". Expert Opin Pharmacother. 4 (6): 999–1001. doi:10.1517/14656566.4.6.999. PMID 12783595.

60. Krueger G, Langley R, Leonardi C, Yeilding N, Guzzo C, Wang Y, Dooley L, Lebwohl M (2007). "A human interleukin-12/23 monoclonal antibody for the treatment of psoriasis". N Engl J Med. 356 (6): 580–592. doi:10.1056/NEJMoa062382. PMID 17287478.

61. Plosker G, Figgitt D (2003). "Rituximab: a review of its use in non-Hodgkin's lymphoma and chronic lymphocytic leukaemia". Drugs. 63 (8): 803–843. doi:10.2165/00003495-200363080-00005. PMID 12662126.

62. Vogel C, Cobleigh M, Tripathy D, Gutheil J, Harris L, Fehrenbacher L, Slamon D, Murphy M, Novotny W, Burchmore M, Shak S, Stewart S (2001). "First-line Herceptin monotherapy in metastatic breast cancer". Oncology. 61. Suppl 2 (Suppl. 2): 37–42. doi:10.1159/000055400. PMID 11694786.

63. LeBien TW (1 July 2000). "Fates of human B-cell precursors". Blood. 96 (1): 9–23. PMID 10891425. Archived from the original on 17 November 2010.

64. Ghaffer A (26 March 2006). "Immunization". Immunology — Chapter 14. University of South Carolina School of Medicine. Archivedfrom the original on 17 November 2010. Retrieved 6 June 2007.

65. Urbaniak S, Greiss M (2000). "RhD haemolytic disease of the fetus and the newborn". Blood Rev. 14 (1): 44–61. doi:10.1054/blre.1999.0123. PMID 10805260.

66. a b Fung Kee Fung K; Eason E; Crane J; Armson A; De La Ronde S; Farine D; Keenan-Lindsay L; Leduc L; Reid GJ; Aerde JV; Wilson RD; Davies G; Désilets VA; Summers A; Wyatt P; Young DC; Maternal-Fetal Medicine Committee; Genetics Committee (2003). "Prevention of Rh alloimmunization". J Obstet Gynaecol Can. 25 (9): 765–73. PMID 12970812.

67. Tini M, Jewell UR, Camenisch G, Chilov D, Gassmann M (2002). "Generation and application of chicken egg-yolk antibodies". Comp. Biochem. Physiol., Part a Mol. Integr. Physiol. 131 (3): 569–574. doi:10.1016/S1095-6433(01)00508-6. PMID 11867282.

68. Cole SP, Campling BG, Atlaw T, Kozbor D, Roder JC (1984). "Human monoclonal antibodies". Mol. Cell. Biochem. 62 (2): 109–20. doi:10.1007/BF00223301. PMID 6087121.

69. Kabir S (2002). "Immunoglobulin purification by affinity chromatography using protein A mimetic ligands prepared by combinatorial chemical synthesis". Immunol Invest. 31 (3–4): 263–278. doi:10.1081/IMM-120016245. PMID 12472184.

70. Brehm-Stecher B, Johnson E (2004). "Single-cell microbiology: tools, technologies, and applications". Microbiol Mol Biol Rev. 68 (3): 538–559. doi:10.1128/MMBR.68.3.538-559.2004. PMC 515252 . PMID 15353569. Archived from the original on 17 November 2010.

71. Williams N (2000). "Immunoprecipitation procedures". Methods Cell Biol. Methods in Cell Biology. 62: 449–453. doi:10.1016/S0091-679X(08)61549-6. ISBN 978-0-12-544164-3. PMID 10503210.

72. Kurien B, Scofield R (2006). "Western blotting". Methods. 38 (4): 283–293. doi:10.1016/j.ymeth.2005.11.007. PMID 16483794.

73. Scanziani E (1998). "Immunohistochemical staining of fixed tissues". Methods Mol Biol. 104: 133–140. doi:10.1385/0-89603-525-5:133. ISBN 978-0-89603-525-6. PMID 9711649.

74. Reen DJ. (1994). "Enzyme-linked immunosorbent assay (ELISA)". Methods Mol Biol. 32: 461–466. doi:10.1385/0-89603-268-X:461. ISBN 0-89603-268-X. PMID 7951745.

75. Kalyuzhny AE (2005). "Chemistry and biology of the ELISPOT assay". Methods Mol Biol. 302: 015–032. doi:10.1385/1-59259-903-6:015. ISBN 1-59259-903-6. PMID 15937343.

76. Saper (2005). "An open letter to our readers on the use of antibodies". Journal of Comparative Neurology. 493: 477–8. doi:10.1002/cne.20839. PMID 16304632.

77. "Implementing Rigor and Transparency in NIH & AHRQ Research Grant Applications".

78. "On the reproducibility of science: unique identification of research resources in the biomedical literature". PeerJ. 2 September 2013. Retrieved 1 September 2014.

79. Bandrowski A, Brush M, Grethe JS, Haendel MA, Kennedy DN, Hill S, Hof PR, Martone ME, Pols M, Tan S, Washington N, Zudilova-Seinstra E, Vasilevsky N (2015). "The Resource Identification Initiative: a cultural shift in publishing". F1000Res. 4: 134. doi:10.12688/f1000research.6555.2. PMC 4648211 . PMID 26594330.

80. "Reporting research antibody use: how to increase experimental reproducibility". F1000. 23 August 2013. Retrieved 1 September 2014.

81. "The Antibody Registry".

82. "Resource Identification Initiative". FORCE11. Retrieved 18 April 2016.

83. Archived 18 November 2010 at WebCite

84. WAM

85. Marcatili P, Rosi A, Tramontano A (2008). "PIGS: automatic prediction of antibody structures". Bioinformatics. 24 (17): 1953–1954. doi:10.1093/bioinformatics/btn341. PMID 18641403. Archivedfrom the original on 17 November 2010. Prediction of Immunoglobulin Structure (PIGS)

86. Archived 18 November 2010 at WebCite

87. RosettaAntibody

88. Park, Hyeongsu. “Written Description Problems of the Monoclonal Antibody Patents after Centocor v. Abbott”. jolt.law.harvard.edu. Retrieved 12 Dec 2014.
89. Lindenmann, Jean (1984). “Origin of the Terms ‘Antibody’ and ‘Antigen’”. Scand. J. Immunol. 19 (4): 281–5. doi:10.1111/j.1365-3083.1984.tb00931.x. PMID 6374880. Archived from the originalon 17 November 2010.
90. Padlan, Eduardo (February 1994). “Anatomy of the antibody molecule”. Mol. Immunol. 31 (3): 169–217. doi:10.1016/0161-5890(94)90001-9. PMID 8114766.
91. “New Sculpture Portraying Human Antibody as Protective Angel Installed on Scripps Florida Campus”. Archived from the original on 17 November 2010. Retrieved 12 December 2008.
92. Pescovitz, David. “Protein sculpture inspired by Vitruvian Man”. Archived from the original on 17 November 2010. Retrieved 12 December 2008.
93. “Emil von Behring — Biography”. Archived from the original on 17 November 2010. Retrieved 5 June 2007.
94. AGN (1931). “The Late Baron Shibasaburo Kitasato”. Canadian Medical Association Journal. 25 (2): 206. PMC 382621 . PMID 20318414.
95. Winau F, Westphal O, Winau R (2004). “Paul Ehrlich—in search of the magic bullet”. Microbes Infect. 6 (8): 786–789. doi:10.1016/j.micinf.2004.04.003. PMID 15207826.
96. Silverstein AM (2003). “Cellular versus humoral immunology: a century-long dispute”. Nat. Immunol. 4 (5): 425–428. doi:10.1038/ni0503-425. PMID 12719732.
97. Van Epps HL (2006). “Michael Heidelberger and the demystification of antibodies” (PDF). J. Exp. Med. 203 (1): 5. doi:10.1084/jem.2031fta. PMC 2118068 . PMID 16523537. Archived (PDF) from the original on 17 November 2010.
98. Marrack, JR (1938). Chemistry of antigens and antibodies (2nd ed.). London: His Majesty’s Stationery Office. OCLC 3220539.
99. “The Linus Pauling Papers: How Antibodies and Enzymes Work”. Archived from the original on 17 November 2010. Retrieved 5 June 2007.
100. Silverstein AM (2004). “Labeled antigens and antibodies: the evolution of magic markers and magic bullets” (PDF). Nat. Immunol. 5 (12): 1211–1217. doi:10.1038/ni1140. PMID 15549122. Archived(PDF) from the original on 18 December 2009.
101. Edelman GM, Gally JA (1962). “The nature of Bence-Jones proteins. Chemical similarities to polypetide chains of myeloma globulins and normal gamma-globulins”. J. Exp. Med. 116 (2): 207–227. doi:10.1084/jem.116.2.207. PMC 2137388 . PMID 13889153.
102. Stevens FJ, Solomon A, Schiffer M (1991). “Bence Jones proteins: a powerful tool for the fundamental study of protein chemistry and pathophysiology”. Biochemistry. 30 (28): 6803–6805. doi:10.1021/bi00242a001. PMID 2069946.
103. Raju TN (1999). “The Nobel chronicles. 1972: Gerald M Edelman (b 1929) and Rodney R Porter (1917–85)”. Lancet. 354 (9183): 1040. doi:10.1016/S0140-6736(05)76658-7. PMID 10501404.

104. Hochman J, Inbar D, Givol D (1973). "An active antibody fragment (Fv) composed of the variable portions of heavy and light chains". Biochemistry. 12 (6): 1130–1135. doi:10.1021/bi00730a018. PMID 4569769.

105. Tomasi TB (1992). "The discovery of secretory IgA and the mucosal immune system". Immunol. Today. 13 (10): 416–418. doi:10.1016/0167-5699(92)90093-M. PMID 1343085.

106. Preud'homme JL; Petit I; Barra A; Morel F; Lecron JC; Lelièvre E (2000). "Structural and functional properties of membrane and secreted IgD". Mol. Immunol. 37 (15): 871–887. doi:10.1016/S0161-5890(01)00006-2. PMID 11282392.

107. Johansson SG (2006). "The discovery of immunoglobulin E". Allergy and Asthma Proceedings. 27 (2 Suppl 1): S3–6. PMID 16722325.

108. Hozumi N, Tonegawa S (1976). "Evidence for somatic rearrangement of immunoglobulin genes coding for variable and constant regions". Proc. Natl. Acad. Sci. U.S.A. 73 (10): 3628–3632. doi:10.1073/pnas.73.10.3628. PMC 431171 . PMID 824647.

109. Gebauer M, Skerra A (June 2009). "Engineered protein scaffolds as next-generation antibody therapeutics". Curr Opin Chem Biol. 13 (3): 245–255. doi:10.1016/j.cbpa.2009.04.627. PMID 19501012.

Biological Membrane

Chapter Contents

Biological membranes are composed of lipid, protein and carbohydrate that exist in a fluid state. Biological membranes are the structures that define and control the composition of the space that they enclose. All membranes exist as dynamic structures whose composition changes throughout the life of a cell. In addition to the outer membrane that results in the formation of a typical cell (this membrane is often referred to as the plasma membrane), cells contain intracellular membranes that serve distinct functions in the formation of the various intracellular organelles, e.g. the nucleus and the mitochondria.

INTRODUCTION

The cell membrane (also known as the plasma membrane or cytoplasmic membrane) is a biological membrane that separates the interior of all cells from the outside environment. The cell membrane is selectively permeable to ions and organic molecules and controls the movement of substances in and out of cells. The basic function of the cell membrane is to protect the cell from its surroundings.

It consists of the phospholipid bilayer with embedded proteins. Cell membranes are involved in a variety of cellular processes such as cell adhesion, ion conductivity and cell signalling and serve as the attachment surface for several extracellular structures, including the cell wall, glycocalyx, and intracellular cytoskeleton. Cell membranes can be artificially reassembled.

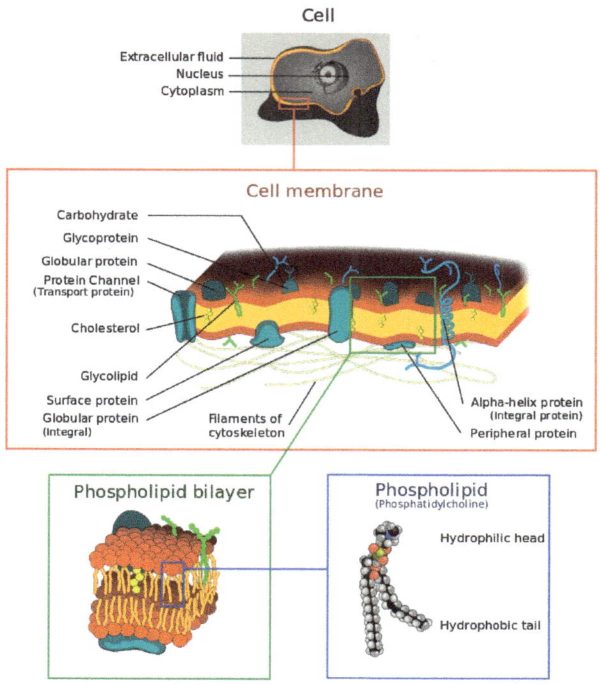

Illustration of a Eukaryotic Cell Membrane

The structure has been variously referred to by different writers as the ectoplast (de Vries, 1885), *Plasmahaut* (plasma skin, Pfeffer, 1877, 1891), *Hautschicht* (skin layer, Pfeffer, 1886; used with a different meaning by Hofmeister, 1867), plasmatic membrane (Pfeffer, 1900), plasma membrane, cytoplasmic membrane, cell envelope and cell membrane.

Some authors that did not believe that there was a functional permeable boundary at the surface of the cell preferred to use the term plasmalemma (coined by Mast, 1924) to the extern region of the cell.

FUNCTION

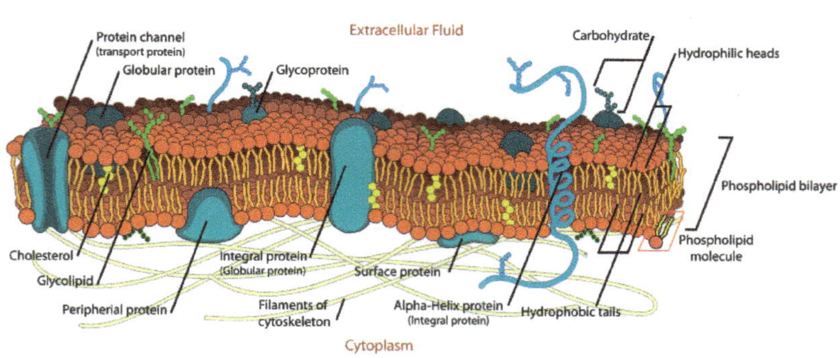

A detailed diagram of the cell membrane

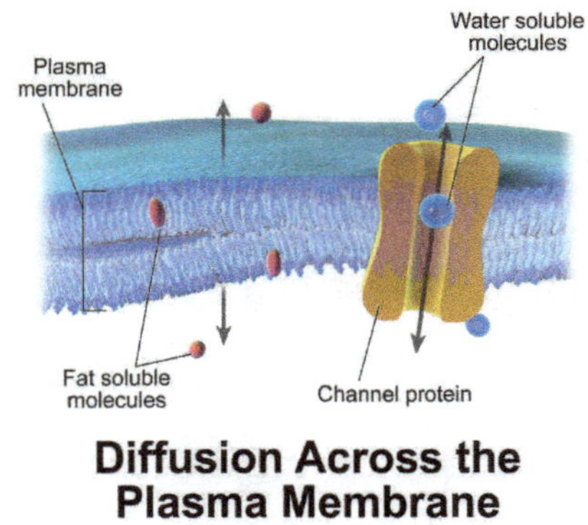

Illustration depicting cellular diffusion

The cell membrane (or plasma membrane or plasmalemma) surrounds the cytoplasm of living cells, physically separating the intracellular components from the extracellular environment. Fungi, bacteria and plants have a cell wall in addition, which provides a mechanical support to the cell and precludes the passage of larger molecules. The cell membrane also plays a role in anchoring the cytoskeleton to provide shape to the cell, and in attaching to the extracellular matrix and other cells to hold them together to form tissues.

The cell membrane is selectively permeable and able to regulate what enters and exits the cell, thus facilitating the transport of materials needed for survival. The movement of substances across the membrane can be either "passive", occurring without the input of cellular energy, or "active", requiring the cell to expend energy in transporting it. The membrane also maintains the cell potential. The cell membrane thus works as a selective filter that allows only certain things to come inside or go outside the cell. The cell employs a number of transport mechanisms that involve biological membranes:

1. Passive osmosis and diffusion: Some substances (small molecules, ions) such as carbon dioxide (CO_2) and oxygen (O_2), can move across the plasma membrane by diffusion, which is a passive transport process. Because the membrane acts as a barrier for certain molecules and ions, they can occur in different concentrations on the two sides of the membrane. Such a concentration gradient across a semipermeable membrane sets up an osmotic flow for the water.
2. Transmembrane protein channels and transporters: Nutrients, such as sugars or amino acids, must enter the cell, and certain products of metabolism must leave the cell. Such molecules diffuse passively through protein channels such as aquaporins (in the case of water (H_2O)) in facilitated diffusion or are pumped across the membrane by transmembrane transporters. Protein channel proteins, also called *permeases,* are usually quite specific, recognizing and transporting only a limited food group of chemical substances, often even only a single substance.

3. Endocytosis: Endocytosis is the process in which cells absorb molecules by engulfing them. The plasma membrane creates a small deformation inward, called an invagination, in which the substance to be transported is captured. The deformation then pinches off from the membrane on the inside of the cell, creating a vesicle containing the captured substance. Endocytosis is a pathway for internalizing solid particles ("cell eating" or phagocytosis), small molecules and ions ("cell drinking" or pinocytosis), and macromolecules. Endocytosis requires energy and is thus a form of active transport.
4. Exocytosis: Just as material can be brought into the cell by invagination and formation of a vesicle, the membrane of a vesicle can be fused with the plasma membrane, extruding its contents to the surrounding medium. This is the process of exocytosis. Exocytosis occurs in various cells to remove undigested residues of substances brought in by endocytosis, to secrete substances such as hormones and enzymes, and to transport a substance completely across a cellular barrier. In the process of exocytosis, the undigested waste-containing food vacuole or the secretory vesicle budded from Golgi apparatus, is first moved by cytoskeleton from the interior of the cell to the surface. The vesicle membrane comes in contact with the plasma membrane. The lipid molecules of the two bilayers rearrange themselves and the two membranes are, thus, fused. A passage is formed in the fused membrane and the vesicles discharges its contents outside the cell.

PROKARYOTES

Gram-negative bacteria have both a plasma membrane and an outer membrane separated by periplasm. Other prokaryotes have only a plasma membrane. Prokaryotic cells are also surrounded by a cell wall composed of peptidoglycan (amino acids and sugars). Some eukaryotic cells also have cell walls, but none that are made of peptidoglycan. The **outer membrane of gram negative microbes** is rich in lipopolysaccharide and thus is different from cell membrane of the microbes. The outer membrane can bleb out into periplasmic protrusions under stress conditions or upon virulence requirements while encountering a host target cell, and thus such blebs may work as virulence organelles.

Structures

Fluid Mosaic Model

According to the fluid mosaic model of S. J. Singer and G. L. Nicolson (1972), which replaced the earlier model of Davson and Danielli, biological membranes can be considered as a two-dimensional liquid in which lipid and protein molecules diffuse more or less easily. Although the lipid bilayers that form the basis of the membranes do indeed form two-dimensional liquids by themselves, the plasma membrane also contains a large quantity of proteins, which provide more structure. Examples of such structures are protein-protein complexes, pickets and fences formed by the actin-based cytoskeleton, and potentially lipid rafts.

Lipid Bilayer

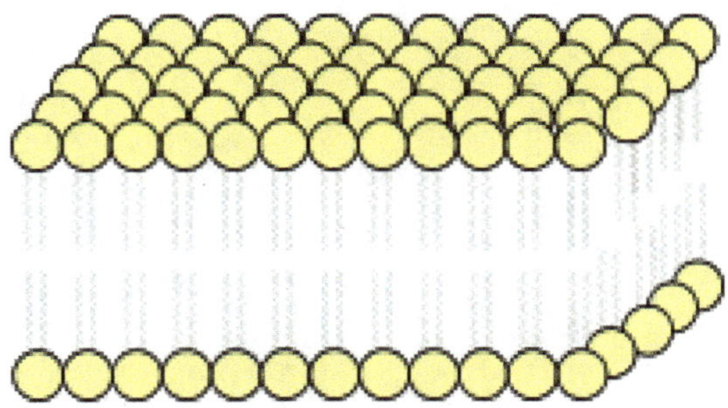

Diagram of the arrangement of amphipathic lipid molecules to form a lipid bilayer. The yellow polar head groups separate the grey hydrophobic tails from the aqueous cytosolic and extracellular environments.

Lipid bilayers form through the process of self-assembly. The cell membrane consists primarily of a thin layer of amphipathic phospholipids that spontaneously arrange so that the hydrophobic "tail" regions are isolated from the surrounding water while the hydrophilic "head" regions interact with the intracellular (cytosolic) and extracellular faces of the resulting bilayer. This forms a continuous, spherical lipid bilayer. Hydrophobic interactions (also known as the hydrophobic effect) are the major driving forces in the formation of lipid bilayers. An increase in interactions between hydrophobic molecules (causing clustering of hydrophobic regions) allows water molecules to bond more freely with each other, increasing the entropy of the system. This complex interaction can include noncovalent interactions such as van der Waals, electrostatic and hydrogen bonds.

Lipid bilayers are generally impermeable to ions and polar molecules. The arrangement of hydrophilic heads and hydrophobic tails of the lipid bilayer prevent polar solutes (ex. amino acids, nucleic acids, carbohydrates, proteins, and ions) from diffusing across the membrane, but generally allows for the passive diffusion of hydrophobic molecules. This affords the cell the ability to control the movement of these substances via transmembrane protein complexes such as pores, channels and gates.

Flippases and scramblases concentrate phosphatidyl serine, which carries a negative charge, on the inner membrane. Along with NANA, this creates an extra barrier to charged moieties moving through the membrane.

Membranes serve diverse functions in eukaryotic and prokaryotic cells. One important role is to regulate the movement of materials into and out of cells. The phospholipid bilayer structure (fluid mosaic model) with specific membrane proteins accounts for the selective permeability of the membrane and passive and active transport mechanisms. In addition, membranes in prokaryotes and in the mitochondria and chloroplasts of eukaryotes facilitate the synthesis of ATP through chemiosmosis.

Membrane Polarity

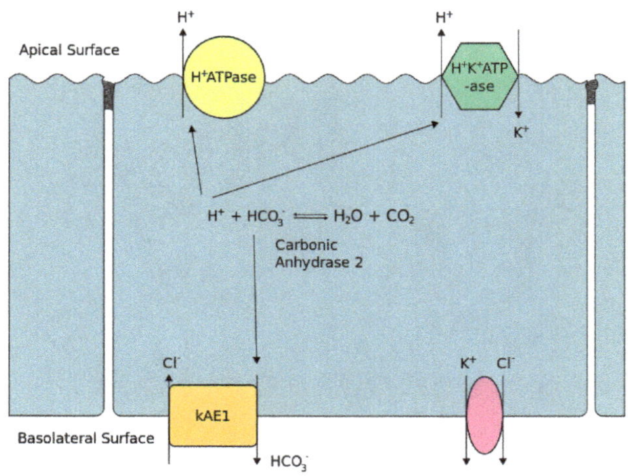

Alpha intercalated cell

The apical membrane of a polarized cell is the surface of the plasma membrane that faces inward to the lumen. This is particularly evident in epithelial and endothelial cells, but also describes other polarized cells, such as neurons. The basolateral membrane of a polarized cell is the surface of the plasma membrane that forms its basal and lateral surfaces. It faces outwards, towards the interstitium, and away from the lumen. Basolateral membrane is a compound phrase referring to the terms "basal (base) membrane" and "lateral (side) membrane", which, especially in epithelial cells, are identical in composition and activity. Proteins (such as ion channels and pumps) are free to move from the basal to the lateral surface of the cell or vice versa in accordance with the fluid mosaic model. Tight junctions join epithelial cells near their apical surface to prevent the migration of proteins from the basolateral membrane to the apical membrane. The basal and lateral surfaces thus remain roughly equivalent to one another, yet distinct from the apical surface.

Membrane Structures

Cell membrane can form different types of "supramembrane" structures such as caveola, postsynaptic density, podosome, invadopodium, focal adhesion, and different types of cell junctions. These structures are usually responsible for cell adhesion, communication, endocytosis and exocytosis. They can be visualized by electron microscopy or fluorescence microscopy. They are composed of specific proteins, such as integrins and cadherins.

Cytoskeleton

The cytoskeleton is found underlying the cell membrane in the cytoplasm and provides a scaffolding for membrane proteins to anchor to, as well as forming organelles that extend from the cell. Indeed, cytoskeletal elements interact extensively and intimately with the cell membrane.Anchoring proteins restricts them to a particular cell surface — for example, the apical surface of epithelial cells that line the vertebrate gut — and limits how far they may diffuse within the bilayer. The cytoskeleton is able to form appendage-like organelles, such as cilia, which are microtubule-based extensions covered by the cell membrane, and filopodia, which are actin-based extensions. These extensions are ensheathed in membrane and project from the surface of the cell in order to sense the external environment and/or make contact with the substrate or other cells. The apical surfaces of epithelial cells are dense with actin-based finger-like projections known as microvilli, which increase cell surface area and thereby increase the absorption rate of nutrients. Localized decoupling of the cytoskeleton and cell membrane results in formation of a bleb.

Composition

Cell membranes contain a variety of biological molecules, notably lipids and proteins. Material is incorporated into the membrane, or deleted from it, by a variety of mechanisms:

- Fusion of intracellular vesicles with the membrane (exocytosis) not only excretes the contents of the vesicle but also incorporates the vesicle membrane's components into the cell membrane. The membrane may form blebs around extracellular material that pinch off to become vesicles (endocytosis).
- If a membrane is continuous with a tubular structure made of membrane material, then material from the tube can be drawn into the membrane continuously.
- Although the concentration of membrane components in the aqueous phase is low (stable membrane components have low solubility in water), there is an exchange of molecules between the lipid and aqueous phases.

Lipids

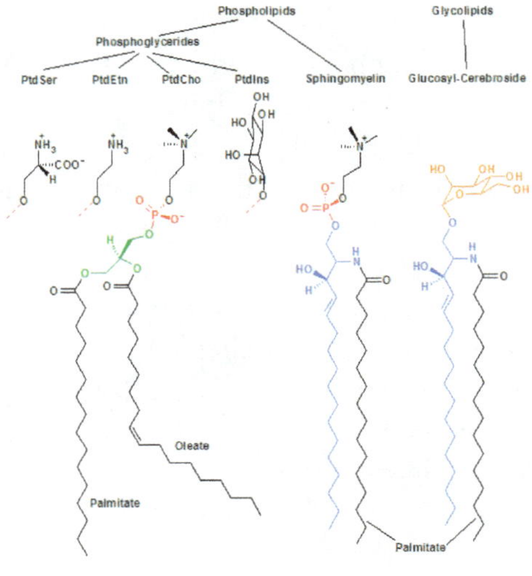

Examples of the major membrane phospholipids and glycolipids: phosphatidylcholine (PtdCho), phosphatidylethanolamine (PtdEtn), phosphatidylinositol (PtdIns), phosphatidylserine (PtdSer).

The cell membrane consists of three classes of amphipathic lipids: phospholipids, glycolipids, and sterols. The amount of each depends upon the type of cell, but in the majority of cases phospholipids are the most abundant. In RBC studies, 30% of the plasma membrane is lipid.

The fatty chains in phospholipids and glycolipids usually contain an even number of carbon atoms, typically between 16 and 20. The 16- and 18-carbon fatty acids are the most common. Fatty acids may be saturated or unsaturated, with the configuration of the double bonds nearly always "cis". The length and the degree of unsaturation of fatty acid chains have a profound effect on membrane fluidity as unsaturated lipids create a kink, preventing the fatty acids from packing together as tightly, thus decreasing the melting temperature (increasing the fluidity) of the membrane. The ability of some organisms to regulate the fluidity of their cell membranes by altering lipid composition is called homeoviscous adaptation.

The entire membrane is held together via non-covalent interaction of hydrophobic tails, however the structure is quite fluid and not fixed rigidly in place. Under physiological conditions phospholipid molecules in the cell membrane are in the liquid crystalline state. It means the lipid molecules are free to diffuse and exhibit rapid lateral diffusion along the layer in which they are present. However, the exchange of phospholipid molecules between intracellular and extracellular leaflets of the bilayer is a very slow process. Lipid rafts and caveolae are examples of cholesterol-enriched microdomains in the cell membrane. Also, a fraction of the lipid in direct contact with integral membrane proteins, which is tightly bound to the protein surface is called annular lipid shell; it behaves as a part of protein complex.

In animal cells cholesterol is normally found dispersed in varying degrees throughout cell membranes, in the irregular spaces between the hydrophobic tails of the membrane lipids, where it confers a stiffening and strengthening effect on the membrane.

MEMBRANE LIPIDS

Membrane lipids are a group of compounds (structurally similar to fats and oils) which form the double-layered surface of all cells (lipid bilayer). The three major classes of membrane lipids are phospholipids, glycolipids, and cholesterol. Lipids are amphiphilic: they have one end that is soluble in water ('polar') and an ending that is soluble in fat ('nonpolar'). By forming a double layer with the polar ends pointing outwards and the nonpolar ends pointing inwards membrane lipids can form a 'lipid bilayer' which keeps the watery interior of the cell separate from the watery exterior. The arrangements of lipids and various proteins, acting as receptors and channel pores in the membrane, control the entry and exit of other molecules and ions as part of the cell's metabolism. In order to perform physiological functions, membrane proteins are facilitated to rotate and diffuse laterally in two dimensional expanse of lipid bilayer by the presence of a shell of lipids closely attached to protein surface, called annular lipid shell.

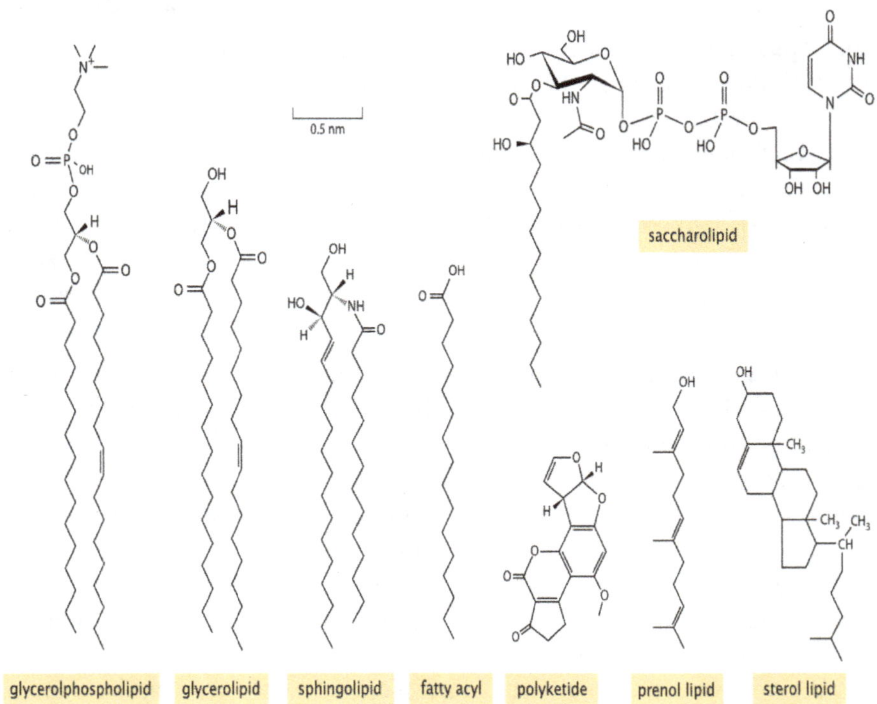

The major membrane lipids

Biological Roles

The bilayer formed by membrane lipids serves as a containment unit of a living cell. Membrane lipids also form a matrix in which membrane proteins reside. Historically lipids were thought to merely serve a structural role. Functional roles of lipids are in fact many: They serve as regulatory agents in cell growth and adhesion. They participate in the biosynthesis of other biomolecules. They can serve to increase enzymatic activities of enzymes.

Non-bilayer forming lipid like monogalactosyl diglyceride(MGDG) predominates the bulk lipids in thylakoid membranes, which when hydrated alone, forms reverse hexagonal cylindrical phase. However, in combination with other lipids and carotenoids/chlorophylls of thylakoid membranes, they too conform together as lipid bilayers.

Membrane Lipid Language

The membrane metabolites of polyunsaturated fatty acids (PUFAs) have an essential role in intercellular biochemical communications. Crawford (2010) in his chapter *Long-chain polyunsaturated fatty acids in human brain evolution* reported, with regard to the *language of lipids,* that the importance of the increased complexity of these lipids was brought about by aerobic metabolism, whereby the simple language of prokaryotes, with only a few words, was developed into a vocabulary of over 1,000 words of eukaryote cells.

About 500 million years ago, some nervous cells and some gut cells of vertebrates, migrated and specialized in a more complex nerrvous system: the *brain*, and in uptake and storage of iodocompounds: the follicular *thyroid*. In the PUFAs, the presence of a double bond between two carbons (or carbon-carbon double bond) provides them with the possibility of changing their molecular structure through enzymes such as phospholipases, cyclooxygenases and lipoxygenases, etc. The resulting substances, called *eicosanoids*: prostaglandins (PG), leukotrienes (LT), lipoxins and tromboxane (TX); and docosanoids: resolvins, protectins, and maresins, are powerful lipid mediators that produce specific actions in the organism; they organize inflammation, hemodynamic, immune response and the repair of tissue.

Many PUFAs cannot be synthesized by animal organisms and are considered *essential*, and therefore should be incorporated into diets. These are: linoleic acid (C18:2 n-6), omega-6 and alpha-linolenic (C18:3 n-3) omega-3, arachidonic acid (AA) - omega - 6 (C20: 4n-6), and docosahexaenoic acid (DHA) - omega -3 (C22:6n-3). These PUFAs are incorporated into the phospholipidic membrane of all the cells of an organism. In parallel, ectodermic cells, differentiated into neuronal cells, became the primitive nervous system and brain. Both these cells synthesized iodolipids, as novel *words* of the chemical *lipid language* developed among cell membranes during the evolution of life. These biochemical signals among cells, since contact and modification of membranes in multicellular organisms formed the bases of adaptation to terrestrial environments, and their alterations are important in the mechanism of apoptosis, carcinogenesis and degenerative diseases, as well as for understand some problems discussed regarding human evolution (as Aquatic ape hypothesis) .

Major Classes

Phospholipids

Phospholipids and glycolipids consist of two long, nonpolar (hydrophobic) hydrocarbon chains linked to a hydrophilic head group.

The heads of phospholipids are **phosphorylated** and they consist of either:

- **Glycerol** (and hence the name **phosphoglycerides** given to this group of lipids), or
- **Sphingosine** (e.g. sphingomyelin and ceramide).

Glycolipids

The heads of glycolipids (glyco- stands for sugar) contain a sphingosine with **one or several sugar units** attached to it. The hydrophobic chains belong either to:

- two fatty acids (FA) - in the case of the phosphoglycerides, or
- one FA and the hydrocarbon tail of sphingosine - in the case of sphingomyelin and the glycolipids.

Galactolipids- monogalactosyl diglyceride(MGDG) and digalactosyl diglycreride(DGDG) form the predominant lipids in higher plant chloroplast thylakoid membranes; liposomal structures formed by total lipid extract of thylakoid membranes have been found sensitive to sucrose as it turns bilayers into micellar structures.

Fatty Acids

The fatty acids in phospho- and glycolipids usually contain an even number, typically between 14 and 24, of carbon atoms, with 16- and 18-carbon being the most common. FAs may be saturated or unsaturated, with the configuration of the double bonds nearly always *cis*. The length and the degree of unsaturation of FAs chains have a profound effect on membranes' fluidity. Plant *thylakoid membranes* maintain high fluidity, even at relatively cold environmental temperatures, due the abundance of 18-carbon fatty acyl chains with three double bonds, *linolenic acid*, as has been revealed by 13-C NMR studies.

Phosphoglycerides

In phosphoglycerides, the hydroxyl groups at C-1 and C-2 of glycerol are esterified to the carboxyl groups of the FAs. The C-3 hydroxyl group is esterified to phosphoric acid. The resulting compound, called phosphatidate, is the simplest phosphoglycerate. Only small amounts of phosphatidate are present in membranes. However, it is a key intermediate in the biosynthesis of the other phosphoglycerides.

Sphingolipids

Sphingosine is an amino alcohol that contains a long, unsaturated hydrocarbon chain. In sphingomyelin and glycolipids, the amino group of sphingosine is linked to FAs by an amide bond. In sphingomyelin the primary hydroxyl group of sphingosine is esterified to phosphoryl choline.

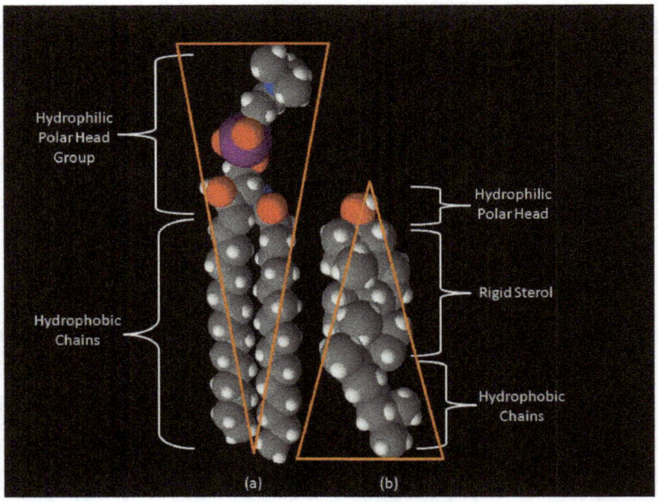

Space-filling models of (a) sphingomyelin and (b) cholesterol.

In glycolipids, the sugar component is attached to this group. The simplest glycolipid is cerebroside, in which there is only one sugar residue, either Glc or Gal. More complex glycolipids, such as gangliosides, contain a branched chain of as many as seven sugar residues.

Sterols

The best known sterol is cholesterol, which is found in humans. Cholesterol also occurs naturally in other eukaryote cell membranes. Sterols have a hydrophobic four-membered fused ring rigid structure, and a small polar head group.

Cholesterol is bio-synthesised from mevalonate via a squalene cyclisation of terpenoids. Cell membranes require high levels of cholesterol - typically an average of 20% cholesterol in the whole membrane, increasing locally in raft areas up to 50% cholesterol (- % is molecular ratio). It associates preferentially with sphingolipids (see diagram) in cholesterol-rich lipid rafts areas of the membranes in eukaryotic cells. Formation of lipid rafts promotes aggregation of peripheral and transmembrane proteins including docking of SNARE and VAMP proteins Phytosterols, such as sitosterol and stigmasterol, and Hopanoids serve a similar function in plants and prokaryotes.

Phospholipids Forming Lipid Vesicles

Lipid vesicles or liposomes are circular pockets that are enclosed by a lipid bilayer. These structures are used in laboratories to study the effects of chemicals in cells by delivering these chemicals directly to the cell, as well as getting more insight into cell membrane permeability. Lipid vesicles and liposomes are formed by first suspending a lipid in an aqueous solution then agitating the mixture through sonication, resulting in a vesicle. By measuring the rate of efflux from that of the inside of the vesicle to the ambient solution, allows researcher to better understand membrane permeability. Vesicles can be formed with molecules and ions inside the vesicle by forming the vesicle with the desired molecule or ion present in the solution. Proteins can also be embedded into the membrane through solubilizing the desired proteins in the presence of detergents and attaching them to the phospholipids in which the liposome is formed. These provide researchers with a tool to examine various membrane protein functions.

Carbohydrates

Plasma membranes also contain carbohydrates, predominantly glycoproteins, but with some glycolipids (cerebrosides and gangliosides). For the most part, no glycosylation occurs on membranes within the cell; rather generally glycosylation occurs on the extracellular surface of the plasma membrane. The glycocalyx is an important feature in all cells, especially epithelia with microvilli. Recent data suggest the glycocalyx participates in cell adhesion, lymphocyte homing, and many others. The penultimate sugar is galactose and the terminal sugar is sialic acid, as the sugar backbone is modified in the Golgi apparatus. Sialic acid carries a negative charge, providing an external barrier to charged particles.

MEMBRANE PROTEIN

Membrane proteins are proteins that interact with, or are part of, biological membranes. They include integral membrane proteins that are permanently anchored or part of the membrane and peripheral membrane proteins that are only temporarily attached to the lipid bilayer or to other integral proteins. The integral membrane proteins are classified as transmembrane proteins that span across the membrane and integral monotopic proteins that are attached to only one side of the membrane. Membrane proteins are a common type of proteins along with soluble globular proteins, fibrous proteins, and disordered proteins. They are targets of over 50% of all modern medicinal drugs. It is estimated that 20–30% of all genes in most genomes encode membrane proteins.

Compared to other classes of proteins, the determination of membrane protein structures has remained a challenge in large part due to the difficulty in establishing experimental conditions where the correct conformation of the protein in isolation from its native environment is preserved.

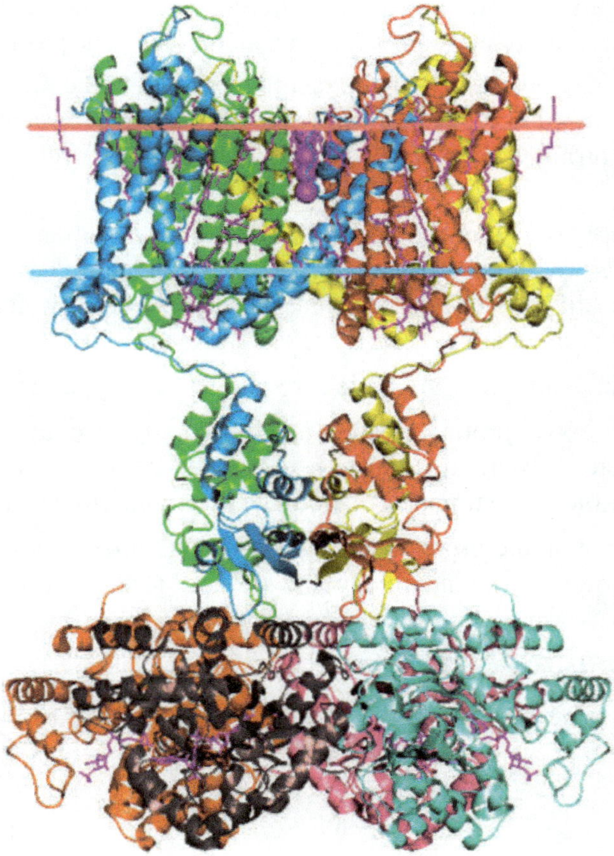

Crystal structure of Potassium channel Kv1.2/2.1 Chimera. Calculated hydrocarbon boundaries of the lipid bilayer are indicated by red and blue dots.

Integral Membrane Proteins

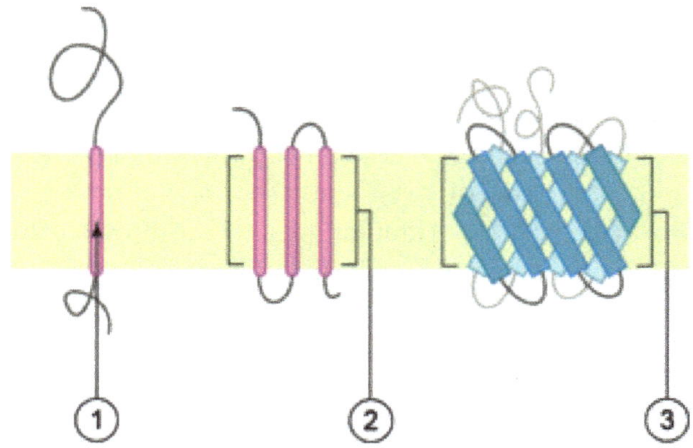

Schematic representation of transmembrane proteins: 1. a single transmembrane α-helix (bitopic membrane protein) 2. a polytopic transmembrane α-helical protein 3. a polytopic transmembrane β-sheet protein. The membrane is represented in light-brown.

Integral membrane proteins are permanently attached to the membrane. Such proteins can be separated from the biological membranes only using detergents, nonpolar solvents, or sometimes denaturing agents. They can be classified according to their relationship with the bilayer:

- Integral polytopic proteins, also known as "transmembrane proteins," are integral membrane proteins that span across the membrane at least once. These proteins may have different transmembrane topology. These proteins have one of two structural architectures:
    - helix bundle proteins, which are present in all types of biological membranes;
    - beta barrel proteins, which are found only in outer membranes of Gram-negative bacteria, lipid-rich cell walls of a few Gram-positive bacteria, and outer membranes of mitochondria and chloroplasts.
- Integral monotopic proteins are integral membrane proteins that are attached to only one side of the membrane and do not span the whole way across.

Peripheral Membrane Proteins

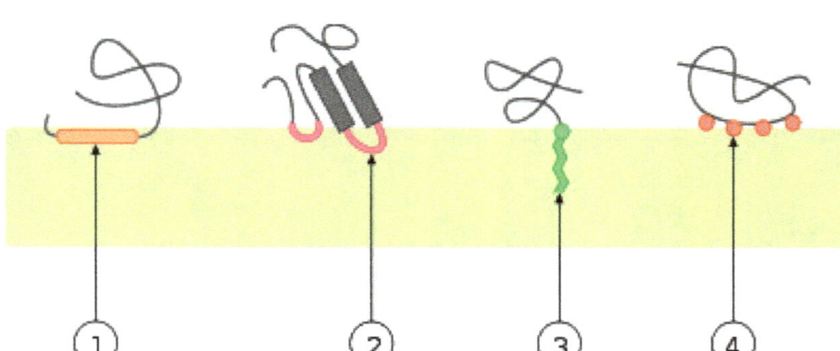

Schematic representation of the different types of interaction between monotopic membrane proteins and the cell membrane: 1. interaction by an amphipathic α-helix parallel to the membrane plane (in-plane membrane helix) 2. interaction by a hydrophobic loop 3. interaction by a covalently bound membrane lipid (*lipidation*) 4. electrostatic or ionic interactions with membrane lipids (*e.g.* through a calcium ion)

Peripheral membrane proteins are temporarily attached either to the lipid bilayer or to integral proteins by a combination of hydrophobic, electrostatic, and other non-covalent interactions. Peripheral proteins dissociate following treatment with a polar reagent, such as a solution with an elevated pH or high salt concentrations.

Integral and peripheral proteins may be post-translationally modified, with added fatty acid or prenyl chains, or GPI (glycosylphosphatidylinositol), which may be anchored in the lipid bilayer.

Polypeptide Toxins

Polypeptide toxins and many antibacterial peptides, such as colicins or hemolysins, and certain proteins involved in apoptosis, are sometimes considered a separate category. These proteins are water-soluble but can aggregate and associate irreversibly with the lipid bilayer and become reversibly or irreversibly membrane-associated.

Membrane Proteins in Genomes

A large fraction of all proteins are thought to be membrane proteins. For instance, about 1000 of the ~4200 proteins of *E. coli* are thought to be membrane proteins. The membrane localization has been confirmed for more than 600 of them experimentally. The localization of proteins in membranes can be predicted very reliably using hydrophobicity analyses of protein sequences, i.e. the localization of hydrophobic amino acid sequences.

Variation

The cell membrane has different lipid and protein compositions in distinct types of cells and may have therefore specific names for certain cell types:

- Sarcolemma in myocytes
- Oolemma in oocytes
- Axolemma in neuronal processes - axons
- Historically, the plasma membrane was also referred to as the plasmalemma

Permeability

The permeability of a membrane is the rate of passive diffusion of molecules through the membrane. These molecules are known as permeant molecules. Permeability depends mainly on the electric charge and polarity of the molecule and to a lesser extent the molar mass of the molecule. Due to the cell membrane's hydrophobic nature, small electrically neutral molecules pass through the membrane more easily than charged, large ones. The inability of charged molecules to pass through the cell membrane results in pH partition of substances throughout the fluid compartments of the body.

REFERENCES

1. Murate, Motohide; Kobayashi, Toshihide. "Revisiting transbilayer distribution of lipids in the plasma membrane". Chemistry and Physics of Lipids. 194: 58–71. doi:10.1016/j.chemphyslip.2015.08.009.
2. Nickels, Jonathan D.; Smith, Jeremy C.; Cheng, Xiaolin. "Lateral organization, bilayer asymmetry, and inter-leaflet coupling of biological membranes". Chemistry and Physics of Lipids. 192: 87–99. doi:10.1016/j.chemphyslip.2015.07.012.
3. Chong, Zhi-Soon; Woo, Wei-Fen; Chng, Shu-Sin (2015-12-01). "Osmoporin OmpC forms a complex with MlaA to maintain outer membrane lipid asymmetry in Escherichia coli". Molecular Microbiology. 98 (6): 1133–1146. doi:10.1111/mmi.13202. ISSN 1365-2958.
4. Forrest, Lucy R. (2015-01-01). "Structural Symmetry in Membrane Proteins". Annual Review of Biophysics. 44 (1): 311–337. doi:10.1146/annurev-biophys-051013-023008. PMID 26098517.
5. Voet, Donald (2012). Fundamentals of Biochemistry: Life at the Molecular Level (4 ed.). Wiley. ISBN 978-1118129180.
6. Dougherty, R. M.; Galli, C.; Ferro-Luzzi, A.; Iacono, J. M. (1987-02-01). "Lipid and phospholipid fatty acid composition of plasma, red blood cells, and platelets and how they are affected by dietary lipids: a study of normal subjects from Italy, Finland, and the USA.". The American Journal of Clinical Nutrition. 45 (2): 443–455. ISSN 0002-9165. PMID 3812343.
7. Lentz, Barry R. (2003-09-01). "Exposure of platelet membrane phosphatidylserine regulates blood coagulation". Progress in Lipid Research. 42 (5): 423–438. doi:10.1016/s0163-7827(03)00025-0. ISSN 0163-7827. PMID 12814644.
8. Lein, Max; deRonde, Brittany M.; Sgolastra, Federica; Tew, Gregory N.; Holden, Matthew A. (2015-11-01). "Protein transport across membranes: Comparison between lysine and guanidinium-rich carriers". Biochimica et Biophysica Acta (BBA) - Biomembranes. 1848 (11, Part A): 2980–2984. doi:10.1016/j.bbamem.2015.09.004.
9. Alberts, Bruce; Johnson, Alexander; Lewis, Julian; Raff, Martin; Roberts, Keith; Walter, Peter (2002-01-01). "The Lipid Bilayer".
10. Daubenspeck, James M.; Jordan, David S.; Simmons, Warren; Renfrow, Matthew B.; Dybvig, Kevin (2015-11-23). "General N-and O-Linked Glycosylation of Lipoproteins in Mycoplasmas and Role of Exogenous Oligosaccharide". PLoS ONE. 10 (11): e0143362. doi:10.1371/journal.pone.0143362. PMC 4657876 . PMID 26599081.
11. Vitrac, Heidi; MacLean, David M.; Jayaraman, Vasanthi; Bogdanov, Mikhail; Dowhan, William (2015-11-10). "Dynamic membrane protein topological switching upon changes in phospholipid environment". Proceedings of the National

Academy of Sciences. 112 (45): 13874–13879. doi:10.1073/pnas.1512994112. ISSN 0027-8424. PMC 4653158 . PMID 26512118.

12. Rojko, Nejc; Anderluh, Gregor (2015-12-07). "How Lipid Membranes Affect Pore Forming Toxin Activity". Accounts of Chemical Research. 48: 3073–3079. doi:10.1021/acs.accounts.5b00403.
13. Kimball's Biology pages, Cell Membranes
14. Singleton P (1999). Bacteria in Biology, Biotechnology and Medicine (5th ed.). New York: Wiley. ISBN 0-471-98880-4.
15. Alberts B, Johnson A, Lewis J, et al. (2002). Molecular Biology of the Cell (4th ed.). New York: Garland Science. ISBN 0-8153-3218-1.
16. Budin, Itay; Devaraj, Neal K. (December 29, 2011). "Membrane Assembly Driven by a Biomimetic Coupling Reaction". Journal of the American Chemical Society. 134 (2): 751–753. doi:10.1021/ja2076873. Retrieved February 18, 2012.
17. Staff (January 25, 2012). "Chemists Synthesize Artificial Cell Membrane". ScienceDaily. Retrieved February 18, 2012.
18. Staff (January 26, 2012). "Chemists create artificial cell membrane". kurzweilai. net. Retrieved February 18, 2012.
19. de Vries, H. 1885. Plasmolytische Studien über die Wand der Vakuolen. Jahrb. wiss. Bot. 16, 465–598
20. Pfeffer, W. 1877. Osmotische Untersuchungen: Studien zur Zell Mechanik. Engelmann, Leipzig.
21. Pfeffer , W. , 1900–1906. The Physiology of Plants, . Translated by A. J. Ewart from the 2nd German ed. of Pflanzenphysiologie, 1897-1904, . Clarendon Press, Oxford.
22. Sharp, L. W. (1921). Introduction To Cytology. New York: McGraw Hill, p. 42.
23. Kleinzeller, A. 1999. Charles Ernest Overton's concept of a cell membrane. In: Membrane permeability: 100 years since Ernest Overton (ed. Deamer D.W., Kleinzeller A., Fambrough D.M.), pp. 1–18, Academic Press, San Diego, .
24. Mast, S. O. 1924. Structure and locomotion in Amoeba proteus. Anat. Rec. 29: 88, .
25. Plowe, J. Q. (1931). Membranes in the plant cell. I. Morphological membranes at protoplasmic surfaces. Protoplasma 12, 196-220, .
26. Wayne, R. 2009. Plant Cell Biology: From Astronomy to Zoology. Amsterdam: Elsevier/Academic Press, p. 17.
27. YashRoy, R.C. (1999). "A structural model for virulence organellae of gram negative organisms with reference to Salmonella pathogenicity in chicken ileum". Indian Journal of Poultry Science. 34 (2): 213–219.
28. Singer SJ, Nicolson GL (Feb 1972). "The fluid mosaic model of the structure of cell membranes". Science. 175 (4023): 720–31. doi:10.1126/science.175.4023.720. PMID 4333397.

29. Doherty GJ, McMahon HT (2008). "Mediation, Modulation and Consequences of Membrane-Cytoskeleton Interactions". Annual Review of Biophysics. 37: 65–95. doi:10.1146/annurev.biophys.37.032807.125912. PMID 18573073.
30. Lodish H, Berk A, Zipursky LS, et al. (2004). Molecular Cell Biology (4th ed.). New York: Scientific American Books. ISBN 0-7167-3136-3.
31. Jesse Gray; Shana Groeschler; Tony Le; Zara Gonzalez (2002). "Membrane Structure" (SWF). Davidson College. Retrieved 2007-01-11.
32. R. B. Gennis. Biomembranes - Molecular Structure and Function. Springer-Verlag, New York (1989).
33. YashRoy R.C. (1990) Lamellar dispersion and phase separation of chloroplast membrane lipids by negative staining electron microscopy. Journal of Biosciences, vol. 15(2), pp. 93-98.https://www.researchgate.net/publication/230820037_Lamellar_dispersion_and_phase_separation_of_chloroplast_membrane_lipids_by_negative_staining_electron_microscopy?ev=prf_pub
34. Venturi, S.; Donati, F.M.; Venturi, A.; Venturi, M. (2000). "Environmental Iodine Deficiency: A Challenge to the Evolution of Terrestrial Life?". Thyroid. 10 (8): 727–9. doi:10.1089/10507250050137851. PMID 11014322.
35. Crawford, M. A.; Bloom, M.; Broadhurst, C. L.; Schmidt, W. F.; Cunnane, S. C.; Galli, C.; Gehbremeskel, K.; Linseisen, F.; Lloyd-Smith, J.; Parkington, J. (1999). "Evidence for the unique function of docosahexaenoic acid during the evolution of the modern hominid brain". Lipids. 34 Suppl: S39–S47. doi:10.1007/BF02562227. PMID 10419087.
36. Venturi, S; Bégin ME (2010). "Thyroid Hormone, Iodine and Human Brain Evolution". In Cunnane S; Stewart K. Environmental Influences on Human Brain Evolution. John Wiley & Sons. pp. 105–124. ISBN 978-0-470-45268-4.
37. Crawford MA (2010). "Long-Chain Polyunsaturated Fatty Acids in Human Brain Evolution". In Cunnane S; Stewart K. Environmental Influences on Human Brain Evolution. John Wiley & Sons. pp. 13–32. ISBN 978-0-470-45268-4.
38. Venturi, Sebastiano (2014). "Iodine, PUFAs and Iodolipids in Health and Disease: An Evolutionary Perspective". Human Evolution-. 29 (1-3): 185–205. ISSN 0393-9375.
39. YashRoy R.C. (1994) Destabilisation of lamellar dispersion of thylakoid membrane lipids by sucrose. Biochimica et Biophysica Acta, vol. 1212, pp. 129-133. https://www.researchgate.net/publication/15042978_Destabilisation_of_lamellar_dispersion_of_thylakoid_membrane_lipids_by_sucrose?ev=prf_pub
40. YashRoy R.C. (1987) 13-C NMR studies of lipid fatty acyl chains of chloroplast membranes. Indian Journal of Biochemistry and Biophysics, vol. 24(6), pp. 177-178.https://www.researchgate.net/publication/230822408_13-C_NMR_studies_of_lipid_fatty_acyl_chains_of_chloroplast_membranes?ev=prf_pub
41. de Meyer F, Smit B. Effect of cholesterol on the structure of a phospholipid bilayer. Proc Natl Acad Sci U S A 2009; 106: 3654-8.

42. Chen, Heshun; Born, Ella; Mathur, Satya N.; Field, F. Jeffrey (December 1, 1993). "Cholesterol and sphingomyelin syntheses are regulated independently in cultured human intestinal cells, CaCo-2: role of membrane cholesterol and sphingomyelin content." (PDF). Journal of Lipid Research. American Society for Biochemistry and Molecular Biology. 34 (12): 2159–67. ISSN 0022-2275.

43. Lang T, Bruns D, Wenzel D, Riedel D, Holroyd P, Thiele C, Jahn R. SNAREs are concentrated in cholesterol-dependent clusters that define docking and fusion sites for exocytosis EMBO J 2001;20:2202-13.

44. Johnson JE, Cornell RB (1999). "Amphitropic proteins: regulation by reversible membrane interactions (review)". Mol. Membr. Biol. 16 (3): 217–235. doi:10.1080/096876899294544. PMID 10503244.

45. Andreeva, A (2014). "SCOP2 prototype: a new approach to protein structure mining". Nucleic Acids Res. 42: D310–4. doi:10.1093/nar/gkt1242. PMC 3964979 . PMID 24293656.

46. Overington JP, Al-Lazikani B, Hopkins AL (December 2006). "How many drug targets are there?". Nat Rev Drug Discov. 5 (12): 993–6. doi:10.1038/nrd2199. PMID 17139284.

47. Krogh, A.; Larsson, B. R.; Von Heijne, G.; Sonnhammer, E. L. L. (2001). "Predicting transmembrane protein topology with a hidden markov model: Application to complete genomes". Journal of Molecular Biology. 305 (3): 567–580. doi:10.1006/jmbi.2000.4315. PMID 11152613.

48. Liszewski, Kathy (1 October 2015). "Dissecting the Structure of Membrane Proteins". Genetic Engineering & Biotechnology News (paper). 35 (17): 1, 14, 16–17.(subscription required)

49. Almén, M.; Nordström, K. J.; Fredriksson, R.; Schiöth, H. B. (2009). "Mapping the human membrane proteome: A majority of the human membrane proteins can be classified according to function and evolutionary origin". BMC Biology. 7: 50. doi:10.1186/1741-7007-7-50. PMC 2739160 . PMID 19678920.

50. Von Heijne, G. (2006). "Membrane-protein topology". Nature Reviews Molecular Cell Biology. 7 (12): 909–918. doi:10.1038/nrm2063. PMID 17139331.

51. Gerald Karp (2009). Cell and Molecular Biology: Concepts and Experiments. John Wiley and Sons. pp. 128–. ISBN 978-0-470-48337-4. Retrieved 13 November 2010.

52. Daley, D. O.; Rapp, M; Granseth, E; Melén, K; Drew, D; von Heijne, G (2005). "Global topology analysis of the Escherichia coli inner membrane proteome". Science. 308 (5726): 1321–3. doi:10.1126/science.1109730. PMID 15919996.

Index

R

S

T

W

X